Poucher's Perfumes, Cosmetics and Soaps – Volume 1
The Raw Materials of Perfumery

Poucher's Perfumes, Cosmetics and Soaps – Volume 1
The Raw Materials of Perfumery

Ninth edition

W.A. Poucher

Edited and revised by A.J. Jouhar
Chief Executive
Information Transfer International

CHAPMAN & HALL
London · New York · Tokyo · Melbourne · Madras

Published by Chapman & Hall, 2–6 Boundary Row, London SE1 8HN

Chapman & Hall, 2–6 Boundary Row, London SE1 8HN, UK

Chapman & Hall, 29 West 35th Street, New York NY10001, USA

Chapman & Hall Japan, Thomson Publishing Japan, Hirakawacho Nemoto Building, 7F, 1-7-11 Hirakawa-cho, Chiyoda-ku, Tokyo 102, Japan

Chapman & Hall Australia, Thomas Nelson Australia, 102 Dodds Street, South Melbourne, Victoria 3205, Australia

Chapman & Hall India, R. Seshadri, 32 Second Main Road, CIT East, Madras 600 035, India

First edition 1923
Ninth edition 1991

© 1991 Chapman & Hall

Typeset in 10/11 Sabon by Thomson Press (India) Ltd, New Delhi
Printed in Great Britain by T.J. Press (Padstow) Ltd, Padstow, Cornwall

ISBN 0 412 27340 3

Apart from any fair dealing for the purposes of research or private study, or criticism or review, as permitted under the UK Copyright Designs and Patents Act, 1988, this publication may not be reproduced, stored, or transmitted, in any form or by any means, without the prior permission in writing of the publishers, or in the case of reprographic reproduction only in accordance with the terms of the licences issued by the Copyright Licensing Agency in the UK, or in accordance with the terms of licences issued by the appropriate Reproduction Rights Organization outside the UK. Enquiries concerning reproduction outside the terms stated here should be sent to the publishers at the London address printed on this page.

The publisher makes no representation, express or implied, with regard to the accuracy of the information contained in this book and cannot accept any legal responsibility or liability for any errors or omissions that may be made.

A catalogue record for this book is available from the British Library

Library of Congress Cataloging-in-Publication data

Poucher, W.A. (William Arthur)
   The raw materials of perfumery / W.A. Poucher; edited and revised by A.J. Jouhar.—9th ed.
     p. cm.—(Poucher's perfumes, cosmetics, and soaps; v. 1)
   ISBN 0–412–27340–3
   1. Perfumes—Dictionaries. I. Jouhar, A.J. (Akbar Jan)
II. Title. III. Series: Poucher, W.A. (William Arthur). Perfumes, cosmetics, and soaps; v. 1.
TP983.P723 vol. 1 1991
668′.5s–dc19
[668′.54′0321]   88-25755
                                                                                         CIP

∞ Printed on permanent acid-free text paper, manufactured in accordance with the proposed ANSI/NISO Z 39.48-199X and ANSI Z 39.48-1984

Contents

Preface   vi
Publisher's note   vii

Dictionary of the raw materials of perfumery and some cosmetic
and toiletry products   1

Appendix: Materials with IFRA guidelines, but not listed in Poucher's   349

Preface

This classic work by Poucher, first published in 1923, was last produced in three volumes titled, respectively *The Raw Materials of Perfumery* (seventh edition, 1974), *The Production, Manufacture and Application of Perfumes* (eighth edition, 1974) and *Modern Cosmetics* (eighth edition, 1974). Its popularity is well demonstrated by there having been three reprints of these editions in 1976, 1979 and 1984, respectively.

The history of events can be traced by reference to the prefaces to earlier editions and those interested should study these with care since they give a fascinating insight into developments in the subject fields covered by *Poucher's Perfumes, Cosmetics and Soaps* over the years. It is not proposed to provide a resumé here.

In this Volume I, the current edition attempts to provide data about raw materials in a more formalized way than before, so that not only the history of some compounds can be checked, but also so that useful reference information can be obtained. It is particularly relevant to do this, since it is not always easy to be certain of nomenclature. Moreover, as we move towards 'ingredient labelling' (a trend not welcomed by some), a high level of uniformity will be needed. Whether this will come from adoption of CTFA terminology, use of CAS numbers or some other system is not clear. Where possible, such data have been included so that readers may identify materials more readily. Where given, CAS numbers are located in the top right-hand corner of each entry.

For many of the compounds listed several alternative names are in use, some of which from the chemist's point of view are either inadequate, ambiguous or occasionally actually misleading. In this edition the compounds have been listed under names which are considered to be chemically satisfactory and which, at the same time, should be reasonably familiar to perfumers; they do not necessarily contain full information as to the structure of the compounds and they make free use of widely accepted trivial names. In most of the entries this is associated with a systematic name which defines the chemical structure, while synonyms which are in use, though sometimes chemically unsatisfactory, also are listed.

Prefixes denoting structural features such as *n-, iso, cis-, trans-, o-, m-, p-* etc. are disregarded in the alphabetical listing. Where appropriate, however, a second entry may be made in the alphabetical position corresponding to the prefix with a cross-reference to the main entry. Prefixes denoting number (for which Greek-derived mono-, di-, tri-, etc. are used rather than their Latin equivalents) are, on the other hand, regarded as forming an integral part of the word and are thus of alphabetical significance. Subsidiary entries with cross-references are made where this is useful.

The systematic naming follows established chemical practice but as this is not completely uniform the following is a brief account of the principles adopted.

Acyclic compounds are usually named as derivatives of the longest unbranched hydrocarbon chain contained within their structure. The carbon atoms in the chain are numbered from one end; if one of the terminal atoms carries a functional group, this is usually designated '1'. The positions of functional groups and other structural features are indicated by the appropriate numeral immediately preceding the verbal element which designates it; thus, *iso*-propanol is named 'propan-2-ol' rather than '2-propanol' or 'propanol-2'. In the case of double and triple bonds, denoted by the elements '-en-' and '-yn-', the lower number of the pair of carbon atoms between which the bond exists is used. The names of normal carboxylic acids are derived from those of the normal hydrocarbons having the same number of carbon atoms by replacing the final 'e' by '-oic acid', e.g. 'n-pentanoic acid' (=*n*-valeric acid); exceptions are acetic and formic acids. In carboxylic acids and aldehydes, the carbonyl carbon is assigned the numeral '1', e.g. lactic acid becomes '2-hydroxypropanoic acid'.

In disubstituted benzene derivatives, the familiar prefixes *o*-, *m*- and *p*- are used to indicate the relative positions of the substituents; with three or more substituents, however, numbers are used to avoid ambiguity. In polycyclic and heterocyclic compounds, the rings are numbered according to current Chemical Abstracts practice.

Melting and boiling points are given in degrees Centigrade. Unless otherwise specified, the boiling points are at atmospheric pressure (760 mmHg); for those determined at other pressures the latter are given in mmHg.

It must be noted that neither the editor nor the publishers can be held responsible for any consequences arising from any errors or omissions in this work.

Thanks are due to Mrs Venetia Oakley of Turville Word Processing Services and the staff of Information TRANSFER International for their patient work in formatting and assembling this volume.

A.J. Jouhar

Publisher's note

Poucher's Perfumes, Cosmetics & Soaps was last available in three volumes. Volume 1 was in its seventh edition and volumes 2 and 3 in its eighth edition.

This volume is the first volume of what will be a two volume set comprising together a ninth edition of the work.

All information given in this volume is correct to the best of our knowledge at the time of writing but it is recommended that the reader consults the IFRA guidelines or the RIFM monographs referred to in the text.

Dictionary of the raw materials of perfumery and some cosmetic and toiletry products

ABIR

Appearance	Powder
Odour	Aromatic
Chemical	Contains curcuma, cardamom, cloves, and sandalwood.
Uses	Used by the Hindus.

ABSOLUTES

Colourless absolutes are known as absoluols, S.I.S., and integral essences.

Source	
NATURAL	Prepared from flowers, leaves, twigs, roots, barks or resins of living or dead plant material.
Isolation	The plant material is extracted by petroleum ether or other volatile solvents in a closed apparatus. After placing these in series, the solvent runs through and is then distilled off at low pressure to be used again. The product or concrete, consisting of the perfume together with the natural insoluble wax and pigments, is shaken in a machine together with a strong alcohol and the insoluble wax is separated. After cooling the alcoholic solution to below zero and filtering off any dissolved wax, the perfume is then isolated from the solution by distillation *in vacuo* at a low temperature. The absolute is left in the retort.
Appearance	Usually coloured or colourless liquids.
Odour	Very fine odours which vary according to the source.
Uses	Fine perfumery.

ACACIA

Source	
NATURAL	An extensive genus of trees and shrubs from the Mimosa section of the N.O. Leguminosae.
Comments	Commercially produced perfumes are based upon the odour of *Robinia pseudacacia*: they are mixtures of synthetics such as anisic aldehyde, methyl anthranilate, iso-butyl benzoate, and phenylacetic aldehyde.

ACACIA GUM [9000–01–5]

CTFA name	Acacia
	Gum senegal, gum arabic.
Source	
GEOGRAPHIC	A plant indigenous to Northern Africa.
NATURAL	Obtained from the bark of the tree *Acacia senegal*, Willd., and other species of N.O. Leguminosae.
Isolation	An incision is made into the bark of the tree and the resulting

	exudate hardens on exposure to the air. The best gum is collected in Kordofan.
Appearance	A hard resinous exudate; it occurs in spheroidal tears up to 32 mm in diameter.
Physical	Insoluble in alcohol but almost completely soluble in twice its weight of water. Also soluble in glycerol and propylene glycol. Aqueous solutions are acidic.
Uses	Preparation of liquid kohol.

ACETALDEHYDE PHENYLETHYL-n-PROPYLACETAL

	Acetal R
	$C_{13}H_{20}O_2$
Isolation	From acetaldehyde-dipropylacetal and phenylethyl alcohol.
Appearance	Colourless liquid.
Odour	Powerful and natural leafy-green note.
Uses	Widely used to add natural green or floral effects to a wide range of fragrance types.

ACETANISOLE [99–93–4]

	p-Methoxyacetophenone
	$C_9H_{10}O_2$
Appearance	White crystals.
Odour	Fragrance, typical for lilac blossom, with warm, heavy and sweet undertones.
Uses	Soaps, particularly mimosa, fougere and trefle.
Physical	m.p. 36–38 °C, practically insoluble in water, soluble in alcohol and oils.
	RIFM Monograph (1974) *FCT*, **12**, 927.

ACETATE C_7 [112–06–1]

	Acetic acid, heptyl ester; heptyl ethanoate; n-heptyl acetate; heptanyl acetate
	$CH_3 \cdot [CH_2]_6 \cdot OCOCH_3$
Source	
NATURAL	Does not occur in nature.
Appearance	A colourless liquid with a slightly floral odour.
Physical	GC RIFM no. 74–3; IRC RIFM no. 74–3.

ACETATE C_{11} [112–19–6]

	Undecenyl acetate; 10-hendecen-1-yl acetate; 10-hendecenyl acetate; undecenylenic acetate
Chemical Name	10-undecen-1-yl acetate
	$CH_2:CH \cdot [CH_2]_9 \cdot OCO \cdot CH_3$

Source	
NATURAL	Not found in nature.
Isolation	Esterification of 10-undecen-1-ol with acetic acid.
Physical	GC RIFM no. 74–6; IRC RIFM no. 74–6.

ACETATE C_{12} [112–66–3]

Acetic acid, dodecyl ester; dodecanyl acetate; n-dodecyl acetate; lauryl acetate

Isolation	Esterification of lauryl alcohol with acetic acid.

ACETIC ACID [64–19–7]

CTFA name	Acetic acid Ethanoic acid CH_3COOH
Source	
NATURAL	Obtained from wood.
CHEMICAL	Produced synthetically from alcohol.
Isolation	Destructive distillation of wood; this produces three fractions: gases, such as methane, which are not condensed; pyroligneous acid; tar, which is the source of creosote. The second fraction contains the acetic acid which may be fixed with lime. Ethyl alcohol may be oxidized into acetic acid by means of an enzyme, *Mycoderma aceti*, which obtains the necessary oxygen from the atmosphere.
Appearance	Colourless liquid.
Odour	Distinctive pungent odour.
Chemical	Acetic acid generally contains 33% of real acid.

ACETIC ACID, GLACIAL [64–19–7]

	Ethanoic acid CH_3COOH
Isolation	Distillation of dried sodium or calcium acetate with sulphuric acid.
Physical	Flammable.
Chemical	The distillate should contain 98.9% weight acetic acid.
Other	The specific gravity rises from 1.058 to 1.075 on the addition of water until the mixture contains 77% acid. Further addition of water lowers the specific gravity. At 46% it has the same specific gravity as the original glaciale and continued dilution causes the specific gravity to decrease.
Uses	Used by perfumers who prefer to dilute it to the strength required.
Comments	May be used in the preparation of smelling salts, and toilet vinegar when it is mixed with various essential oils.

ACETONE [67-64-1]

Chemical name	2-Propanone
CTFA name	Acetone
	Dimethyl ketone; beta-ketopropane; pyroacetic ether CH_3COCH_3
Appearance	Colourless volatile liquid.
Odour	Characteristic pungent, sweetish taste.
Physical	b.p. 56.3 °C, m.p. −94.8 °C: highly flammable miscible with water and alcohol, chloroform, ether and most alcohols.
Uses	In the preparation of nail varnishes (useful solvent for celluloid and nitro cotton).
Comments	Prolonged or repeated topical use may cause erythema or dryness.

ACETOPHENONE [98-86-2]

Chemical name	Methyl phenyl ketone
	Phenyl methyl ketone; 1-phenylethanone; acetylbenzene; 'hypnone' C_8H_8O
Source	
NATURAL	Occurs in the oils of labdanum resin and *Stirlingia latifolia*.
Appearance	Colourless liquid, solidifying in the cold.
Odour	Pungent–sweet, reminiscent of hawthorn with floral undertones.
Physical	b.p. 202 °C, m.p. 21 °C, slightly soluble in water but freely soluble in alcohol, chloroform and ether, fatty oils and glycerol.
Uses	In perfumes such as hawthorn, mimosa and foin coupe and in low cost fragrances for soaps, detergents, household and industrial products.
Comments	Blends well with anisic aldehyde, terpineol, and heliotropin. RIFM Monograph (1973) *FCT*, **11**, 99 and (1973) *FCT*, **11**, 1079.

ACETYL CEDRENE [68867-57-2]

	Vertofix
	$C_{17}H_{26}O$
Isolation	Obtained from cedar wood oil by the acetylation of terpenes.
Appearance	Slightly viscous light yellow liquid.
Odour	Warm and woody, reminiscent of cedar and vetiver.
	RIFM Monograph (1978) *FCT*, **16**, 639.

4-ACETYL-6-TERT-BUTYL-1,1-DIMETHYLINDAN [3848-24-6]

	4-acetyl-1,1-dimethyl-6-tert-butylindan; celestolide; ethanone, 1-[6-, 1-dimethylethyl-2,3-dihydro-1,1-dimethyl-]H-indane
Source	
NATURAL	does not occur in nature.
Appearance	White crystals.
Physical	GC RIFM no. 72–96; IRC RIFM no. 72–96.

7-ACETYL-1,1,3,4,4,6-HEXAMETHYL-TETRAHYDRO-NAPHTHALENE [21145-77-7]

	Tonalid
	$C_{18}H_{26}O$
Isolation	Synthesized from dimethyl butene and p-cymene with subsequent acetylation.
Appearance	Colourless to white crystals.
Odour	Sweet and musky–woody.
Physical	b.p. 248 °C, m.p. 46 °C.
Uses	Wide range of applications from toiletries to household products.
Comments	Synthetic tetralin-type musk.
	RIFM Monograph (1983) *FCT*, **21**, 645.

ALCOHOL C_6 [111-27-3]

	Caproic acid; 1-hexanol; n-hexyl alcohol
Chemical name	Hexan-1-ol
	$CH_3 \cdot [CH_2]_4 \cdot CH_2OH$
Source	
NATURAL	Is a constituent of several essential oils and aromas—especially apple, strawberry, tea, violet and other flowers.
Isolation	Reduction of ethyl caproate with sodium alcoholate.
Physical	GC RIFM no. 74–125; IRC RIFM no. 74–125.

ALCOHOL C_{10} See n-DECYL ALCOHOL

ALCOHOL C_{11} [112-43-6]

	Undecylenic alcohol
Chemical name	10-undecen-1-ol
	$CH_2 \cdot CH[CH_2] \cdot CH_2OH$
Source	
NATURAL	Found in the leaves of Litsea odorifera.
Isolation	Sodium reduction of undecylenic acid esters.
Physical	GC RIFM no. 71–10; IRC RIFM no. 71–10.
	RIFM Monograph (1978) *FCT*, **16**, 641.

ALDEHYDE C_6 [111-71-7]

	Caproaldehyde; caproic aldehyde; hexaldehyde; hexoic aldehyde
Chemical name	1-hexanal
Source	
NATURAL	Found in a number of essential oils.
Isolation	Oxidation of n-hexanol.
Appearance	A colourless mobile liquid with a strong fatty odour.
Physical	GC RIFM no. 71–12; IRC RIFM no. 71–12.

ALDEHYDE C_7 See n-HEPTANAL

ALDEHYDE C_8 See n-OCTYL ALDEHYDE

ALDEHYDE C_9 See n-NONYL ALDEHYDE

ALDEHYDE C_{10} See DECYL ALDEHYDE

ALDEHYDE C_{11}, SATURATED See UNDECYLIC ALDEHYDE

ALDEHYDE C_{11}, UNSATURATED See UNDECYLENIC ALDEHYDE

ALOE-WOOD OILS [84837-08-1]

	Agar-attar; Oriental lignaloes
History	Was valued by the Egyptians and later by Arabs and Jews in the Middle Ages.
Source	
GEOGRAPHIC	Trees native to Bengal, Burma, Assam and Java.
NATURAL	Distillation of the wood of *Aquilaria agallocha*.
Odour	Compares with ambergris and sandalwood.

ALLYL AMYL GLYCOLATE [67634-00-8]

Chemical name	Glycolic acid, 2-pentyloxy allyl ester
	$C_{10}H_{18}O_3$
Appearance	Colourless liquid.
Odour	Strong, fruity, pineapple with a green galbanum nuance.
Uses	To reinforce modern green and oriental fragrances with strength and tenacity.

ALLYL CAPROATE [123-68-2]

	Allyl hexanoate; hexanoic acid, 2-propenyl ester
Chemical name	2-propenyl hexanoate
	$CH_2 \cdot CH \cdot CH_2 \cdot OOC[CH_2]_4 \cdot CH_3$
Source	
NATURAL	Does not occur in nature.
Isolation	Esterification of allyl alcohol with caproic acid.
Physical	GC RIFM no. 71–20; IRC RIFM no. 71.20.

ALLYL CAPRYLATE [4230-97-1]

Allyl octanoate; octanoic acid, 2-propenyl ester; 2-propenyl octanoate; 2-propenyl octylate
$CH_2:CH \cdot CH_2 \cdot OCO \cdot [CH_2]_6 \cdot CH_3$

Source
NATURAL — Not thought to occur in nature.
Isolation — Esterification of allyl alcohol with octanoic acid.
Appearance — A colourless oily liquid.
Physical — GC RIFM no. 75–1; IRC RIFM no. 75–1.

ALLYL CYCLOHEXYL ACETATE [4728-82-9]

Allyl cyclohexaneacetate; allyl cyclohexylacetate; allyl hexahydrophenylacetate; cyclohexaneacetic acid, 2-propenyl ester; 2-propen-1-yl cyclohexaneacetate.
$C_6H_{11} \cdot CH_2 \cdot OCO \cdot CH_2 \cdot CH:CH_2$

Source
NATURAL — Does not occur in nature.
Chemical — Esterification of allyl alcohol with cyclohexaneacetic acid.
Appearance — A colourless liquid.
Physical — GC RIFM no. 74–276; IRC RIFM no. 74–276.

ALLYL CYCLOHEXYL PROPIONATE [2705-87-5]

Chemical name — Allyl-3-cyclohexyl propionate
$C_{12}H_{20}O_2$
Isolation — By direct esterification of allyl alcohol with cyclohexanepropionic acid.
Appearance — Colourless, slightly oily liquid.
Odour — Powerful, sweet-fruity, resembling pineapple.
Uses — As a topnote ingredient for fruity and fruity–floral compositions and in combination with citrus notes to provide a fruity characteristic to the fragrance.
RIFM Monograph (1973) *FCT*, **11** 491 and (1973) *FCT*, **11**, 1081 (Binder, p. 66).

ALLYL IONONE [79-78-7]

Chemical name — alpha-Allyl ionone
Hexalon, Cetone V
$C_{16}H_{24}O$
Isolation — By citral condensation with allyl acetone followed by cyclization.
Appearance — Pale yellow oily liquid.
Odour — Peculiar oily-sweet, slightly sweet-nutty and woody character.
Physical — b.p. approx. 265 °C.

Uses	In combination with other ionones in modern aldehydic creations and in fragrances with fruity-aldehydic topnotes.
Comments	Its odour tends to change with ageing, particularly under poor storage conditions. RIFM Monograph (Allyl alpha-ionone) (1973) *FCT*, **11**, 493 and *FCT*, **11**, 1081 (Binder, p. 67).

ALLYL PHENOXYACETATE [7493-74-5]

Acetate PA; acetic acid, phenoxy, 2-propenyl ester; 2-propenyl phenoxyacetate

$CH_2:CH \cdot CH_2 \cdot OCO \cdot CH_2 \cdot O \cdot C_6H_5$

Source	
NATURAL	Does not occur in nature.
Isolation	Esterification of allyl alcohol with phenoxyacetic acid.
Physical	GC RIFM no. 74–19; IRC RIFM no. 74–19.

ALUMINIUM NAPHTHOSULPHONATE [1300-81-8]

Chemical name	Aluminium beta-naphthodisulphonate 2-hydroxynaphthalenedisulphonic acid aluminium salt $Al_2(C_{10}H_5(OH)(SO_3)_2)_3$
Source	
CHEMICAL	Prepared from barium beta-naphtholodisulphonate and aluminium sulphate.
Appearance	Fine almost white powder.
Physical	Soluble in water and glycerol; practically insoluble in ether.
Uses	Has a mild antiseptic and astringent action and has been used as a deodorant.

AMBERGRIS (TINCTURE) [8038-65-1]

History	One of the most valuable materials used by the perfumer, prized for many centuries for its curious perfume and reputed aphrodisiac properties, and thought at one time to be the excrement of a bird, a form of congealed gum or bitumen or even a marine fungus.
Source	
GEOGRAPHIC	Collected from the shores of Australia, New Zealand and the Indian Ocean.
NATURAL	Pathological growth formed in the stomach of male sperm whales (*Physeter macrocephalus*/catadon) when it is feeding on squid or cuttle fish and is caused by the irritation of indigestible beaks in the whale's stomach. It is found either in the stomach or lodged in the bowel, although the most common source has traditionally been as collected flotsam rather than the slaughter of whales.

Isolation	Ambergris is powdered and 30 g are added to 1 l of alcohol. The solution is agitated continuously for several days at 25–30 °C and then occasionally for as long as maturing can be afforded. A few days before use it is filtered.
Appearance	Ten different and distinct types of ambergris exist and the colours range from pure white, silver grey (from New Zealand), golden (from the North African coast), golden grey (from the Gulf of Aden), pale yellow (from Australia), dark grey with golden striations, black with some golden streaks inside (from the Azores), hard black (from all parts of the world), dry and dark grey (from the Persian Gulf), and dark reddish-brown (from Madagascar). The best quality is light grey.
Odour	Varies and, although characteristic is difficult to define, has been described as 'musky', 'musty', 'earthy', and reminiscent of the sea.
Physical	True ambergris has a specific gravity between 0.780 and 0.920 and softens about 60 °C. It liquefies at higher temperatures and is melted by boiling water. It is flammable, and soluble in ether and certain volatile and fixed oils.
Chemical	The three major components are the triterpene alcohol ambrein, epicoprostanol and coprostanone.
Other	A hot needle should enter the mass easily, without sticking, and a characteristic odour is given off. An amber-coloured molten drop should appear on removing the point. Of the four animal extracts it has the least animal character. It has the longest duration of evaporation.
Uses	Ambergris extract is confined exclusively to the preparation of expensive perfume oils.
Comments	Substitutes are often used as genuine ambergris is difficult to obtain. RIFM Monograph (1976) *FCT*, **14**, 675.

AMBRETTE SEED OIL AND ABSOLUTE [8015-62-1]

	Ambrette oil
Source	*Hibiscus abelmoschus*, L. (musk seed), N. O. Malvaceae (mallow family).
GEOGRAPHIC	Central and South America, Indonesia and India.
Isolation	The essential oil is obtained from the dried and powdered seeds by steam distillation; fatty acids can be removed by solvent extraction (usually alcohol) to give the so-called absolute.
Yield	0.3–0.5% depending upon source of the plant material.
Appearance	Liquid.
Odour	Musky, aromatic, very persistent, with a distinct cognac note.
Chemical	Sesquiterpene alcohol farnesol has been identified as a constituent.
Uses	Exalting agent in expensive, fine fragrances. RIFM Monograph (Ambrette seed oil; *Hibiscus abelmoschus*) (1975) *FCT*, **13**, 705.

AMBRETTOLIDE [123-69-3]

Cyclohexadecen-7-olide
$C_{16}H_{28}O_2$

Source	
NATURAL	In ambrette seed oil.
Isolation	Synthesized from dihydroxypalmitic acid or from aleuritic acid.
Physical	b.p. approx. 300 °C.
Appearance	Colourless viscous liquid.
Odour	Extremely tenacious floral-musky with a sandalwood note.
Uses	One of the best fixatives used in perfumes.

RIFM Monograph (1975) *FCT*, **13**, 707.

AMBROSIA OIL [89997-47-7]

Source	Chenopodium oil
GEOGRAPHIC	Native plant of South America.
NATURAL	Seed of *Chenopodium ambrosioides*, N. O. Chenopodiaceae.
Isolation	The oil is distilled from the seeds.
Appearance	Red oil.
Odour	Fine odour of geraniol, also of boldo leaf oil, possibly due to traces of ascaridol.

RIFM Monograph (1976) *FCT*, **14**, 713.

IFRA GUIDELINE

The Committee recommends that the following materials should not be used as fragrance ingredients:

—Allylisothiocyanate
—Chenopodium oil
—3,7-Dimethyl-2-octen-1-ol (6,7-Dihydrogeraniol)
—Furfurylideneacetone
—Methyl methacrylate
—Phenylacetone (Methyl benzyl ketone)
—Esters of 2-octynoic acid, except those covered elsewhere in these Guidelines i.e. methyl and allyl heptine carbonate
—Esters of 2-nonynoic acid, except methyl octine carbonate
—Thea sinensis absolute

These recommendations are based on the absence of reports on the use of these materials as fragrance ingredients and inadequate evaluation of possible physiological effects resulting from their use in fragrances.

March 1988, last amendment July 1990

AMIDOL [95-86-3]

Diaminophenol hydrochloride

Source	
CHEMICAL	Derived from 2,4-diaminophenol.

Uses	Hair dye which yields fairly permanent brown and black shades.
Comments	First made by Gauche and patented as a photographic developer, it does not stain the skin and has, therefore, advantages over other types of dyes.

o-AMINOBENZOIC ACID [118-92-3]

	Anthranilic acid; 2-aminobenzoic acid; vitamin L
Appearance	White to pale yellow crystals.
Odour	Sweetish taste.
Physical	m.p. 145 °C, freely soluble in hot water and alcohol; solutions in alcohol, ether and particularly glycerol show an amethyst fluorescence.
Comments	Some of its esters are widely distributed in nature, although it does not appear to exist in any plant.

AMMONIACUM [9000-03-7]

	Gum ammoniac
History	The plant *Ferula marmarica*, Aschers et Taub, which grows in Cyrenia, was used by the ancients for fumigation; mentioned by Dioscorides and Pliny.
Source	
GEOGRAPHIC	Iran.
NATURAL	The flowering and fruiting stem of *Dorema ammoniacum*, Don, N. O. Umbelliferae, which grows to about 6 ft (1.8 m).
Isolation	The plant contains a large quantity of latex, exuded when punctured by a species of beetle. Some of the milky juice hardens in the hot sun and falls to the ground. It is collected at the end of July.
Appearance	Hard gum-resin in the shape of irregular, rounded, yellowish to brownish tears.
Odour	Persistent, and recalling castor: slightly sweetish, bitter and somewhat acrid taste.
Physical	m.p. 45–55 °C; brittle when cold but soft when warm; partly soluble in water, alcohol, ether and can form emulsions with water.
Uses	Can be used in small quantities of strong alcoholic tincture in preparing opoponax perfumes.

AMMONIUM CARBONATE [10361-29-2]

CTFA name	Ammonium carbonate Hartshorn
Isolation	Obtained from the two salts of ammonia: ammonium bicarbonate, $H(NH_4)CO_3$, and ammonium carbamate, $(NH_4)CO_2(NH_2)$.
Physical	Occurs in translucent fibrous pieces, but on exposure to air becomes much effloresced owing to loss of NH_3 with formation of the bicarbonate.

Ammonium chloride

Uses	In the preparation of smelling salts and occasionally in stearin creams.
Comments	Must be kept tightly closed in a cool place.

AMMONIUM CHLORIDE [12125-02-9]

CTFA name	Ammonium chloride
	Ammonium muriate; sal ammoniac; salmiac
	NH_4Cl
Source	
CHEMICAL	Prepared by the combination of hydrochloric acid and ammonia.
Appearance	White colourless crystalline powder.
Odour	None.
Physical	Readily soluble in water, less so in alcohol; almost insoluble in acetone, ether and ethyl acetate.
Uses	Lotions of 5% solutions are used as skin bleaches.
Comments	Has medicinal uses.

AMMONIUM HYDROXYDE [7664-41-7]

COLIPA name	Ammonia
	'Spirit of hartshorn' Ammonia solution
	NH_4OH
Isolation	Ammonia liquid contains 27–30% by weight of ammonia.
Appearance	Colourless liquid.
Odour	Pungent, very diffusive lachrymatory.
Physical	s.g. 0.880.
Chemical	Strongly alkaline.
Uses	Used in the preparation of stearin creams, yielding a soft product.
Comments	Reaction with sulphuric or other strong mineral acids is exothermic.

iso-AMYL ACETATE [123-92-2]

CTFA name	Isoamyl acetate
	3-Methylbut-1-yl acetate; 3-methyl-1-butanol acetate; amylacetic ester; pear oil; banana oil.
	$CH_3COOCH_2CH_2CH(CH_3)_2$
Source	
NATURAL	Constituent of angophora, apple, banana, and coconut oils.
Isolation	Obtained by fractionation of fused oil.
Appearance	Colourless liquid.
Odour	Powerful: fresh-fruity reminiscent of pear, banana and apple.
Physical	b.p. 142 °C; soluble in water and miscible with alcohol, ethyl acetate and amyl alcohol.
Uses	Occasionally, in fine perfumery it is used in traces to 'lift' the initial odour note of chypre and heavy-type bouquets. Technical

	quality material is much used in the production of nail enamels and cellulose lacquers, as it is a middle boiling solvent.
Comments	Used in the composition of artificial pear flavourings in boiled sweets.
	RIFM Monograph (1975) *FCT*, **13**, 551.

iso-AMYL ALCOHOL [123-51-3]

	iso-Butyl carbinol; 3-methylbutan-1-ol
	$(CH_3)_2CHCH_2CH_2OH$
Source	
NATURAL	Constituent of the oils of certain species of eucalyptus, geranum Bourbon and of other oils.
Isolation	Product of fermentation and obtained by purifying fusel oil
Appearance	Clear liquid.
Odour	Characteristic choking, disagreeable.
Physical	b.p. 132 °C.
Uses	Solvent for gums and resins in the lacquer industry and also has wide application in the essence industry.
	RIFM Monograph (1978) *FCT*, **16**, 785.

iso-AMYL ANISATE [27739-29-3]

	iso-Amyl-*p*-methoxybenzoate
	$CH_3OC_6H_4COOCH_2CH_2CH(CH_3)_2$
Isolation	Fractionation of fusel oil.
Appearance	Colourless liquid.
Odour	Soft fragrant, suggestive of neliotrope.
Physical	b.p. 312 °C.
Uses	As a fixative in floral perfumes.

iso-AMYL BENZOATE [94-46-2]

	3-Methyl-but-1-yl benzoate
	$C_6H_5COOCH_2CH_2CH(CH_3)_2$
Isolation	Esterification of fusel oil with anisic acid.
Appearance	Colourless oil.
Odour	Balsamic, sweet, slightly ambergris.
Other	Possesses good fixative properties.
Uses	Useful in compounding oriental and amber perfumes.
	RIFM Monograph (1973) *FCT*, **11**, 495 and (1973) *FCT*, **11**, 1081.

iso-AMYL BENZYL ETHER [122-73-6]

	Amyl benzyl oxide; 3-methylbut-1-yl benzyl ether
	$C_6H_5CH_2OCH_2CH_2CH(CH_3)_2$
Isolation	Fractionation of fusel oil.

Odour   Reminiscent of gardenia.
Physical   b.p. 235 °C.
Uses   In heavy floral types for soaps, gardenia, tuberose, ylang-ylang.
           RIFM Monograph (1978) FCT, **16**, 647.

iso-AMYL-n-BUTYRATE [106-27-4]

3-Methylbut-1-yl butanoate; iso-amyl-n-butyrate
$CH_3CH_2CH_2COOCH_2CH_2CH(CH_3)_2$

Source
NATURAL   In Cocoa oil (with the propionic acid ester).
CHEMICAL   Consists mainly of iso-amyl butyrate together with other isomers.
Appearance Colourless liquid.
Odour   Ethereal fruity, reminiscent of apricot, pineapple and banana.
Physical   b.p. 186 °C.
Uses   In the manufacture of apricot, banana, pineapple, raspberry, strawberry, and cider essences.
            RIFM Monograph (1979) FCT, **17**, 833.

iso-AMYL CAPROATE [2198-61-0]

Amyl hexylate; 3-methylbut-1-yl hexanoate; fusel-amyl caproate; amyl hexoate.
$CH_3(CH_2)_4COOCH_2CH_2CH(CH_3)_2$

Isolation   Fractionation of fusel oil.
Appearance   Colourless liquid.
Odour   Powerful banana-type.
Physical   b.p. 220 °C.
Uses   Preparation of artificial fruit essences, and occasionally in perfumery work to give a new note to bouquet perfumes.
             RIFM Monograph (1979) FCT, **17**, 825 and (1988) FCT, **26**, 285.

iso-AMYL CAPRYLATE [2035-99-6]

Amyl octoate; 3-methylbut-1-yl octanoate
$CH_3(CH_2)_6COOCH_2CH_2CH(CH_3)_2$

Isolation   Fractionation of fusel oil.
Appearance   Colourless liquid.
Odour   Oily, fruity odour reminiscent of elderflower with orris undertones.
             RIFM Monograph (1979) FCT, **17**, 827.

iso-AMYL CINNAMATE [7779-65-9]

3-Methylbut-1-yl 3-phenylacrylate
$C_6H_5CH= CHCOOCH_2CH_2H(CH_3)_2$

Isolation   Fractionation of fusel oil.

Odour	Balsamic, mild amber, cocao bean-like.
Uses	Useful fixative in soap compounds and fancy bouquets.
	RIFM Monograph (Amyl(iso) Cinnamate) (1975) *FCT*, **13**, 709.

alpha-AMYL CINNAMIC ALDEHYDE [122-40-7]

	2-Benzylidene-heptan-1-al; *alpha-n*-amylcinnamic aldehyde
	$C_{14}H_{18}O$
History	First marketed as jasmine aldehyde.
Isolation	Synthesized by the condensation of benzaldehyde and *n*-heptanal.
Appearance	Pale yellow oily liquid.
Odour	Mild oily herbaceous floral, mainly jasmine impure samples have a by-odour of benzaldehyde.
Physical	b.p. 285 °C
Uses	Extensively used to provide a jasmine-floral character to a wide range of floral compositions for soaps, detergents and household products.
	RIFM Monograph (1973) *FCT*, **11**, 855.

o-tertiary-AMYLCYCLOHEXYL ACETATE [67874-72-0]

	2-(1,1-Dimethylprop-1-yl)cyclohexyl acetate; 'Coniferan'-tert-amyl-cyclohexyl acetate; 2-tert-pentyl cyclohexanyl acetate
Source	
CHEMICAL	Commercial samples consist predominantly of the *cis* isomer.
Appearance	Colourless liquid.
Odour	Balsamic green, camphoraceous.
Other	Excellent stability and does not discolour.
Uses	In pine-type compositions and for fine quality bouquet perfumes.
	RIFM Monograph (1976) *FCT*, **14**, 679 (Amylcyclohexyl acetate [mixed isomers])

iso-AMYL ETHYL CARBINOL [18720-65-5]

	6-Methylheptan-3-ol; Ethyl-*iso*-Amylcarbinol
	$C_2H_5CH(OH)CH_2CH_2CH(CH_3)_2$
Source	
NATURAL	Occurs naturally in Japanese peppermint oil.
CHEMICAL	Isomeric with octyl alcohol.
Physical	b.p. 166 °C

iso-AMYL ETHYL CARBINYL ACETATE [32764-34-4]

	6-Methylhept-3-yl acetate; Ethyl-*iso*-amyl carbinyl acetate
	$C_2H_5CH(OOCCH_3)C_5H_{11}$
Appearance	Colourless liquid.
Odour	Harsh, oily-fruity, herbaceous.
Uses	Used as a low cost masking odour.

n-AMYL ETHYL KETONE [106–68–3]

Octan-3-one; Ethyl-n-Amyl Ketone
$C_2H_5CO(CH_2)_4CH_3$

Source
NATURAL — French lavender oil.
Appearance — Colourless liquid.
Odour — Slightly pungent, herbaceous fruity smell.
Physical — b.p. 172 °C
Uses — As a topnote in lavender and other fresh herbaceous compounds.

AMYL-iso-EUGENOL [10484–36–3]

1-pentoxy-2-methoxy-4-propenyl-benzene; amyl oxy-iso-eugenol; iso-Eugenol amylether
$C_{15}H_{22}O_2$

Appearance — Liquid.
Odour — Heavy sweet balsamic slightly spicy.
Physical — b.p. approx. 300 °C
Uses — Makes exquisite bouquets combined with ylang, heliotropin and rose. Useful modifier for carnation tuberose and many other floral types.
RIFM Monograph (1979) FCT, **17**, 513.

iso-AMYL FORMATE [110–45–2]

3-Methylbut-1-yl formate; iso-pentyl-formate
$HCOOCH_2CH_2CH(CH_3)_2$

Appearance — Colourless liquid.
Odour — Pungent fruity green.
Physical — b.p. 123.5 °C, m.p. −93.5 °C
Uses — In the production of fruit essences (plum); sometimes used in perfumes of the Peau d'Espagne type.
RIFM Monograph (1979) FCT, **17**, 829.

n-AMYL FUROATE [4996–48–9]

n-Pentylfuran-2-carboxylate; Amyl pyromucate
$C_4H_3OCOOC_5H_{11}$

Appearance — Colourless or pale yellow oily liquid.
Odour — Sweet caramel-winey character.
Uses — In imitation rum, maple, caramel and butterscotch.

iso-AMYL HEPTIN CARBONATE

Amyl heptin carboxylate; 3-methyl-1-yl oct-2-ynoate 'vert de violette'; iso-amyl octanoate
Isolation — Similar to methyl heptin carbonate.

Appearance	Colourless liquid.
Odour	Powerful sweet leafy green reminiscent of violets; sometimes sold as 'vert de violette'.
Physical	b.p. 250 °C
Uses	Traces are used in violet, sweet-pea, and lilac perfumes.
Comments	Should be kept to a 10% alcoholic solution and used with discretion. IFRA GUIDELINE The Committee recommends that the following materials should not be used as fragrance ingredients: —Allylisothiocyanate —Chenopodium oil —3,7-Dimethyl-2-octen-1-ol (6,7-Dihydrogeraniol) —Furfurylideneacetone —Methyl methacrylate —Phenylacetone (Methyl benzyl ketone) —Esters of 2-octynoic acid, except those covered elsewhere in these Guidelines i.e. methyl and allyl heptine carbonate —Esters of 2-nonynoic acid, except methyl octine carbonate —Thea sinensis absolute These recommendations are based on the absence of reports on the use of these materials as fragrance ingredients and inadequate evaluation of possible physiological effects resulting from their use in fragrances. March 1988, last amendment July 1990

n-AMYL HEPTYLATE [7493–82–5]

n-Pentyl heptanoate; Amyl heptanoate
$CH_3(CH_2)_5COO(CH_2)_4CH_3$

Appearance	Colourless liquid.
Odour	Powerful green fruity note.
Uses	In the preparation of fruit essences; traces may be used for modifying the odours of perfume compounds.

alpha-AMYLHYDROCINNAMIC ALDEHYDE

2-Benzylheptan-1-al; Dihydro amylcinnamal
$C_6H_5CH_2CH(CHO)(CH_2)_4CH_3$

Comments	Has similar jasmine character to amyl cinnamic aldehyde.

iso-AMYL LAURATE [6309–51–9]

CTFA name	Isoamyl laurate
	3-Methylbutyl dodecanoate; amyl laurinate
Appearance	Colourless mobile liquid.
Odour	Faint smell recalling that of peach.
Uses	Useful for giving a new note to a common odour.

n-AMYL METHYL KETONE See METHYL-n-AMYLKETONE

iso-AMYL MYRISTATE [10484–35–2]

Amyl myristinate; 3-methylbut-1-yl tetradecanoate
$CH_3(CH_2)_{12}COOCH_2CH(CH_3)_2$
Isolation — Fractionation of fusel oil.
Appearance — Colourless oily liquid.
Odour — Mild waxy note.
Physical — b.p. 320 °C.
Uses — Fixative in fancy compounds and violet types.

iso-AMYL PHENYLACETATE [102–19–2]

3-Methylbut-1-yl phenylacetate
$C_6H_5CH_2COOCH_2CH_2CH(CH_3)_2$
Appearance — Colourless liquid.
Odour — Powerful; a decided animal note of castoreum type.
Physical — b.p. 268 °C.
Uses — In heavy perfumes of chypre type and also in soap perfumes: used in conjunction with ethyl cinnamate, coumarin, musk, heliotropin, rose, jasmine, and tuberose as a powder perfume.
Comments — Should be used with discretion.
RIFM Monograph (1978) FCT, **16**, 791.

iso-AMYL PROPIONATE [105–68–0]

3-Methylbut-1-yl propanoate
$C_2H_5COOCH_2CH_2CH(CH_3)_2$
Appearance — Colourless liquid.
Odour — Fruity apricot, pineapple type.
Physical — b.p. 160.2 °C.
Uses — In traces to impart a new note to a perfume compound; also employed in fruity compositions of the apricot, banana, cherry, tutti-fruity, raspberry type etc.
RIFM Monograph (1975) FCT, **13**, 809.

iso-AMYL PYRUVATE [7779–72–8]

3-Methylbut-1-yl 2-ketopropanoate
$CH_3COCOOCH_2CH_2CH(CH_3)_2$
Appearance — Colourless liquid.
Odour — Fresh ethereal rum, caramel with fruity undertones.
Physical — b.p. 185 °C.
Uses — In fancy compositions and as a topnote in citrus colognes.

iso-AMYL SALICYLATE [87-20-7]

3-Methylbut-1-yl o-hydroxybenzoate; Trefle
$HOC_6H_4COOCH_2CH_2CH(CH_3)_2$

Isolation Synthesized by the action of hydrochloric acid on a solution of salicyclic acid in amyl alcohol.
Appearance Colourless amber–yellow liquid.
Odour Herbaceous green, slightly floral.
Physical b.p. 277 °C.
Other Blends well with clary sage oil, oakmoss resin, geraniol and terpineol.
Uses Extensively used as the basis of many floral and non-floral perfume types: useful as a modifier and fixer of perfumes such as carnation, chypre, trefle and orchid.
RIFM Monograph (1973) *FCT*, **11**, 859.

iso-AMYL UNDECYLENATE [5421-27-2]

3-Methylbut-1-yl hendec-10-enoate
$CH_2\!=\!CH(CH_2)_8COOC_5H_{11}$

Appearance Colourless liquid.
Odour Oily rose-type.
Physical b.p. 160 °C at 3 mm Hg.
Uses Useful in the preparation of mimosa and rose compounds; a good modifier for bouquets.

iso-AMYL iso-VALERATE [659-70-1]

Amyl valerianate; 3-methylbut-1-yl 3-methyl-butanoate
$(CH_3)_2CHCH_2COOCH_2CH_2CH(CH_3)_2$

Appearance Colourless liquid.
Odour Ripe apples.
Physical b.p. 190 °C.
Uses Used in compounding fruit essences, and as a modifier in perfumes of the crab apple type.
RIFM Monograph (1978) *FCT*, **16**, 789.

AMYROL

Source
NATURAL The name given to a mixture of alcohols occurring in West Indian sandalwood oil.

ANETHOLE [104-46-1]

CTFA name Anethole
iso-Estragole; 1-methoxy-4-(1-propenyl)-benzene; *p*-propenylanisole; anise camphor
$C_{10}H_{12}O$

Source	
NATURAL	Oils of anise, star-anise, fennel, *Magnolia kobus*, and *Osmorrihiza longistylis*: may also occur in the oil distilled from *Artemisia caudata*.
Isolation	Synthetically from anisole and propionaldehyde.
Appearance	Colourless oily liquid.
Odour	Typical of star anise; very sweet, warm herbaceous.
Physical	b.p. 236 °C, m.p. 21.4 °C; practically insoluble in water, miscible with ether and chloroform.
Chemical	On oxidation yields aubepine (anisic aldehyde).
Uses	A flavour in dental preparations and liqueurs; occasionally used in perfumes for soaps and household products.
Comments	*cis* and *trans* isomers exist.
RIFM Monograph (1973) *FCT*, **11**, 863.

ANGELICA OIL [8015–64–3]

Angelica root oil

Source	
GEOGRAPHIC	This biennial plant is cultivated in Northern Europe and Asia. The roots are raised from seed in the first year, planted out in the autumn, hoed repeatedly during the ensuing spring, and dug up in the autumn.
NATURAL	*Archangelica officinalis*, L., N. O. Umbelliferae.
Isolation	Steam distillation from the fresh or dried comminuted roots.
Yield	0.3–0.4%.
Appearance	Colourless to light brown oil.
Odour	Pleasantly aromatic, a mixture of musk and pepper with a spicy top note.
Physical	Almost insoluble in water.
Chemical	The somewhat musky odour is conditioned by the presence of a musk-like lactone which is converted from pentadecanal acid of angelica root oil.
Uses	A remarkably good fixative is obtained when the resinoid is extracted from the root with volatile solvents. The oil is largely used in flavouring liqueurs and both root and seed oils are very valuable blenders in dental flavours. In perfumery used occasionally in traces in masculine notes.
Comments	Should be kept well closed, cool and away from light.
RIFM Monograph (Root oil) (1975) *FCT*, **13**, 713, and (Seed oil) (1974) *FCT*, **12**, 821.
IFRA GUIDELINE
The Committee recommends for applications on areas of skin exposed to sunshine, excluding bath preparations, soaps and other products which are washed off the skin, to limit Angelica root oil to 3.9% in the compound (see remark on phototoxic ingredients in the introduction).

This recommendation is based on the results of RIFM on the phototoxicity of Angelica root oil (*Food Cosmet. Toxicol.*, **13**, 713 (1975) Special issue II) and on the observed no-effect level of a sample in a test using the hairless mouse. (Private communication to IFRA).

June 1975, amended October 1978

ANISALDEHYDE [123-11-5]

p-Methoxybenzaldehyde; anisic aldehyde; aubepine
$C_8H_8O_2$

Source	
NATURAL	Oils of Russian anise, bitter fennel, star anise leaf, Chinese aniseed, and buchu, and also in the absolutes of *Acacia cavenia* and *A. farnesiana* (possibly also wallflower), and of the extract of vanilla.
CHEMICAL	Synthesized by oxidizing anethole or *p*-cresyl methyl ether and methylating *p*-hydroxy-benzaldehyde at normal temperatures.
Appearance	Pale yellow liquid.
Odour	Powerful sweet-floral, recalling hawthorn.
Physical	b.p. 248 °C
Chemical	Readily oxidizes to anisic acid on exposure: sold as 'crystallized aubepine' in combination with sodium bisulphite.
Uses	In soaps, perfumes such as acacia, cassie, new-mown hay, heliotrope, lilac, and wallflower, and for blending in face powder perfumes.

RIFM Monograph (1974) *FCT*, **12**, 823.

ANISEED OIL [8007-70-3]

CTFA name	Anise oil
	Anise oil, star anise oil
Source	
GEOGRAPHIC	*Illicium verum* and probably other varieties of N. O. Magnoliaceae, cultivated in China. An oil of more delicate odour and flavour is obtained from the fruit of *Pimpinella anisum*, L., N. O. Umbelliferae, grown in Spain, Russia and Germany.
CHEMICAL	Chief constituent is anethole, together with traces of pinene, phellandrene, dipentene, *l*-limonene, hydroquinone ethyl ether, methyl chaicol, anisic analdehyde, anisic acid, anisic ketone, *p*-cymene, cineol, safrole and terpineol.
Isolation	Steam distilled from the comminuted seeds: output can be increased by 50% if the leaves are also used.
Yield	2–3% according to whether the fruit is wet or dry; with modern apparatus it is possible to obtain 5%. The oil distilled from the leaves may be used as an adulterant.
Appearance	Colourless or pale yellow oil.
Odour	Powerful and typical.

Anisic acid

Physical	Slightly soluble in water.
Uses	As a toner in many different perfume types; popular flavour in dental preparations, especially if blended with peppermint.
Comments	Should be kept well-closed, cool, away from light. RIFM Monograph (1973) *FCT*, **11**, 865, (Anis oil) and (1975) *FCT*, **13**, 715 (Anis Star oil).

ANISIC ACID [579-75-9]

p-Methoxybenzoic acid
$CH_3OC_6H_4COOH$

Appearance	White crystals.
Odour	Practically odourless.
Physical	b.p. approx. 280 °C.
Uses	Has little value as a fragrance ingredient.

ANISOLE [100-66-3]

Methyl phenyl ether; Methoxybenzene
$C_6H_5OCH_3$

Appearance	Colourless liquid.
Odour	Powerfully sweet aromatic.
Physical	b.p. 155 °C.
Chemical	Resinifies with formaldehyde.
Uses	In low cost fragrances for soaps, detergents and household products. RIFM Monograph (1979) *FCT*, **17**, 243.

ANISYL ACETATE [104-21-2]

p-Methoxybenzyl acetate. It has the misnomer 'Cassie Ketone' in some countries.
$CH_3COOCH_2C_6H_4OCH_3$

Appearance	Colourless liquid.
Odour	Mild fruity-floral, slightly balsamic.
Physical	b.p. 235 °C.
Uses	Excellent in sweet florals such as lilac, acacia, honeysuckle.

ANISYL ALCOHOL [105-13-5]

p-Methoxybenzyl alcohol; anisic alcohol; anise alcohol; anisol
$CH_3OC_6H_4CH_2OH$

Source	
NATURAL	Constituent of Tahiti vanilla.
Appearance	Colourless liquid.
Odour	Mild floral sweet odour reminiscent of lilac and hawthorn.
Physical	b.p. 259 °C.

| Uses | Useful in lilac, apple blossom, sweet pea, gardenia etc. |
| | RIFM Monograph (1974) *FCT*, **12**, 825. |

ANISYL FORMATE [122–91–8]

	p-Methoxybenzyl formate
	$HCOOCH_2C_6H_4OCH_3$
Appearance	Colourless liquid.
Odour	Sweet herbaceous green reminiscent of seaweed.
Uses	Useful in small quantities in lilac, heliotrope, and tuberose compounds.
	RIFM Monograph (1976) *FCT*, **14**, 685.

ANYME OIL

Source	
GEOGRAPHIC	*Myoporum crassefolium*, a tree native to Australia.
Isolation	Distillation of the wood.
Appearance	Vivid yellowish liquid, with a slight fluorescence.
Odour	Persistent pinaceous woody-type.

APOPIN OIL [8008–51–3]

History	Known to natives as 'Shuyu'.
Source	
NATURAL	A Lauraceous tree, usually *Cinnamonum camphora*.
CHEMICAL	Contains camphor, eugenol, safrole, and cineol.
Isolation	Distilled from the tree in Taiwan.
Uses	Used mixed with camphor oil.

ARABIS

	Rock Cress
Source	
NATURAL	A large genus belonging to N. O. Cruciferae. *Arabis albida*, L., found in rock gardens. Bears snow-white flowers in spring.
Odour	The flowers have a delicate odour suggestive of heliotropin.

ARAUCARIA OIL

Source	
GEOGRAPHIC	Principally from the island of New Caledonia.
NATURAL	*Callitropsis araucarioides*, Compton, a tree of N. O. Pinaceae; grows at some 800 ft (240 m) above sea-level to a height of about 30 ft (9 m).
CHEMICAL	Contains almost 60% eudesmol, also eudesmene, geraniol as ester, and about 1% unidentified phenols.

24 Areca nuts

Isolation	Distilled from the wood.
Yield	About 6%.
Appearance	Viscous oil.
Odour	Pleasant rose-type.

ARECA NUTS [89957–52–8]

Betel nuts; pinang

Source	
GEOGRAPHIC	A palm cultivated in India and Java.
NATURAL	*Areca catechu*, L., N. O. Palmae.
Isolation	The nuts are powdered.
Appearance	Powder.
Uses	In tooth powders and pastes.

ARNICA FLOWER OIL [8057–65–6]

CTFA name	Arnica extract
	Extract of Arnica
Source	
NATURAL	*Arnica montana*.
Isolation	The oil can be distilled. The flowers can be extracted with petroleum ether.
Yield	0.95% is obtained by extraction.
Appearance	An olive-brown oil is distilled. A yellow wax-like substance of characteristic odour is extracted.
Odour	The distilled oil has a persistent 'heady' odour.

AROMADENDRAL [5150–32–3]

$C_9H_{12}O$

Source	
NATURAL	Different species of eucalyptus.
Isolation	Distillation from eucalyptus oils.
Odour	Reminiscent of cummin.

AROMANTHEMES

A name given to natural flower perfumes of high concentration. They are stated to be pure without the addition of any synthetic or other substance.

ARROWROOT

Source	
GEOGRAPHIC	South Africa, St Vincent, and Bermuda.
NATURAL	Obtained principally from the rhizome of *Maranta arundinacea*, L., N. O. Marantaceae.

ARTEMISIA OILS [8008-93-3]

Source
GEOGRAPHIC   North America, Morocco, Algeria, Yugoslavia, and France.
NATURAL   The most important sources are: *Artemisia annua*, yellow, with an odour like basil; *A. herba alba*, fragrant, greenish yellow; *A. caudata*, anise-like odour; *A. vulgaris*, bright green with a cineol-like odour; *A. glutinosa*, a very fragrant odour reminiscent of a mixture of borneol, spike and rosemary oils.
Yield   0.3%.
Odour   Herbaceous and spicy.
Uses   In balsamic complexes and nature-based perfumes.
RIFM Monographs (1975) *FCT*, **13**, 712 (Wormwood and Absinthium oil)
(1974) *FCT*, **12**, 709 (Dracunculus L.)
(1976) *FCT*, **14**, 737 (Pallens oil)
(1975) *FCT*, **13**, 719 (Vulgaris oil).

ASAFETIDA [9000-04-8]

CTFA name   Asafoetida extract
Asafoetida resinoid; Devil's dung; food of the Gods; asant
Source
GEOGRAPHIC   Iran and Afghanistan.
NATURAL   *Ferula foetida*, Regel, N. O. Umbelliferae.
CHEMICAL   Vanillin is a constituent.
Isolation   The dried, resinous gum is extracted by an incision in the root. The impurities are removed by macertion with alcohol and filtration.
Yield   Depends upon raw material and extraction procedure.
Appearance   Milky exudate, which dries to a resinous gum. After filtration the liquid contains resin and essential oil.
Odour   Intense and garlic-like.
Chemical   Contains three sulphur-containing compounds.
Uses   Limited due to its intensity: valued, in traces, as a component of certain types of oriental perfumes and is also used as a toner.

ASPARAGUS OIL [84775-79-1]

Source
NATURAL   The small pale rose flowerets of *Asparagus sprengeri*, Regel, N. O. Liliaceae, known as Smilax.
Odour   Intense, and narcotic, resembling the fatty aldehydes.

ATLAS CEDARWOOD OIL [8000-27-9]

History   This balsamic liquid has been used by natives for many years for medicinal purposes.

Source	
GEOGRAPHIC	Northern Africa.
NATURAL	*Cedrus atlantica*, Manetti, N. O. Pinaceae.
CHEMICAL	Odour due to a mixture of isomeric ketones called atlantone: also contains considerable quantities of casinene.
Isolation	Steam distillation of waste wood, such as chippings, sawdust etc.
Yield	3–5%.
Appearance	Viscous yellow balsamic liquid.
Odour	Mimosa-type.
Uses	A double rectified oil is used in soap perfumery. The terpeneless oil and the resinoid extracted by volatile solvents are now commercial products in perfumes and soaps: blends well with a variety of bouquets.

AURANTIENE

Source	
CHEMICAL	The residual terpenes in the refining of orange oil.
Uses	Soap perfume.

AURANTIOL [89–43–0]

Aurcol; Aurantine, Anturalal, Auranol

Source	
CHEMICAL	Condensation product of methyl anthranilate with hydroxy-citronellal (see Schiff's Base).
Appearance	Highly viscous greenish-yellow liquid.
Odour	Intense sweet heavy floral of orange blossom.
Uses	Widely in delicate floral compositions such as mugnet, honeysuckle, jasmine and orange blossom types. RIFM Monograph (1975) *FCT*, **13**, 549.

AVOCADO OIL [8024–32–6]

CTFA name	Avocado oil Alligator pear oil
History	Mexican and South American women are known to have used it for about four centuries as protection for the epidermis against dry winds.
Source	
GEOGRAPHIC	Tropical America.
NATURAL	*Persea gratissima*, trees of N. O. Lauraceae (avocado pear).
Isolation	Prepared from damaged fruits: dehydration at 130 °C in an atmosphere of either nitrogen or carbon dioxide is followed by hydraulic expression of the crude runnings: the clear oil is refined after the runnings have been settled in tanks.
Appearance	Deep green liquid by transmitted light and deep red by reflected light thus having a fluorescent appearance.
Physical	Not chilled for cosmetic purposes as many of the valuable

	constituents would be precipitated: has a tendency to reduce the surface tension of liquids thus forming finer emulsions.
Chemical	Can be bleached; contains 77% oleic acid and 11% linoleic acid, and is rich in lecithin and phytosterine, with a high oil-soluble vitamin content.
Other	One of the most penetrating oils.
Uses	Valuable in tissue and massage creams and has exhibited some skin healing properties; can be adapted in any cosmetic formula in which oil is an ingredient.

AZALEA

Source	
NATURAL	N. O. Ericaceae
GEOGRAPHIC	A genus of shrubs native to North America and the East.
Appearance	Five-stamened flowers in various colours, resembling rhododendrons.

AZULENE [275–51–4]

CTFA name	Azulene
Source	
NATURAL	Colouring matter occurring in the volatile oils of camomile, milfoil, and absinthe.
Appearance	Intensely blue.
Other	Considered to have a healing and anti-inflammatory effect.
Uses	In skin creams, face pads, and cosmetic lotions.

BACKHOUSIA OIL [84775–80–4]

History	Had been a raw material used in the manufacture of ionine but the process is no longer profitable as a result of competition with lemongrass and *Litsea cubeba*.
Source	
GEOGRAPHIC	The tree grows in the coast area between Brisbane and Gympie.
NATURAL	*Backhousia citriodora*, F. v. Muell, N. O. Myraceae, Queensland, Australia.
Isolation	Distillation from the leaves.
Odour	Similar to lemongrass.
Uses	Soap perfumery instead of lemon oil to produce a new sweet note.

BALSAMO BLANCO [8007–00–9]

	White Peru balsam, bauma blanc de son sonate.
Source	
NATURAL	Fruit-pods of *Myroxylon pereira*, Klotzsch, N. O. Leguminosae.
Isolation	Obtained by pressure.
Appearance	Resembles turbid honey with a yellowish-white colour.

Odour	Recalls coumarin or melilot; alters slightly on exposure, to approach that of cedarwood.
Chemical	Constituents include cinnamic acid, cinnamic alcohol and phenyl propyl alcohol.
Comments	This balsam is entirely different from that obtained from the trunk of the tree.
Uses	No longer used as a fragrance ingredient due to its skin sensitizing properties and has now been replaced by synthetic alternatives.

Peru Balsam: RIFM Monograph (1974) *FCT*, **12**, 951, Special Issue I (Binder, p. 618).

IFRA GUIDELINE

The Committee recommends that Peru balsam (the exudation from *Myroxylon pereirae* (Royle) Klotzsch) should not be used as a fragrance ingredient. Only preparations produced by methods which give products not showing a potential for sensitization (e.g. distillation) should be used.

This recommendation is based on test results of RIFM on the sensitizing potential of Peru balsam (D. L. Opdyke (1974), *Food Cosmet. Toxicol.*, **12**, 951) and absence of sensitizing reactions in tests with several samples of Peru balsam oil (D. L. Opdyke (1974), *Food Cosmet. Toxicol.*, **12**, 953), as well as with compounds containing Peru balsam oil (private communication to IFRA).

October 1974, amended May 1976

BARBERRY [8054-40-8]

CTFA name	Berberis extract
	Berberis vulgaris, L., also known as jaundice berry, woodsour, sowberry, pepperidge bush, sourspine.
Source	
GEOGRAPHIC	Europe and North America.
NATURAL	A spring shrub of N. O. Berberidaceae. It attains a height of a few feet; its red fruit appears in autumn following yellow flowers which bloom earlier in the year. One of the best species is *B. bealii*; its nine-petalled flowers are yellowish green and open to about 0.5 in. (13 mm) in diameter.
Isolation	Extracted from the fresh root bark.
Odour	The yellow flowers have an odour suggestive of citronellyl acetate: those of *B. Bealii* recall the odour of lily of the valley. However, after standing in water for 24 h the latter changes to that of lily-rose with suggestions of anisic aldehyde and indole.

BARIUM SULPHATE [7727-43-7]

CTFA name	Barium sulfate
	Blanc fixe; CI 77120
	$BaSO_4$

Source	
NATURAL	The mineral barytes.
CHEMICAL	Can be obtained from a soluble barium salt and a soluble sulphate.
Isolation	Obtained by precipitation.
Appearance	Soft, fine, white powder when dried.
Odour	Odourless, tasteless.
Physical	Practically insoluble in water.
Uses	Manufacture of grease paints, face powders, and compacts.
Comments	Barytes is not suitable for use in cosmetics.

BASIL OIL [8015-73-4]

CTFA name	Basil extract
	Ocimum basilicum extract; extract of basil
Source	
GEOGRAPHIC	Yugoslavia, France, Italy, USA, Eastern Europe, Madagascar, Réunion and the Seychelles.
NATURAL	*Ocimum basilicum*, L., N.O. Labiateae.
Isolation	Steam distilled from the dried herb.
Yield	0.1–0.2%.
Appearance	Limpid yellow liquid.
Odour	Fresh, somewhat herbaceous, penetrating fragrance.
Chemical	Principal constituents are methyl chavicol and linalol.
Uses	Commonly used in balsamic complexes and as a trace constituent in nature-based perfumes.
	RIFM Monograph (1973) *FCT*, **11**, 867.

BAY OIL [8006-78-8]

CTFA name	Bay oil
	Myrica oil; Oil of Bay
Source	
GEOGRAPHIC	West Indies and South and Central America.
NATURAL	*Myrcia acris*, D.C., N.O. Myrtacae and *Pimenta racemosa*.
Isolation	Bay leaves are collected when the shrub is about 5 years old; after adding common salt or sea-water, steam distillation results in light and heavy oil fractions, which are mixed.
Yield	A 10-year-old tree provides 60 to 100 lb (27 to 45 kg) of leaves per year; distillation yields 2%. Although sea-water increases the rate of distillation, fresh-water gives higher yields.
Odour	Powerful, spicy, and sweet.
Uses	Imparts freshness to soaps; as a trace in the preparation of carnation oils.
	RIFM Monograph (1973) *FCT*, **11**, 869 (*Pimenta racemosa*) and (1976) *FCT*, **14**, 337 (sweet: *Laurus nobilis* L.).

BAY PINE (OYSTER) OIL

History	The tree which yields this oil has been cultivated in the Nilgiris district of India for 100 years.
Source	
GEOGRAPHIC	India.
NATURAL	*Callitris rhomboidea*, N. O. Coniferae.
Isolation	Distilled from the leaves.
Chemical	Contains about 30% esters when calculated as geranyl acetate.
Uses	Particularly suitable for perfuming soaps.

BDELLIUM

History	Bdellium was mentioned 6000 years ago (*Genesis* ii 12) and 2000 years later (*Numbers* xi 7).
Source	
GEOGRAPHIC	India, Africa.
NATURAL	The name given to several gum resins exuded by species of Buseraceae. The Museum of the Pharmaceutical Society of Great Britain has four specimens: Indian Bdellium (a product of *Balsamodendron mukul*, Hooker) which occurs in irregular dark reddish-brown masses covered with characteristic 'pinheads' of lustrous resin; its odour is reminiscent of cedarwood: Opaque Bdellium which occurs in hard spherical yellow ochre-coloured lumps; this variety is attributed to *B. playfairii*, Hooker: African Bdellium which occurs in hard rough lumps red to transmitted light; it has an unpleasant odour and smooth-fractured surfaces have a slate-like appearance: Gafal (attributed to *Boswellia papyfera* or *Commiphora africana*) which comprises miscellaneous yellow and black translucent masses with a slightly aromatic odour.
Appearance	Irregular masses of hardened gum resins.
Physical	Resembles myrrh.
Comments	Should not be confused with bissabol or perfumed bdellium which has a quite different odour.

BEESWAX   [8006–40–4 (white); 8012–89–3 (yellow)]

CTFA name	Beeswax
Source	
GEOGRAPHIC	Spain, Morocco, Jamaica and California.
NATURAL	Secreted from the underside of the honey bee (*Apis mellifica*) and used to form the walls of honeycomb cells.
Isolation	Extracted by treating the honeycombs with hot water and straining off the separated wax. Prolonged exposure of the raw material to air and sunlight or to the bleaching action of chromic and other acids results in white beeswax.

Yield	0.1–0.2%.
Odour	Mild, oily, and honey-like: slightly balsamic taste.
Physical	Yellow waxy appearance; soft to brittle: practically insoluble in water.
Chemical	Complex mixture of about 80% ceryl myristate (myricin), esters from alcohols and acids of similar molecular size, about 13% free acids (mainly cerotic acid), and a small percentage of hydrocarbons: reacts with alkalis to form emulsion complexes.
Uses	In floral fine fragrances, water-in-oil emulsions and cosmetic creams.

RIFM Monograph (1976) *FCT*, **14**, 691.

BENTONITE [1302-78-9]

CTFA name	Bentonite Hydrated aluminium silicate; wilkinite; soap clay $Al_2O_3 4(SiO_2).H_2O$
History	Name derived from the area where it was first discovered – near Fort Benton, Wyoming, USA.
Source	
GEOGRAPHIC	Principally found in Otay, but deposits have also been discovered in Dakota, New Jersey, Utah, Nevada, Montana, Mississippi, and in Alberta, Canada.
NATURAL	A montmorillonite type of clay consisting mainly of silicon dioxide and alumina together with oxides of iron, lime, magnesium soda, and potash.
Appearance	A clay which occurs in white, rose, green and brown forms; suspensions are dark coloured.
Physical	Sodium bentonite can absorb up to 30 times its own weight of water to form a gelatinous mass. The property is much increased by the addition of small quantities of alkaline substances such as magnesium oxide.
Other	Two main types exist: sodium bentonite and calcium bentonite.
Uses	In face masks and, because of its remarkable properties as a suspending agent, in liquid make-up preparations where the pigment content is high.

BENZALDEHYDE [100-52-7]

CTFA name	Benzaldehyde Benzoic aldehyde; artificial bitter almond oil C_6H_5CHO
Source	
NATURAL	Found in the oils of macassar seeds, in cinnamon root and bark, champaca, patchouli and cassia and in the concretes prepared from cassie and orange blossom: may also occur in elderflowers.
CHEMICAL	Can be prepared synthetically (e.g. from benzal chloride and lime or by oxidation of toluene).

Appearance	Colourless liquid.
Odour	Powerful sweet character reminiscent of bitter almonds.
Physical	Strongly refractive and becomes yellowish on keeping; soluble to a small extent in water, miscible with alcohol, ether and oils; b.p. 179 °C.
Chemical	Oxidizes in air to benzoic acid.
Other	The artificial product is prone to contain some chlorine.
Uses	Traces in violet- and heliotrope-type perfume; and as a soap perfume. Benzaldehyde, free from chlorine (designated benzaldehyde FFC), is sold as artificial bitter almond oil.

RIFM Monograph (1976) *FCT*, **14**, 693.

BENZALDEHYDE DIMETHYLACETAL [1125-88-8]

$C_6H_5COOCH\ C_6H_5CH(OMe)_2$

Odour	Sweet-green and warm note. Softer than benzaldehyde.
Other	Similar properties to benzaldehyde.
Physical	b.p. 198 °C.
Uses	In violet compounds.

RIFM Monograph (1979) *FCT*, **17**, 711.

BENZOIC ACID [65-85-0]

Benzenecarboxylic acid; phenylformic acid; dracylic acid
C_6H_5COOH

Source	
NATURAL	In numerous essential oils and gum resins, e.g. benzoin, Peru balsam, neroli and champaca oils: occurs in nature in free and combined forms.
Isolation	Manufacturing processes include air oxidation of toluene, the hydrolysis of benzotrichloride and the decarboxylation of phthalic anhydride.
Appearance	Monoclinic tablets, plates and leaflets.
Physical	Sublimes at about 100 °C: mixtures of excess benzoic acid and water form two liquid phases and water solubility is increased by alkaline substances such as borax or trisodium phosphate.
Uses	A preservative active against fungi but pH-dependent because of dissociation.

RIFM Monograph (1979) *FCT*, **17**, 715.

BENZOIN (EXTRACT) [9000-05-9]

CTFA name	Benzoin (and Benzoin extract) Benzoin Siam resinoid; Benzoin Sumatra resinoid
History	The botanical source of Sumatra benzoin was determined in 1787 by Jonas Dryander.

Source	
GEOGRAPHIC	Thailand, Laos, Cambodia and Sumatra.
NATURAL	Resinous exudation from *Styrax benzoin*, Dryander, and other species of N. O. Styraceae; the tree (which resembles birch) from which Siam benzoin is obtained is thought to be *Styrax tonkinensis*; it is common in almost every altitude of northern Indo-China and grows rapidly, reaching maturity in about 10 years. Sumatra benzoin is a pathogenic product, only exuded when the plant has been wounded. This tree, which flourishes in the north-eastern part of Sumatra, grows rapidly and reaches 20 ft (6 m) in a few years.
Isolation	Solvent extraction of the resin.
Yield	85–95% depending on source of the plant material.
Appearance	Thailand (Siam) benzoin occurs in tears, which are yellowish brown to white and brittle. Sumatra benzoin is imported in rectangular blocks, which display numerous yellowish-white 'almonds' when broken. Sumatra benzoin is a milky resinous sap, which hardens in the air and is then scraped off the bark.
Odour	Siam benzoin has a sweet, balsamic, chocolate-like odour. Sumatra benzoin has a warm, sweet, powdery odour.
Physical	Siam benzoin assumes a reddish-brown translucent appearance when exposed to air.
Chemical	Thailand (Siam) benzoin contains up to 40% benzoic acid and esters of benzoic acid, with up to 5% vanillin.
Uses	Benzoin is the balsamic resin; benzoin extract is an extract from this resin. Siam benzoin is used to give body to almost any perfume and as a fixative. Sumatra benzoin is used in cheaper perfumery and toilet preparations: also widely used in medicine in skin lotions and tinctures with or without tragacanth.
Comments	Siam benzoin is almost completely soluble in its own weight of warm 90% alcohol. If the solution is filtered bright and the alcohol recovered at low temperature *in vacuo*, an article is produced similar to those sold under such names as benzoin resinodor, 'clairs' etc. It is the only kind used in fine perfumery. Penang and Palembang benzoins are the poorest quality, and occur in red–brown porous masses. Note: not to be confused with benzoin (2-hydroxy-1,2-diphenylethanone). RIFM Monograph (1973) *FCT*, **11**, 871 (resinoid).

BENZOPHENONE [119-61-9]

	Diphenylmethanone; benzoylbenzene; diphenyl ketone $C_{13}H_{10}O$
Isolation	Distillation from benzoyl chloride and benzene.
Appearance	White crystalline solid.
Odour	Sweet, powdery floral (geranium-like).
Physical	b.p. 305.9 °C.

Uses   Fixative for heavy perfumes. Extensively in sweet woody geranium notes for low cost soap fragrances and household products.
RIFM Monograph (1973) *FCT*, **11**, 873.

BENZOYL ACETONE [93-91-4]

$C_6H_5COCH_2COCH_3$

Appearance	White crystals.
Odour	Balsamic vanilla-opoponax odour.
Physical	m.p. 61 °C, b.p. 261 °C.
Uses	In oriental, amber, chypre and heather compounds.

BENZYL ACETATE [140-11-4]

CTFA name	Benzyl acetate
	Acetic acid phenylmethyl ester; acetic acid benzyl ester
	$C_9H_{10}O_2$
Source	
NATURAL	In oils of jasmine, gardenia, hyacinth, and ylang-ylang.
Isolation	Synthesized by acetylating benzyl alcohol.
Appearance	Colourless liquid.
Odour	Fresh light-fruity jasmine-like.
Physical	Practically insoluble in water, miscible with alcohol or ether; b.p. 215 °C.
Uses	One of the most useful synthetics: widely employed in soap perfumery: used as the basis of all jasmine perfumes, being modified with methyl anthranilate and linalol.
Comments	Cheap to manufacture.

RIFM Monograph (1973) *FCT*, **11**, 875.

BENZYL ACETONE [2550-26-7]

4-Phenylbutan-2-one
$C_6H_5CH_2CH_2COCH_3$

Appearance	Oil.
Odour	Floral green.
Physical	b.p. 235 °C.
Uses	As a modifier for benzyl acetate in soap fragrances.

RIFM Monograph (1983) *FCT*, **21**, 647.

BENZYL ALCOHOL [100-51-6]

CTFA name	Benzyl alcohol
	Benzenemethanol; phenylcarbinol; phenylmethanol; alpha-hydroxytoluene
	C_7H_8O

Source	
NATURAL	Occurs in the oils of tuberose, ylang-ylang, champaca, wallflower, jasmine, clove, hyacinth, robinia pseudo-acacia, and cassie flowers.
Isolation	Synthesized by heating benzyl chloride over a long period with a solution of potassium carbonate.
Appearance	Colourless liquid.
Odour	Faintly aromatic, becoming 'almondy' on exposure (by auto-oxidation), due to the formation of benzaldehyde and dibenzyl oxide: sharp, burning taste.
Physical	b.p. 205 °C. 4% soluble in water.
Uses	As a blender and fixative for some florals.

RIFM Monograph (1973) FCT, **11**, 1011.

BENZYL BENZOATE [120-51-4]

CTFA name	Benzyl benzoate
	Benzoic acid phenylmethyl ester; benzoic acid benzyl ester; benzylbenzenecarboxylate; phenylmethylbenzoate; benylate
	$C_{14}H_{12}O_2$
Source	
NATURAL	In the balsams of Tolu and Peru and in the oils of tuberose, ylang-ylang, and hyacinth.
Isolation	Synthesized from benzyl chloride and sodium benzoate.
Appearance	Colourless oily liquid solidifying in the cold.
Odour	Faint sweet-balsamic character.
Physical	b.p. 324 °C: insoluble in water, miscible with alcohol, chloroform, ether and oils.
Uses	As a fixative for many floral and oriental types and as a solvent.

RIFM Monograph (1973) FCT, **11**, 1015.

BENZYL n-BUTYL ETHER [588-67-0]

	Rose-oxide; iso-butyl benzyl ether
	$C_6H_5CH_2O(CH_2)_3CH_3$
Appearance	Clear, colourless liquid.
Odour	Sweet, fresh floral and somewhat chemical.
Physical	b.p. 212 °C.
Chemical	Stable to acids and alkali.
Uses	In small proportions in rose compounds.

BENZYL n-BUTYRATE [103-37-7]

	Aldehyde C19 (so called)
	$CH_3CH_2CH_2COOCH_2C_6H_5$
Appearance	Colourless liquid.
Odour	Characteristic heavy, fruity.

Benzyl chloroacetate

Physical	b.p. 240 °C.
Uses	In modifying jasmine perfumes.

RIFM Monograph (1974) *FCT*, **12**, 827.

BENZYL CHLOROACETATE [140-18-1]

Benzyl monochloracetate
$ClCH_2COOCH_2C_6H_5$

Appearance	Colourless liquid.
Odour	Mild jasmine-type.

BENZYL CINNAMATE [103-26-4]

3-Phenyl-2-propenoic acid methyl ester; *trans*-cinnamic acid benzyl ester; cinnamein; benzyl 3-phenylacrylate
$C_6H_5CH\!=\!CHCOOCH_2C_6H_5$

Source	
NATURAL	In the balsams of Tolu and Peru.
CHEMICAL	Synthesized by refluxing a mixture of sodium cinnamate, benzyl chloride and alcohol.
Appearance	Crystalline solid.
Odour	Mild, sweet, balsamic.
Physical	m.p. 35 °C, insoluble in water, soluble in alcohol, ether and oils.
Uses	Mainly as a fixative: especially useful in preparing oriental-type perfumes: also employed in compounding ambre colognes and lavender waters.

RIFM Monograph (1973) *FCT*, **11**, 1017.

BENZYL CYANIDE [140-29-4]

Phenylacetonitrile; benzeneacetonitrile; alpha-tolunitrile
$C_6H_5CH_2CN$

Isolation	Product of a reaction between benzyl chloride and sodium cyanide and formed during the synthesis of phenylacetic acid.
Appearance	Colourless oily liquid.
Odour	Herbaceous green with sweet floral undertones.
Physical	b.p. 234 °C.

RIFM Monograph (1982) *FCT*, **20**, 803.

BENZYL ETHYL CARBINOL [701-70-2]

1-Phenylbutan-2-ol; Ethyl benzyl carbinol
$C_6H_5CH_2CH(OH)C_2H_5$

Appearance	Oily, colourless liquid.
Odour	Floral, earthy, resembling hyacinth.
Physical	b.p. 220 °C.
Uses	As a modifier in floral and herbaceous fragrance types.

BENZYL ETHYL ETHER [539–30–0]

Benzyl ethyl oxide; ethoxymethylbenzene; Ethyl benzyl ether
$C_6H_5CH_2OC_2H_5$

Isolation	Prepared from sodium ethoxide and benzyl bromide.
Appearance	Colourless oily liquid.
Odour	Powerful, fruity fragrance.
Physical	b.p. 187 °C, practically insoluble in water, miscible with alcohol or ether.
Uses	In soap perfumery.

BENZYL FORMATE [104–57–4]

Formic acid phenylmethyl ester; formic acid benzyl ester
$HCOOCH_2C_6H_5$

Appearance	Colourless liquid.
Odour	Powerful, earthy, green-herbaceous character.
Physical	m.p. 4 °C, b.p. 202 °C, practically insoluble in water, soluble in alcohol.
Uses	In jasmine and tuberose compounds.
Comments	Oils containing this substance require long maturing.

RIFM Monograph (1973) *FCT*, **11**, 1019.

BENZYL HEPTINE CARBONATE

Benzyl oct-2-ynoate
$CH_3(CH_2)_4C\equiv CCO_2CH_2C_6H_5$

History	Similar properties to methyl heptine carbonate.
Physical	b.p. 255 °C.
Comment	This material is no longer used as a fragrance ingredient.

IFRA GUIDELINE
The Committee recommends that the following materials should not be used as fragrance ingredients:

—Allylisothiocyanate
—Chenopodium oil
—3,7-Dimethyl-2-octen-1-ol (6,7-Dihydrogeraniol)
—Furfurylideneacetone
—Methyl methacrylate
—Phenylacetone (Methyl benzyl ketone)
—Esters of 2-octynoic acid, except those covered elsewhere in these Guidelines i.e. methyl and allyl heptine carbonate
—Esters of 2-nonynoic acid, except methyl octine carbonate
—Thea sinensis absolute

These recommendations are based on the absence of reports on the use of these materials as fragrance ingredients and inadequate evaluation of possible physiological effects resulting from their use in fragrances.

March 1988, last amendment July 1990

BENZYL METHYL CARBINOL [698-87-3]

1-Phenylpropan-2-ol; Methyl benzyl carbinol
$C_6H_5CH_2CH(OH)CH_3$

Appearance	Colourless liquid.
Odour	Sweet floral-green odour.
Physical	b.p. 220 °C.
Chemical	Isomeric with phenyl propyl alcohol.
Uses	A modifier.

BENZYL PHENYLACETATE [102-16-9]

$C_6H_5CH_2COOCH_2C_6H_5$

Appearance	Colourless, slightly viscous liquid.
Odour	Mild, sweet, honey-like.
Physical	b.p. 318 °C.
Uses	A fixative for jasmine, rose and eglantine compounds. RIFM Monograph (1973) FCT, 11, 1027.

BENZYL PROPIONATE [122-63-4]

$C_2H_5COOCH_2C_6H_5$

Appearance	Colourless liquid.
Odour	Fruity-sweet, floral, jasmine type.
Other	When mixed with the acetic acid ester in a proportion of about 20% produces an odour recalling rum.
Physical	b.p. 222 °C.
Uses	Gives a freshness to jasmine fragrances. RIFM Monograph (1975) FCT, 13, 723.

BENZYL SALICYLATE [118-58-1]

CTFA name	Benzyl salicylate
	Benzyl o-hydroxybenzoate; 2-hydroxybenzoic acid phenylmethyl ester; salicylic acid benzyl ester
	$C_{14}H_{12}O_3$
Isolation	Synthesized from methyl salicylate and benzyl alcohol by transesterification, or from sodium salicylate and benzyl chloride.
Appearance	Viscous colourless liquid.
Odour	Faint floral, balsamic.
Physical	b.p. 300 °C, slightly soluble in water, miscible with alcohol or ether.
Uses	Widely used in carnation, jasmine, lily, lilac and bouquet compounds; also used as a fixative and solvent and can be employed as an oil-soluble sun-screening agent in anti-sunburn oils and lotions.

Comments	If applied to sensitive areas of the skin, it may cause slight smarting. RIFM Monograph (1973) *FCT*, **11**, 1029.

BENZYL *iso*-VALERATE [103–38–8]

Benzyl 3-methylbutanoate; benzyl valerianate
$(CH_3)_2CHCH_2COOCH_2C_6H_5$

Appearance	Colourless liquid.
Odour	Powerful fruity-herbaceous.
Physical	b.p. 246 °C.
Uses	A modifier in the preparation of artificial rose and opoponax compounds.

BENZYLIDENE ACETONE [122–57–6]

4-Phenylbut-3-en-2-one
$C_6H_5CH=CHCOCH_3$

Isolation	Obtained when benzaldehyde and acetone are condensed with the aid of dilute caustic soda.
Appearance	Crystalline solid.
Odour	Recalls sweet-pea.
Physical	b.p. 262 °C.
Uses	Once used, in traces, in eau-de-cologne, jonquille, violet, fern, musk and lavender.
Comments	Now no longer used as a fragrance ingredient due to its skin sensitizing properties. RIFM Monograph (1973) *FCT*, **11**, 1021 (Binder, p. 135). IFRA GUIDELINE The Committee recommends that benzylidene acetone should not be used as a fragrance ingredient. This recommendation is based on the findings of RIFM on the sensitizing potential of this material (*Food Cosmet. Toxicol.*, **11**, 1021 (1973)) and on information on sensitizing effects in tests on guinea pigs and humans. (Private communication to IFRA). <div style="text-align:right">June 1974</div>

BERGAMOT MINT OIL [68917–15–7]

Source	Mentha citrata oil
GEOGRAPHIC	Florida at Arlington Farm, near Washington (D.C.), far Western USA.
NATURAL	Obtained from the fresh plants of *Mentha citrata*, which is a hybrid between *Mentha aquatica* and *Mentha viridis*, Ehrh, N.O. Labiate.
Isolation	Steam distillation of the fully grown plant.
Yield	1%.

Bergamot oil

Appearance	Yellowish oil.
Odour	Bright and refreshing, resembling bergamot and rosewood oil, reminiscent of lavender oil.
Chemical	Contains about 50% of linalyl acetate.
Uses	In perfumery to achieve freshness.

BERGAMOT OIL [8007-75-8]

Source		
	GEOGRAPHIC	Southern part of the province of Reggio-Calabria (in Italy), South America and Western Africa.
	NATURAL	*Citrus bergamia.*
Isolation		Obtained by machine cold expression from the fresh peel of the nearly ripe fruit.
Yield		0.5%.
Appearance		Yellow to brownish-green liquid: if bright green, the oil is either old and contains copper, or has been artificially tinted.
Odour		Fresh, fruity, citrus.
Other		Open to skilful adulteration, principally by use of terpinyl acetate, glyceryl acetate, ethyl citrate and ethyl laurate.
Uses		Used extensively in eau-de-cologne and lavender water; blends well with almost any synthetic compound. Low or furocoumarine free oils are generally now used due to the phototoxicity of normally extracted bergamot oil.
Comments		The leaves of the bergamot tree are sometimes distilled yielding oil containing linalyl acetate and some methyl anthranilate.

RIFM Monographs (1973) *FCT*, **11**, 1031 (Expressed)
(1973) *FCT*, **11**, 1035 (Rectified).

IFRA GUIDELINE

The Committee recommends for applications on areas of skin exposed to sunshine, excluding bath preparations, soaps and other products which are washed off the skin, not exceeding a level of 5-methoxypsoralen of 75 ppm in the compound. Assuming a 5-methoxypsoralen content of 0.35%, this level of bergamot oil expressed in the compound would approximate to 2.0% (see remark on phototoxic ingredients in the introduction).

This recommendation is based on the published literature on the phototoxicity of this material, summarized by D. L. Opdyke, *Food Cosmet. Toxicol.*, **11**, 1031-3 (1973), and on more recent investigations published in *Contact Dermatitis* (1977) **3**, 225-39.

October 1974, amended October 1978
RIFM Monograph Bergamot oil expressed (1973) *FCT*, **11**, 1031 (Binder, p. 143).

BERGAPTEN(E) [484-20-8]

5-Methoxypsoralen; bergaptan; heraclin; majudin; 5-MOP
$C_{12}H_{18}O_4$

Source	
NATURAL	Bergamot oil.
Isolation	Isolated from the distillation residues of bergamot oil.
Appearance	Crystalline material.
Physical	Melts and sublimes at 188 °C: soluble in absolute alcohol.
Chemical	A naturally occurring analogue of psoralen.
Comments	Has been used to promote tanning in suntan preparations but is photo-mutagenic. Its use is forbidden in the EC except as a component (at normal levels) of essential oils.

BETULA LENTA

Sweet birch oil

Source	
GEOGRAPHIC	Tennessee and Virginia, USA.
NATURAL	Obtained from the bark of *Betula lenta*, L., N.O. Betulaceae, known locally as 'black birch': obtained from the trees when they are in sap and 1–2 ft (30–60 cm) in diameter.
Isolation	Distillation of the wood.
Appearance	Liquid.
Physical	Almost identical to the oil from *Gaultheria procumbens*, L., N.O. Ericaceae.
Chemical	Consists almost entirely of methyl salicylate.
Other	By adding a few crystals of tartaric or citric acid, shaking and then filtering, the oil can be decolourized.
Uses	Compounding new-mown hay perfumes: may be used in the preparation of synthetic cassie oil.

RIFM Monograph (1979) *FCT*, **17**, 907.

BIGNONIA

Source	
GEOGRAPHIC	Humid areas of Central America and India.
NATURAL	Bignonia is an indigenous genus of climbing plant. The principal species are *B. suaveolens* and *B. suberosa*. The flowers are trumpet-shaped and have an exquisite fragrance.

BIRCH BUD OIL [84012–15–7]

Source	
NATURAL	*Betula alba*, L. (European white birch), N.O. Betulaceae.
Isolation	Distillation from the resinous leaf buds.
Yield	approx. 5%.
Appearance	Viscid yellow liquid.
Odour	Balsamic.
Chemical	Contains a sesquiterpene alcohol (betulenol) and some paraffin, which often crystallizes out at normal temperatures.
Uses	The manufacture of hair washes.

BIRCH TAR OIL [84012-15-7]

'Essential oil of birch wood', Rectified Birch Tar.

Source	
GEOGRAPHIC	Russia.
NATURAL	*Betulae alba.*
Isolation	Destructive distillation of the wood (which produces the empyreumatic oil) followed by rectification.
Yield	Varies depending upon quality of the wood and production procedure.
Appearance	Dark brown tarry oil.
Odour	Woody and tarry.
Physical	Soluble in benzene, chloroform, ether, glacial acetic acid, carbon disulphide and oil of turpentine.
Uses	In traces in masculine fragrances and as a perfume for leather. RIFM Monograph (1973) *FCT*, **11**, 1037.

BISABOL [100084-96-6]

Perfumed bdellium, sweet myrrh, opoponax, habbak hadi

History	Collected in the same regions as herrabol or true myrrh, but packed separately in goat-skin bags.
Source	
GEOGRAPHIC	From the Ogaden region.
NATURAL	A resinous gum exudation from *Commiphora erythraea*, var. *glabrescens*, Engl., N.O. Burseraceae.
Appearance	Occurs in yellowish-brown lumps, with a waxy fracture, striated and interspersed with white markings.
Odour	Reminiscent of olibanum.
Uses	Used in religious ceremonies in China.
Comments	Differs from the somewhat foetid medicinal gum-resin previously obtained from *Balsamodendron kataf*, Kunth, N.O. Umbelliferae.

BISMUTH OXYCHLORIDE [7787-59-9]

CTFA name	Bismuth oxychloride
	Bismuth chloride oxide; bismuth subchloride; basic bismuth chloride; bismuthyl chloride; synthetic pearl essences; pearl white; 'blanc d'Espagne; 'blanc de perle'; CI 77163; Pigment White 14 BiOCl
Appearance	A pure crystalline or powder solid.
Physical	Has greater opacity and better slip properties than natural pearl essences: special pearling agents can be produced with various crystal sizes and shapes to simulate the brilliance and pearlescent effect obtained with natural pearl essences based on guanine.

Chemical	Darkens on exposure to light and an ultraviolet light absorber improves light stability: can be obtained as 100% dry powder or as a dispersion in water, alcohol, castor oil or mineral oil.
Uses	Often used in combination with other pearling agents to give special effects in products such as lipsticks, eye cosmetics, and pressed powders.

BITHIONOL [97-18-7]

2,2'-Thiobis(4,6-dichlorophenol); TBP
$C_{12}H_6O_2SCl_4$

Isolation	Prepared from 2,4-dihalo substituted phenol with S halides.
Appearance	White or greyish-white crystalline powder.
Odour	Faint phenolic-type.
Physical	Insoluble in water, soluble in alcohol, acetone and ether.
Uses	Soaps and deodorant preparations as an effective bactericide.
Comments	Banned by the FDA from use in cosmetics.

BITTER ORANGE OIL See ORANGE OIL, BITTER

BIXINE [1393-63-1]

Annatto

Source	
GEOGRAPHIC	Central America and the East and West Indies.
NATURAL	Red colouring matter obtained from the pulp of the fruit of *Bixa orellana*, L., N.O. Bixaceae.
Appearance	Red liquid.

BLACK-CURRANT CONCRETE [97676-19-2]

Source	
NATURAL	Obtained from the buds of *Ribes nigrum*, L., a common bush belonging to N.O. Glossularieae.
Isolation	Steam cohobation.
Yield	2.4–3%.
Appearance	Dark-green semi-crystalline essence.
Odour	Strong and characteristic.
Chemical	Constituents include terpenes and sesquiterpenes such as nopinene, *p*-sabinene, *d*-caryophyllene and *d*-cadinene.

BLES DES PAGODES

Source	
NATURAL	An oil from an unknown botanical species grown in Annam.
Odour	Recalls palmarosa and ginger-grass with a spice of cumic aldehyde.

BOG MYRTLE [90064-18-8]

Candleberry myrtle

Source	
GEOGRAPHIC	Europe and Northern America.
NATURAL	*Myrica gale*, L., N.O. Myricaceae; bears berry-like fruit covered with a greenish-yellow waxy secretion.
Isolation	When the berries are placed in hot water, the wax floats to the surface.
Appearance	Yellow solid.
Odour	A fragrant odour when burned.
Uses	In the manufacture of certain candles.

BOIS D'OLHIO

Source	
GEOGRAPHIC	The tree is a native of Japan, but is also found in Brazil.
NATURAL	*Olhio vehermilho* (Ole Vermelho), *Myrospermum erythroxylon*, Fr. Allem.
Isolation	The oil is distilled from the wood of the tree using volatile solvents.
Yield	1%.
Appearance	Yellowish mobile oil.
Odour	A mixture of santal, cedar and rose.
Uses	A fixative for soap perfumery.

BOIS DE ROSE FEMELLE [8015-77-8]

Bois de Rose, Rosewood oil
Cayenne linaloe oil

History	Bois de rose oil is an important export from the northern states of Brazil.
Source	
GEOGRAPHIC	French Guiana and northern parts of the states of Para and Amazonas.
NATURAL	Burseraceae or Lauraceae, of which there are several varieties. Occurs most commonly in *Aniba rosaeodora*.
Isolation	Steam distilled from chipped wood, previously softened in hot water.
Yield	0.8–1.6%.
Appearance	Yellowish oily liquid (the yellower the wood, the more oil it contains). The true oil is characteristically viscous.
Odour	Floral, reminiscent of rose, orange and mignonette.
Chemical	Contains 60–90% linalol, 5% terpineol, 2% geraniol, and probably some nerol together with traces of cineol and methyl heptenol.
Uses	In lilac and lily types of perfume, and can also be employed as a modifier on numerous other types such as May blossom, corylopsis, and rose perfumes; invaluable in soaps.

BOLDO LEAF OIL [8022-81-9]

Source	
GEOGRAPHIC	Chile.
NATURAL	*Peumus boldus*, Mol., N.O. Monimiaceae.
Isolation	Distilled from the leaves.
Yield	2%.
Appearance	Golden-yellow volatile oil.
Odour	Reminiscent of cineol and ascaridol.
Chemical	Contains cymene.
	RIFM Monograph (1982) *FCT*, **20**, 643.

BORAX [1303-96-4 (hydrous); 1330-43-4 (anhydrous)]

CTFA name	Sodium borate
	Sodium tetraborate, sodium borate, sodium biborate
	$Na_2B_4O_7 \cdot 10H_2O$
Source	
GEOGRAPHIC	The calcium and magnesium salts are found in heavy deposits in Chile and Peru.
NATURAL	Borax forms the bed of dried up lakes in California and Nevada.
Isolation	Prepared by boiling the calcium and magnesium salts with sodium carbonate and crystallizing the liquid which is run off. Naturally occurring borax is purified by recrystallization.
Physical	A weak alkali.
Uses	In the preparation of cold creams and stearin creams, as a saponifying agent.

BORIC ACID [10043-35-3]

CTFA name	Boric acid
	Boracic acid, orthoboric acid
	H_3BO_3
Source	
NATURAL	Occurs in nature as the mineral sassolite.
CHEMICAL	Obtainable from borax.
Isolation	Prepared by adding sulphuric acid to borax.
Appearance	Lustrous scales, colourless transparent crystals, white granules or powder.
Odour	None.
Physical	m.p. 171 °C.

Comments: The terpeneless oil is about three times the concentration of the natural product.
RIFM Monograph (1978) *FCT*, **16**, 653.

Uses	As a neutralizing agent in shampoos. A soothing eye lotion may be prepared with a weak solution of boracic acid in rose water.

BORNEOL [507–70–0]

Sumatra camphor; Borneo camphor; Malayan camphor; 1,7,7-trimethyl-bicyclo [2.2.1] heptan-2-ol; 2-hydroxybornane
$C_{10}H_{17}OH$

History	Borneo camphor has been collected for hundreds of years but the synthetic d- and l-borneols have largely replaced it.
Source	
GEOGRAPHIC	A tree indigenous to Borneo and Sumatra is a source of the essential oil.
NATURAL	*Dryobalanops camphora*, Colebr., N.O. Dipterocarpaceae (grow to a height of over 100 ft (30 m) and have at the top a fine canopy of white flowers); also occurs in citronella, thyme, lavender, spike and rosemary, Canadian snake root, African ginger and nutmeg.
CHEMICAL	A constituent of numerous volatile oils. Can be obtained by synthesis.
Isolation	Obtained from oils distilled from other pinus species and can be obtained by separation from turpentine.
Appearance	Appears as a solid crystalline deposit in the cracks and fissures of the tree.
Odour	Reminiscent of patchouli and labdanum (or pepper and camphor).
Physical	m.p. 204 °C, b.p. 214 °C.
Uses	In religious ceremonies; also occasionally used in soap perfumery in camphoraceous odours of the rosemary–lavender type.

RIFM Monograph (1978) *FCT*, **16**, 655 (1-Borneol).

iso-BORNEOL [124–76–5]

1,7,7-trimethyl-bicyclo [2.2.1] heptan-2-ol; 2-hydroxybornane; Borneo camphor
$C_{10}H_{17}OH$

Isolation	The camphor is obtained on oxidation of iso-borneol.
Appearance	Crystalline solid.
Chemical	Stereoisomeric with *d*-borneol.
Physical	m.p. 212 °C, b.p. 214 °C.

RIFM Monograph (1979) *FCT*, **17**, 531.

BORNYL ACETATE [76–49–3]

l-Bornyl acetate; 1,7,7-trimethylbicyclo [2.2.1]-heptan-2-ol acetate; laevo-Bornylacetate
$CH_3COOC_{10}H_{17}$

Source	
NATURAL	Occurs naturally in the oil in pine needles, where the odour is much finer.

Boronia absolute 47

Isolation	Can be synthesized by heating borneol with anhydrous acetic acid in the presence of sulphuric acid; or from turpentine or pinene by the Wagner–Meerwein rearrangement.
Physical	m.p. 29 °C, b.p. 226 °C.
Appearance	Colourless liquid or crystalline mass.
Odour	Sweet herbaceous reminiscent of pine needles.
Uses	In theatre and cinema sprays, where a powerful, persistent odour is required; also suitable for many fougere and lavender soap compounds.
Comments	The butyric, formic, iso-valeric, and propionic acid esters differ only slightly in commercial odour value.

RIFM Monograph (1973) *FCT*, **11**, 1041 (laevo-Bornyl acetate).

iso-BORNYL ACETATE [125-12-2]

$CH_3COOCH_{10}H_{17}$

Odour	Mild, balsamic, camphoraceous, reminiscent of pine needles.
Physical	b.p. 227 °C.
Uses	Very extensively used in low cost household and soap fragrances.

RIFM Monograph (1975) *FCT*, **13**, 552.

iso-BORNYL FORMATE [1200-67-5]

$HCOOCH_{10}H_{17}$

Odour	Green-earthy, herbaceous–camphoraceous.
Physical	b.p. 212 °C.
Uses	Rarely used in fragrances.

BORONIA ABSOLUTE [8053-33-6]

Boronia Oil

History	It was not possible to obtain the true perfume until 1924, when specimens of a concrete otto and essential oil of boronia were shown at the British Empire Exhibition.
Source	
GEOGRAPHIC	Australian swampy forests.
NATURAL	Small, bell-like flowers (sometimes the leaves) of a species of boronia, probably *B. megastigma* (a wild shrubby tree 3–4 ft (0.9–1.2 m) high, belonging to N.O. Rutaceae).
Isolation	The flowers are combed off the shrubs, sieved to remove leaflets and sepals, then treated with volatile solvents to produce the concrete. Extraction produces the absolute.
Yield	0.1–0.2%.
Appearance	Green oily liquid.
Odour	Spicy and herbaceous-type.
Chemical	Contains some beta-ionone.
Uses	In high class chypre and fougere perfumes.
Comments	Oils have been distilled from other species of boronia.

BOUVARDIA

History	The flowers were cultivated by Baron Humboldt who discovered them in Central America in the course of his journeys.
Source	
NATURAL	Various species of N.O. Rubiaceae such as President Garfield, Bridesmaid, Dazzler, Alfred Neuner, Humboldtii, Corymbeflora, and Jasmaflora, var. *odorata*. The flowers of these plants are of the jasmine type and range in colour from whites and yellows to brilliant scarlet.
Odour	Delicious jasmine-type fragrance.
Comments	Artificial bouvardia oils are compounded on the basis of benzyl acetate, formate and propionate, with the use of modifiers such as linalol, methyl-anthranilate, and hydroxy-citronellal.

trans-ω-BROMSTYRENE [588-72-7]

trans-1-Brom-2-phenylethylene; omega-bromostyrene; bromstyrol(e); bromstyrolene; hyacinthine

$C_6H_5CH\!=\!CHBr$

Isolation	Manufactured from cinnamic acid and bromine: soda transforms the intermediate dibromophenyl propionic acid to bromostyrene.
Appearance	Pale yellow mobile liquid.
Odour	Harsh, hyacinth odour.
Physical	b.p. 219 °C.
Chemical	Oxidized in air to bromacetophenone.
Uses	Once widely in soap perfumes: little used in fine perfumery, where it is replaced by phenylacetic aldehyde.
Comments	Has to be blended carefully.

RIFM Monograph (1973) *FCT*, **11**, 1043.

BROOM [8023-80-1]

Broom absolute, Genet absolute

Source	
GEOGRAPHIC	A small shrub which grows wild throughout the whole of the Mediterranean, but is cultivated for production purposes in Southern France and Morocco.
NATURAL	Known as genet in Provence, the shrub, *Cytisus scoparius*, Link., belongs to N.O. Leguminosae.
Isolation	Available first as a concrete; both absolute and concrete are extracted from the blossoms by means of volatile solvents.
Yield	Approximately 0.2% concrete which yields some 20–40% absolute.
Odour	Sweet and honey-like.
Uses	Frequently in fine fragrances of floral type.
Comments	In artificial broom ottos methyl-*p*-cresol is often an important constituent.

Genet absolute RIFM Monograph (1976) *FCT*, **14**, 779.

BRUYERE d'ANNAM

Source	
GEOGRAPHIC	The hills of Southern and Cochin China.
NATURAL	Comes from the cultivated shrub *Cathetus fasciculata*, Lour, N.O. Euphorbiaceae.
Isolation	Distillation of the plants.
Yield	Varies depending upon locality and whether the plants are fresh or dry.
Appearance	Oil, either green or yellow depending upon nature of the soil.
Odour	Reminiscent of cajuput.
Chemical	Contains 31% cineol.
Uses	Soap perfumery.

BUCHU OIL [68650-46-4]

Buchu leaf oil; bucco; bucku; buku

Source	
GEOGRAPHIC	Several species of herb indigenous to Cape Colony, South Africa.
NATURAL	*Barosma betulina* and *Barosma crenulata*, N.O. Rutaceae.
Isolation	Steam distilled from the fresh leaves.
Yield	0.8–2.5%.
Appearance	Oil.
Odour	Pleasant, but intense and camphoraceous reminiscent of blackcurrants.
Chemical	2% has been obtained from *B. venusta*: this contained myrcene, estragol, linalol, aldehydes, sesquiterpenes, small quantities of esters and phenols.
Uses	Only used in small quantities because of the intense odour.

BUPLEUROL [57197-03-2]

$C_{10}H_{20}O$

Source	
GEOGRAPHIC	Sardinia.
NATURAL	Obtained from the essential oil of *Bupleurum fruticosum*, L., N.O. Umbelliferae.
Appearance	Liquid.
Odour	Rose-like.

iso-BUTYL ACETATE [110-19-0]

2-Methyl propyl acetate
$CH_3COOCH_2CH(CH_3)_2$

Appearance	Colourless liquid.
Odour	Sweet and recalling raspberry and pear.
Uses	In traces for modifying artificial rose ottos.

n-BUTYL ACETATE [123-86-4]

CTFA name	Butyl acetate
	Acetic acid butyl ester
	$C_6H_{12}O_2$
Isolation	Obtained from n-butyl alcohol and either acetic acid or acetic anhydride.
Appearance	Colourless liquid.
Odour	Ethereal-fruity, pungent.
Physical	b.p. 126 °C.
Uses	A solvent in the lacquer industry.
	RIFM Monograph (1979) *FCT*, **17**, 515.

iso-BUTYL ALCOHOL [78-83-1]

	2-Methylpropan-1-ol
	$(CH_3)_2CHCH_2OH$
Comments	Has a number of aromatic esters.

n-BUTYL ALCOHOL [71-36-3]

CTFA name	n-Butyl alcohol
	1-butanol; propyl carbinol
	$CH_3(CH_2)_3OH$
Source	
CHEMICAL	Several synthetic routes.
Appearance	A highly refractive liquid which burns with a strongly luminous flame.
Odour	Similar to fusel oil but weaker; the vapour is irritant and causes coughing.
Physical	b.p. 108 °C, miscible with many organic solvents.
Chemical	Similar properties to the acetate.

iso-BUTYL ANTHRANILATE [7779-77-3]

	2-Methylprop-1-yl o-aminobenzoate
	$NH_2C_6H_4COOCH_2CH(CH_3)_2$
Appearance	Colourless or pale straw-coloured liquid.
Odour	Fruity sweet warm odour.
Physical	b.p. 270 °C.
Uses	As a modifier for jasmine, neroli and orange blossom fragrances.

iso-BUTYL BENZOATE [120-50-3]

	2-Methylpropyl benzoate
	$C_6H_5COOCH_2CH(CH_3)_2$
Appearance	Colourless oil.

Odour	Pleasant 'eglantine' odour.
Physical	b.p. 242 °C.
Uses	Of great value in the preparation of carnation, orange blossom, sweet-pea and trefle perfumes.
	RIFM Monograph (1975) *FCT*, **13**, 553.

n-BUTYL BENZOATE [136-60-7]

	$C_6H_5COOC_4H_9$
Appearance	Colourless oily liquid.
Physical	b.p. 250 °C.
Uses	As a fixative in heavy floral.
	RIFM Monograph (1983) *FCT*, **21**, 651.

iso-BUTYL n-BUTYRATE [539-90-2]

	2-Methylpropyl n-butanoate
	$CH_3CH_2CH_2COOCH_2CH(CH_3)_2$
Appearance	Colourless liquid.
Odour	Fruity, pungent reminiscent of pear and banana.
Physical	b.p. 157 °C.
Uses	Used in modifying floral bouquets.
	RIFM Monograph (1979) *FCT*, **17**, 521.

n-BUTYL n-BUTYRATE [109-21-7]

	Butanoic acid butyl ester, butyric acid butyl ester
	$C_8H_{16}O_2$
Isolation	Synthesized from n-butyl alcohol and n-butyric acid.
Appearance	Colourless liquid.
Odour	Fruity, pineapple odour.
Physical	b.p. 165 °C, practically insoluble in water, miscible with alcohol or ether.
Uses	In pineapple flavourings.
	RIFM Monograph (1979) *FCT*, **17**, 521.

iso-BUTYL CINNAMATE [122-67-8]

	2-Methylprop-1-yl 3-phenylacrylate
	$C_6H_5CH{=}CHCHOOCH_2CH(CH_3)_2$
Appearance	Colourless slightly viscous liquid.
Odour	Herbaceous-balsamic, sweet amber.
Physical	b.p. 271 °C.
Uses	A useful constituent of oriental bouquets; blends well with oakmoss, vetivert, patchouli, coumarin, vanillin, and heliotropin.
	RIFM Monograph (1976) *FCT*, **14**, 799.

alpha-n-BUTYL CINNAMIC ALDEHYDE [7492-44-6]

2-Benzylidene-pentan-1-al; Butylcinnamal
$C_6H_5CH{=}(C_4H_9)CHO$

Isolation	Condensation of normal hexyl aldehyde with benzaldehyde.
Appearance	Pale yellow liquid.
Odour	Reminiscent of green, oily-herbaceous.
Physical	b.p. 252 °C.
Uses	As a modifier for the more popular amyl and hexyl derivatives.

RIFM Monograph (1980) *FCT*, **18**, 657.

p-tert-BUTYL-m-CRESOL [2219-72-9]

4-tert-Butyl-3-methylphenol; p-tertiary-butyl-m-cresol
$(CH_3)_3CC_6H_3(OH)(CH_3)$

Appearance	Colourless oily liquid.
Odour	Powerful odour of Russian leather.
Physical	m.p. 23 °C.
Uses	Superior to birch tar oil as a base for Cuir compounds.

n-BUTYL p-CRESYL ETHER [10519-06-9]

n-Butyl p-tolyl ether; p-cresol butyl ether
$CH_3(CH_2)_3OC_6H_4CH_3$

Appearance	Colourless liquid.
Odour	Powerful ylang-ylang type; more flowery than the methyl ether.
Physical	b.p. 220 °C.
Uses	Similar to those of the methyl ether.

p-tertiary-BUTYL CYCLOHEXANOL [98-52-2]

Padoryl, Patchone, PTBCH
$C_{10}H_{20}O$

Isolation	From p-tertiary-butyl phenol by hydrogenation.
Appearance	White crystalline powder or needles.
Odour	Powerful dry, woody-camphoraceous with leather-like undertones.
Uses	In woody, pine, camphoraceous and herbaceous fragrances for soap, detergents and household products where it gives power and diffusive character to the compositions.

RIFM Monograph (1974) *FCT*, **12**, 833.

o-tertiary-BUTYL CYCLOHEXYL ACETATE [88-41-5]

Verdox, OTBCHA
$C_{12}H_{22}O_2$

Isolation	Hydrogenation of o-tert-butyl-phenol followed by acetylation.
Appearance	Colourless to pale yellow liquid; may solidify at lower temperatures.
Odour	Piney-woody, fruity, fresh camphoraceous.

Chemical	Stable to acids and alkalis.
Physical	b.p. 230 °C.
Uses	Widely used to reinforce topnotes of pine, herbaceous, woody and some floral compositions mainly for soaps, detergents, airfresheners, masking agents and household products.
Comments	Exhibits remarkable stability in many hostile products. The commercial product is usually a mixture of the *cis-* and *trans-* isomers.

p-tertiary-BUTYLCYCLOHEXYL ACETATE [32210-23-4]

'Vertenex'
$C_{12}H_{22}O_2$

Appearance	Colourless liquid.
Odour	A fine quality woody odour.
Physical	b.p. 232 °C.
Chemical	Stable and does not discolour.
Uses	Widely used in rose compositions and also in the perfumes of soap and detergents.
	RIFM Monograph (1978) *FCT*, **16**, 657.

p-tertiary-BUTYL-alpha-METHYL DIHYDROCINNAMIC ALDEHYDE
[80-54-6]

'Lilial'; Lily Aldehyde
$C_{14}H_{20}O$

Isolation	Synthesized by the condensation of *p*-tert-butyl-benzaldehyde with propionaldehyde followed by catalytic hydrogenation.
Appearance	Colourless liquid.
Odour	Reminiscent of lily of the valley, lily and tulip.
Physical	b.p. 258 °C.
Uses	Widely used in soap, detergent and cosmetic perfumes.
	RIFM Monograph (1978) *FCT*, **16**, 659.

iso-BUTYL FORMATE [542-55-2]

2-Methylprop-1-yl formate
$HCOOCH_2CH(CH_3)_2$

Appearance	Colourless liquid.
Odour	Raspberry-like.
Physical	b.p. 98 °C.
Uses	Used as a modifier in perfumery.

n-BUTYL FORMATE [592-84-7]

$HCOOC_4H_9$

Appearance	Colourless liquid.
Odour	Ethereal-rum-like.

54 *n*-Butyl furoate

Physical	b.p. 107 °C.
Uses	In perfumes of the peau d'espagne type.

n-BUTYL FUROATE [583-33-5]

n-Butylfuran-2-carboxylate
$C_4H_3OCOOC_4H_9$

Appearance	Colourless liquid.
Odour	Sweet, brandy-like, with a burnt woody note.
Physical	b.p. approx. 216 °C.

n-BUTYL LACTATE [138-22-7]

n-Butyl 2-hydroxy propanoate
$CH_3CHOHCOOC_4H_9$

Physical	b.p. 200 °C.
Uses	In the lacquer industry as a plasticizer and solvent. RIFM Monograph (1979) *FCT*, **17**, 727.

iso-BUTYL β-NAPHTHYL ETHER [2173-57-1]

2-(2-Methyl propoxy) naphthalene; beta-naphthol butyl ether
$C_{10}H_7OC_4H_9$

Appearance	White crystalline leaves.
Odour	Sweet, fruity, orange blossom floral.
Physical	b.p. 307 °C, m.p. 33 °C.
Uses	As a fixative in some cologne and orange blossom compounds.

iso-BUTYL PHENYLACETATE [102-13-6]

2-Methylprop-1-yl phenylacetate
$C_6H_5CH_2COOCH_2CH(CH_3)_2$

Appearance	Colourless liquid.
Odour	Powerful and diffusive rosy-musk.
Physical	b.p. 247 °C.
Uses	Extensively used in floral perfumes, in particular tobacco, carnation, tuberose and white rose compounds; up to 20% can be used in fancy bouquets.
Comments	Named 'eglantine' like the benzoate. RIFM Monograph (1975) *FCT*, **13**, 811.

n-BUTYL PHENYLACETATE [122-43-0]

$C_6H_5CH_2COO(CH_2)_3CH_3$

Appearance	Colourless liquid.
Odour	Sweet, honey-like with a slight floral note.

Physical	b.p. 258 °C.
Uses	Used in a similar way to iso-butyl phenylacetate.
	RIFM Monograph (1983) *FCT*, **21**, 657.

iso-BUTYL PROPIONATE [540-42-1]

	2-Methylprop-1-yl propanoate
	$C_2H_5COOCH_2CH(CH_3)_2$
Appearance	Liquid.
Odour	Intense, fruity, rum-like.
Physical	b.p. 137 °C.
Uses	In fruit essences.

2-iso-BUTYLQUINOLINE [93-19-6]

	2-iso-Butylquinoline
	$(CH_3)CHCH_2C_9H_6N$
Appearance	Viscous oily liquid.
Odour	Powerful woody-earthy, mossy note resembling oakmoss.
Physical	b.p. approx. 255 °C.
Uses	In fancy perfumes of the chypre type.
	RIFM Monograph (1976) *FCT*, **14**, 311
	(1978) *FCT*, **16**, 661.

6-sec-BUTYL QUINOLINE [68141-26-4]

	iso-Butyl quinoline
	$C_{13}H_{15}N$
Appearance	Liquid: varies from straw yellow to green or dark brown.
Odour	Intense earthy-tarry-mossy resembling that of oakmoss and vetivert.
Physical	b.p. approx. 252 °C.
Uses	Unusual effects may be obtained only when used in small proportions.
Comments	A proportion of 8-sec-butyl quinoline may occur in the commercial product.
	RIFM Monograph (1976) *FCT*, **14**, 311
	(1978) *FCT*, **16**, 661.

iso-BUTYL SALICYLATE [87-19-4]

	2-Methyprop-1-yl o-hydroxybenzoate
	$HOC_6H_4COOCH_2CH(CH_3)_2$
Appearance	Colourless liquid.
Odour	Harsh-herbaceous floral resembling clover and orchid.
Physical	b.p. 260 °C.

Uses	In trefle and orchidee perfumes; blends well with carnation compounds and has some advantage over methyl salicylate for compounding synthetic cassie. RIFM Monograph (1975) *FCT*, **13**, 813.

n-BUTYL SALICYLATE [2052-14-4]

	n-Butyl o-hydroxybenzoate $HOC_6H_4COO(CH_2)_3CH_3$
Appearance	Colourless liquid.
Odour	Persistent trefle-type with a rough herbaceous character.
Physical	b.p. 268 °C, m.p. 6 °C.
Chemical	Blends well with iso-eugenol.
Uses	As a base for fancy perfumes. RIFM Monograph (1978) *FCT*, **16**, 663.

n-BUTYL STEARATE [123-95-5]

CTFA name	Butyl stearate
	n-Butyl n-octadecanoate $CH_3(CH_2)_{16}COO(CH_2)_3CH_3$
Appearance	Colourless liquid.
Physical	m.p. 28 °C, b.p. 315 °C.
Uses	Used in conjunction with dibutyl phthalate as a plasticizer in nail enamels: can also be used in lipstick formulations as a solvent for fluorescent dyestuffs and for waxes and fats.

n-BUTYL UNDECYLENATE [109-42-2]

	n-Butyl hendec-10-enoate $CH_2{=}CH(CH_2)_8COO(CH_2)_3CH_3$
Appearance	Resembles oil.
Odour	Cognac.
Physical	b.p. 252 °C.
Uses	In cognac compounds.

iso-BUTYL n-VALERATE [10588-10-0]

	iso-Butyl valerianate $CH_3(CH_2)_3COOCH_2CH(CH_3)_2$
Appearance	Colourless oily liquid.
Odour	Fruity, ethereal, apple–raspberry like.
Physical	b.p. 170 °C.
Uses	Traces are used as a modifier in fruity perfumes.

n-BUTYL n-VALERATE [591-68-4]

	n-Butyl n-pentanoate; butyl valerianate
	$CH_3(CH_2)_3COO(CH_2)_3CH_3$
Appearance	Colourless liquid.
Odour	Ethereal-fruity somewhat choking character.
Physical	b.p. 186 °C.

BUTYLATED HYDROXYANISOLE [25013-16-5]

CTFA name	BHA
	A mixture of isomers of tertiary butyl-substituted 4-methoxy-phenols, e.g. 2-tert-butyl-4-methoxyphenol and 3-tert-butyl-4-methoxyphenol
	$C_{11}H_{16}O_2$
Appearance	White crystalline powder or a waxy solid.
Odour	Phenolic.
Physical	Insoluble in water, soluble in organic solvents and in oils and propylene glycol; miscible with waxes.
Uses	As an antioxidant in glyceride oils, fats, essential oils and cosmetic products to stabilize and prevent rancidity.
Comments	Used in concentrations of 0.02–0.03%.

BUTYRALDEHYDE [123-72-8]

	n-Butanal; butyric aldehyde
	$CH_3(CH_2)_2CHO$
Appearance	Colourless liquid.
Odour	Sharp, penetrating, irritating.
Physical	b.p. 76 °C, a flammable liquid, miscible with many organic solvents and partially soluble in water.
Uses	In traces in modern sophisticated perfume compounds.
	RIFM Monograph (1979) FCT, 17, 731.

CABREUVA OIL [68188-03-4]

Source	
GEOGRAPHIC	Brazil, Paraguay and Argentina.
NATURAL	*Myocarpus frondosus* and *Myocarpus fastigiatus*.
Isolation	Steam distillation of sawdust and wood chippings from saw mills.
Yield	1.6–1.8%.
Odour	Woody and fatty.
Uses	A woody fragrance component in perfumes and as a source of nerolidol.
	RIFM Monograph (1982) FCT, 20, 645 (*Myrocarpus frondosus* and *M. fastigiatus*).

CACAO BUTTER [8002-31-1]

COLIPA name	Cocoa butter
	Theobroma oil
Source	
NATURAL	Seeds of *Theobroma cacao*, L., N.O. Sterculiaceae.
Appearance	Yellowish-white solid, brittle below 25 °C.
Odour	A chocolate odour and taste.
Physical	m.p. 30–35 °C; insoluble in water, slightly soluble in alcohol, very soluble in chloroform, ether, petroleum ether and benzene.
Chemical	Chief constituents are glycerides of stearic, palmitic, oleic, arachidic and linoleic acids.
Uses	A grease paint remover; about 0.5% makes application of lipsticks to the lips smooth and easy: used as the base of suppositories and pessaries.

CACTUS

History	The original name given by Theophrastus to spiny plants.
Source	
GEOGRAPHIC	Most cacti are indigenous to tropical America.
NATURAL	Some species of the family Cactaceae grow 40–50 ft (12–15 m) high. The plants are succulent, spiny and leafless. The fruit is often a vivid red in colour with a somewhat acid flavour.
Odour	Reminiscent of vanilla, rose and jasmine.

CADE OIL [8013-10-3]

CTFA name	Juniper tar
Source	
GEOGRAPHIC	Europe.
NATURAL	The shrub, *Juniperus oxycedrus*.
Isolation	Destructive distillation of the wood.
Yield	1.4–1.6%.
Odour	Smoky and tar-like.
Uses	Used in small quantities in leather bases and in fougeres or pine compositions.
	RIFM Monograph (1975) *FCT*, **13**, 773 (Rectified).
	IFRA GUIDELINE
	The Committee recommends that crude Cade oil should not be used as a fragrance ingredient. Only rectified oils obtained by pyrolysis of the wood and twigs of Juniperus oxycedrus L., followed by fractional distillation under vacuum should be used.
	Such rectified oils are characterized by a maximum boiling point of 260 °C (at atmospheric pressure) and the lack of significant evaporation residue.

This recommendation is based on the fact that fractional distillation is effective in removing the underdesirable polynuclear aromatic hydrocarbon content of crude Cade oil (K. Bouhlal, J.M. Meynadier et al. (1988), Parf. Cosm. Arômes No 83, 73–82).

July 1990

CADINENE

β Cadineni [523–47–7]
α Cadinene [24406–05–1]

Chemical name	Hexahydro-4-iso-propyl-1,6-dimethylnaphthalene $C_{15}H_{24}$
Source	
NATURAL	Found in the essential oils of cedarwood, santal, sassafras, wormwood, and patchouli.
Isolation	From oil of cade, obtained from destructive distillation of wood from different species of conifers.
Physical	b.p. 275 °C.
Uses	In synthetic ylang-ylang oil up to 5%; not extensively used in perfumery.

RIFM Monograph (1973) *FCT*, **11**, 1045.

CAJUPUT OIL

Source	
GEOGRAPHIC	Indonesia, Australia and Malaysia.
NATURAL	Wild *Melaleuca minor* and other species of melaleuca.
Isolation	Steam distillation of the fresh leaves and twigs.
Yield	0.8–1% depending upon production procedure and source of the plant material.
Appearance	Colourless or yellowish liquid.
Odour	Powerful and eucalyptus-like with a bitter aromatic taste.
Physical	Very slightly soluble in water; miscible with alcohol, chloroform, ether and carbon disulphide.
Uses	Not used extensively as it is difficult to obtain.
Comments	Must be kept cool and unexposed to air or light.

RIFM Monograph (1976) *FCT*, **14**, 701 (*Malaleuca leucadendron*, L.).

CALAMINE

[8011–96–9]

COLIPA name	Calamine
Source	
CHEMICAL	A basic zinc carbonate coloured with a small amount (5%) of ferric oxide.
Appearance	Amorphous pink or reddish-brown powder.
Uses	With zinc oxide in sunburn lotions; has a mild astringent and soothing action on the skin.

CALAMUS OIL [8015-79-0]

History	Mentioned in the Bible, and by Theophrastus, Dioscorides and Pliny; also described by Pomet in his 'Histoire des Drogues': highly esteemed in India, where it is sold in nearly every bazaar.
Source	
GEOGRAPHIC	A plant cultivated along river banks in Korea, India and the USSR.
NATURAL	*Acorus calamus*, L., N.O. Aroideae.
Isolation	Steam distillation of the fresh or dried roots.
Yield	1.2–1.5% depending upon source of the raw material.
Odour	Earthy and aromatic; reminiscent of patchouli.
Chemical	Contains pinene, camphene, camphor, calamene, a tertiary alcohol, eugenol, and asarone.
Uses	As a toner and in chypre compositions. RIFM Monograph (1977) *FCT*, **15**, 623 (*Acorus calamus*, L.).

CALANTAS WOOD

Source	
NATURAL	The tree *Toona calantas*, N.O. Meliaceae.
Odour	Very strong.
Chemical	The principal constituent of the essential oil obtained from the wood is cadinene.
Uses	The manufacture of cigar boxes.

CALCIUM PHOSPHATE [7758-87-4]

	Tricalcium orthophosphate; tribasic calcium phosphate; tricalcium phosphate $Ca_3(PO_4)_2$
Source	
NATURAL	Occurs naturally as the mineral osteolite.
Isolation	Obtained by adding ordinary sodium phosphate to a solution of calcium chloride in the presence of ammonia.
Appearance	Amorphous powder.
Odour	Odourless and tasteless.
Physical	Practically insoluble in water, alcohol or acetic acid.
Uses	Used with chalk as the base for some dental creams.

CALCIUM STEARATE [1592-23-0]

CTFA name	Calcium stearate Calcium octadecanoate $C_{36}H_{70}CaO_4$
Appearance	White, granular, fatty powder.
Physical	Insoluble in water, slightly soluble in hot alcohol.
Other	Special fine grades of this powder have good covering power with adhesion and good 'slip' properties and are used in cosmetic work.

Uses As a tabletting lubricant and also in talcum and face powders.
Comments Commercial preparations may contain palmitate.

CALENDULA CONCRETE/CALENDULA ABSOLUTE [84776–23–8]

CTFA name	Calendula oil/extract
Source	
GEOGRAPHIC	Europe and North America.
NATURAL	The cultivated marigold, *Calendula officinalis*, N.O. Compositae.
Isolation	The concrete is obtained by extraction of the flowers and then the absolute.
Yield	0.3–0.4% concrete from which about 30% absolute is obtained.
Odour	A typical, bitter and herbaceous odour.
Chemical	The oil contains calendulin, carotenoid pigments and a saponin.
Comments	The supply of this oil is inconsistent.

CALIFORNIAN LAUREL OIL [90131–78–5]

The tree has several names: 'Spice', 'Californian olive', 'Californian baytree', 'pepper tree', 'mountain laurel'.

Source	
GEOGRAPHIC	A tree indigenous to California and Oregon.
NATURAL	*Umbellularia califonica*, Nutt, N.O. Lauraceae.
Isolation	Distillation of the leaves.
Appearance	Pale yellow liquid.
Odour	Aromatic; can be irritating if strongly inhaled.
Chemical	Identified constituents are umbellulone, cineol, methyleugenol, l-pinene, eugenol, and safrole.

CALIFORNIAN POPPY

White bush poppy, Californian poppy wort

Source	
NATURAL	A herbaceous perennial of N.O. Papaveraceae, with large white flowers from June to September.
Odour	Delicate and magnolia-like.

CAMELLIA [68647–73–4]
[84650–60–2]

CTFA name	Camellia oil
	Thea sinensis oil, Tea oil
Source	
GEOGRAPHIC	China and Japan.
NATURAL	The source shrub belongs to N.O. Ternstroemiaceae; the most attractively perfumed variety is *C. sasanqua*.
Isolation	A fixed oil can be expressed from the seeds. Young leaves can be distilled to produce an essential oil.
Appcarance	Oil.

Odour	Sweet.
Chemical	The essential oil contains 97% eugenol and an ester with an odour reminiscent of geraniol.
Comments	Synthetic ottos are usually compounds of linalol, ylang-ylang, and iso-eugenol, blended with jasmine, orange flower and musk. The leaves are used as a hair tonic.

IFRA GUIDELINE

The Committee recommends that the following materials should not be used as fragrance ingredients:

- Allylisothiocyanate
- Chenopodium oil
- 3,7-Dimethyl-2-octen-1-ol (6,7-Dihydrogeraniol)
- Furfurylideneacetone
- Methyl methacrylate
- Phenylacetone (Methyl benzyl ketone)
- Esters of 2-octynoic acid, except those covered elsewhere in these Guidelines i.e. methyl and allyl heptine carbonate
- Esters of 2-nonynoic acid, except methyl octine carbonate
- Thea sinensis absolute

These recommendations are based on the absence of reports on the use of these materials as fragrance ingredients and inadequate evaluation of possible physiological effects resulting from their use in fragrances.

March 1988, last amendment July 1990

CAMOMILE OIL, GERMAN [8002-66-2]

History	Piesse named the pigment, azulene, which causes the colour of the oil.
Source	
GEOGRAPHIC	Cultivated in Hungary, Germany, Egypt and the USSR.
NATURAL	*Matricaria chamomilla*, L.
Isolation	Steam distillation of two parts of the inflorescence.
Yield	4 kg pure flowers yield 14 g oil (0.35%); 1 kg pure calices yields 5.1 g oil (0.51%).
Appearance	The oil from the flowers is blue with a butter-like consistency when cooled; that obtained from the calices is faintly green turning to yellow after a few days.
Odour	Typically sweet and herbaceous like.
Physical	Very soluble in alcohol.
Uses	In perfumery as a blender in oriental compounds and sometimes in combination with patchouli, lavender, and oakmoss.
Comments	In the glass and porcelain industries camomile oil has been used as a solvent for platinum chloride, to coat vessels with platinum. It should be kept well closed, cool and protected from light.

RIFM Monograph (1974) *FCT*, **12**, 851.

CAMOMILE OIL, MOROCCAN [92202-02-3]

Source	
GEOGRAPHIC	A common annual which covers vast areas of western Morocco.
NATURAL	*Ormenis mixta*, L., and *O. multicaulis*.
Isolation	Distilled from the plants.
Yield	0.1%.
Appearance	Bluish-green oil, which turns yellow with ageing.
Odour	Similar to other camomiles, except for a rosy–honeyed character.

CAMOMILE OIL, ROMAN [8015-92-7]

COLIPA name	Chamomile oil
	Anthemis nobilis oil; camomille oil; English camomille oil
Source	
GEOGRAPHIC	Wild plants which grow in Western and Southern Europe, but are cultivated in Great Britain, Germany, Belgium, Bulgaria, Yugoslavia, France and Hungary.
NATURAL	*Anthemis nobilis*, L., N.O. Compositae.
Isolation	Steam distillation of the flowers, or whole plants.
Yield	1% from the flowers; reduced to 0.3% when the whole plant is distilled.
Appearance	Pale blue to greenish-blue liquid, which becomes brownish on exposure to light or air.
Odour	Characteristic odour of the flowers, fresh, sweet and herbaceous, with a burning taste.
Physical	Slightly soluble in water.
Chemical	Principal constituents are angelic acid, and its iso-amyl and iso-butyl esters.
Uses	As a perfume in shampoo powders; is also believed to tint the hair: has limited use as a tobacco flavour.
Comments	Should be kept well closed, cool and protected from light. RIFM Monograph (1974) *FCT*, **12**, 853.

CAMPHENE [79-92-5]

Chemical name	2,2-Dimethyl-3-methylenebicyclo[2.2.1]heptane
	2,2-Dimethyl-3-methylenenorbornane; 3,3-dimethyl-2-methylene-norcamphane
	$C_{10}H_{16}$
Source	
NATURAL	Occurs in many essential oils such as turpentine (*l*- and *d*-forms), in cypress oil (*d*-form), in camphor oil, bergamot oil, oil of citronella, neroli, ginger and valerian.
Appearance	*dl*-form: cubic crystals.
Odour	Mild, oily-camphoraceous.
Physical	b.p. 159°C, m.p. 51–52°C; volatilizes on exposure to air:

moderately soluble in alcohol, soluble in ether or chloroform, insoluble in water.
RIFM Monograph (1975) *FCT*, **13**, 735.

CAMPHOR [8008–51–3] (oils)
[76–22–2] (chemical)

CTFA name	Camphor
	Gum camphor; 1,7,7-trimethylbicyclo[2.2.1]heptan-2-one; 2-bornanone
	$C_{10}H_{16}O$
Source	
GEOGRAPHIC	A tree indigenous to China, Japan and the island of Taiwan: also cultivated in Sri Lanka and California.
NATURAL	Obtained from the essential oil of the wood of *Cinnamomum camphora*, L., N.O. Lauraceae.
Isolation	The camphor and volatile oil are obtained by steam distillation from the wood, the former being separated from the latter by pressure. Purified camphor is prepared by sublimation from a mixture of crude camphor, charcoal or lime. Camphor may be synthesized by the oxidation of borneol.
Appearance	Transparent crystalline mass; can have a crystalline form.
Odour	Distinctive and medicinal; a slightly bitter, cooling taste.
Physical	Sublimes appreciably at room temperature and cannot be powdered in a mortar unless moistened with an organic solvent.
Uses	Mainly in the manufacture of celluloid: in cosmetics it is mixed with hard paraffin and sold as 'camphor ice': also used as a commercial source of safrole.

RIFM Monograph (1976) *FCT*, **14**, 703 (Brown).
RIFM Monograph (1973) *FCT*, **11**, 1047 (White).
RIFM Monograph (1975) *FCT*, **13**, 739 (Yellow).
RIFM Monograph (1978) *FCT*, **16**, 665 (USP).

3-(*iso*-CAMPHYL-5)CYCLOHEXANOL [69621–68–7]

	$C_{16}H_{28}O$
Isolation	Condensation of camphene and guaiacol, with subsequent hydrogenation.
Appearance	Colourless, viscous liquid.
Odour	Tenacious, warm, woody, balsamic, similar to sandalwood.
Chemical	Mixture of isomers.

RIFM Monograph (Isocamphyl cyclohexanol, mixed isomers) (1976) *FCT*, **14**, 801.

CANADA BALSAM [85085–34–3]

Canada turpentine

Source	
GEOGRAPHIC	From trees widely distributed throughout northern America.

Cananga oil 65

NATURAL	The balsam fir, *Abies balsamea*, Miller, and the hemlock spruce, both N.O. Coniferae.
Isolation	The oleo-resin forms in cavities in the trees and is collected by puncturing the trunk with the pointed spout of a can into which the liquid runs.
Appearance	Clear, transparent liquid: becomes viscous and eventually dries to a hard resin.
Chemical	Contains about 20% volatile oil, mainly l-pinene.
Uses	Excellent fixative for soap perfumes containing a high percentage of citrus oils; may be used up to as much as 10% in cologne and verbena compounds.

CANADIAN SNAKE ROOT OIL [89957–73-3]

Source	
GEOGRAPHIC	A plant which grows on rocky ground or in old woods in Canada and the United States.
NATURAL	*Asarum canadense*, L., N.O. Aristolochiaceae.
Isolation	The rhizome, known as wild ginger, is deprived of its roots, cleaned and dried.
Appearance	Yellowish oil.
Odour	Pungent, recalling patchouli and ginger.
Chemical	Isolated constituents include a phenol of creosote-like odour, pinene, d-linalol, l-borneol, l-terpineol, geraniol and methyl eugenol.
Uses	In traces in eau-de-cologne.

CANANGA OIL [68606–83-7]

	Ylang-Ylang oil
Source	
GEOGRAPHIC	Wild plants found in Java, Réunion, and Southern Asia: cultivated on Madagascar, Java, the Philippines and the Comoro Islands.
NATURAL	*Cananga odorata*, Hook, and probably also *C. latifolia*, N.O. Anonaceae.
Isolation	Obtained by distillation of the flowers, which appear when the plants are 3 years old.
Yield	About 1%; can be increased by 50% with prolonged treatment. The first distillate yields the so-called 'ylang-ylang extra' (the finest oil); of the following three fractions, the first two are of no commercial value.
Appearance	Light yellow fragrant oil.
Odour	Sweet and floral.
Chemical	Among the important constituents identified are linalol, geraniol, p-cresol methyl ether, cadinene, safrole, nerol, farnesol, eugenol, iso-eugenol, methyl-eugenol, benzyl acetate and benzoate, methyl salicylate and anthranilate.
Uses	In floral fragrances such as jasmine, violet and lilac: a valuable

soap perfume. The terpeneless oil is sometimes sold as synthetic ylang-ylang.
RIFM Monograph (1973) FCT, **11**, 1049.

CANDELILLA WAX [8006-44-8]

COLIPA name	Candelilla wax
Source	
NATURAL	Obtained from various Euphorbiaceae species.
Isolation	Separated by immersion in boiling water. The wax is skimmed off the surface and purified by treatment with sulphuric acid.
Appearance	Brownish to yellowish-brown, hard, brittle but easily pulverized lumps.
Chemical	Soluble in chloroform and turpentine; practically insoluble in water.
Uses	In lipstick formulations.

CANELLA OIL

Source	
GEOGRAPHIC	West Indies.
NATURAL	*Canella alba*, Murray, N.O. Canellaceae.
Isolation	Distilled from the bark.
Appearance	Pale yellow liquid.
Odour	Aromatic, recalling cinnamon and clove.
Uses	In Eastern bouquets.

CANTHARIDIN [56-25-7]

$C_{10}H_{12}O_4$

Source	
NATURAL	Obtained from the dried beetle, *Cantharis vesicatoria*.
Appearance	Crystallizable solid.
Odour	Odourless and tasteless.
Physical	Sparingly soluble in water; sublimes at about 110 °C.
Uses	In hair preparations, after dissolution in ether or chloroform.
Comments	A vesicant.

CAPARRAPI BALSAM

Source	
GEOGRAPHIC	United States of America.
NATURAL	*Laurus giganteus*, a large evergreen forest tree of N.O. Lauraceae.
Isolation	Obtained by an incision in the trunk of the tree.
Appearance	Varies according to age of the tree.
Odour	Aromatic.
Uses	Used locally for medicinal purposes.

CAPE OIL

	Capé oil
History	First distilled in 1911 by Roure–Bertrand Fils.
Source	
GEOGRAPHIC	The Ivory Coast of West Africa.
NATURAL	*Popowia capea*, Camus, N.O. Anonaceae.
Isolation	Distillation of the dried leaves.
Yield	0.59%.
Appearance	Greenish-yellow mobile oil.
Odour	Resembles patchouli; becomes more powerfully aromatic on evaporation.
Comments	Used by the natives as a perfume: also used as a bath by macerating the crumbled leaves and stalks in cold water.

CARAWAY OIL [8000–42–8]

CTFA name	Caraway oil
	Carum carvi oil
History	Caraway was first cultivated in Holland about 1815.
Source	
GEOGRAPHIC	A biennial herb cultivated throughout Europe, but especially in Holland and to some extent England.
NATURAL	*Carum carvi*, L., N.O. Umbelliferae.
Isolation	Obtained from the dried, ripe seeds which are crushed and steam distilled.
Yield	3–5%; more if the seeds are particularly good.
Appearance	Oily liquid.
Odour	Intense and spicy.
Chemical	Consists of carvone and limonene.
Uses	Useful in the manufacture of toilet soap, and, in combination with other oils, for the flavouring of mouth washes: used only in small quantities in perfumery due to its intensity, but in combination with cassia oil it yields pleasant brown Windsor perfumes.
	RIFM Monograph (1973) *FCT*, **11**, 1051 (*Carum carui*, L.).

CARBON TETRACHLORIDE [56–23–5]

	Tetrachloromethane
	CCl_4
Source	
CHEMICAL	Obtained from carbon disulphide and chlorine in the presence of a catalyst.
Appearance	Colourless, clear, non-inflammable, heavy liquid.
Odour	Sweet and nauseating.
Physical	b.p. 76.7 °C; miscible with organic solvents; almost insoluble in water.
Uses	Can be used as a liquid hair and scalp cleanser.

Cardamom oil

Comments	Toxic. The use of this material in cosmetics is forbidden in most countries.

CARDAMOM OIL [8000-66-6]

CTFA name	Cardamom oil *Elettaria cardamomum* oil
History	Cardamoms have been cultivated from earliest times.
Source	
GEOGRAPHIC	Reed-like perennial plant: grows wild and is also cultivated in India, Tanzania and Guatemala.
NATURAL	*Elettaria cardamomum*, Maton, which belongs to the ginger family: there are two varieties cultivated in India – Mysore and Malabar; the former yields a higher percentage of oil. The plants are grown placed between tea and rubber plants.
Isolation	The fruits are collected in September or October, dried in the open air, trimmed by machines, sorted to tint and bleached by sulphur vapour.
Yield	1–5% of essential oil. The roots yield about 0.6% of an oil with a very aromatic odour.
Appearance	Oily liquid.
Odour	The oil obtained from the seeds has a characteristic, balsamic odour.
Uses	A spicy toner in both masculine and feminine perfumes; recommended as a useful ingredient of lily of the valley perfumes. The powdered seeds are occasionally used in the manufacture of incense and fumigating powders with liberal quantities of cascarilla and sandalwood. RIFM Monograph (1974) *FCT*, **12**, 837 (*Elletaria cardamom* (L.), Maton).

CARMINE [1390-65-4]

CTFA name	Carmine Carminic acid; CI 75470; CI Natural Red 4 $C_{22}H_{20}O_{13}$
Source	
CHEMICAL	The aluminium lake of the colouring agent cochineal.
Isolation	Prepared by precipitating decoctions of cochineal with alum, gelatine, etc.; the best results are said to be dependent upon fine dry weather.
Appearance	Bright red solid pieces, easily reduced to powder.
Chemical	Readily soluble in alkaline liquids: contains about 50% carminic acid.
Other	Has a deep red colour in water and is yellow to violet in acid solutions.
Uses	In toilet preparations such as lip salves and rouges.

CARNATION (*Dianthus caryophylius*) [89958-29-2]

Oeillet

Source	
NATURAL	One of our most beautiful garden flowers.
CHEMICAL	Can be duplicated synthetically from clove oil.
Isolation	Carnation absolute can be obtained by steam distillation of the flowers.
Yield	0.003%.
Appearance	Light green concrete oil.
Odour	Powerful fragrance.
Physical	m.p. 52–55 °C (oil).
Uses	Only in the most exclusive perfumes.
Comments	Extremely costly to obtain carnation absolute.

CAROB SEED GUM [9000-40-2]

CTFA name	Locust bean gum
	Algaroba; carob flour; ceratonia; Saint John's Bread
History	It is believed that the carobs from which this gum is obtained are what are referred to in the New Testament as the 'husks' eaten by the prodigal son, and the 'locusts' that St John lived on in the wilderness.
Source	
GEOGRAPHIC	The countries bordering the Mediterranean, Syria and Western Asia.
NATURAL	From the bean of *Ceratonia siliqua*, a small branched tree of N.O. Leguminosae.
Appearance	Appears commercially as a white powder.
Physical	When the powder is combined with water it swells and forms a mucilage on heating. The enzyme associated with the mother substance is thus destroyed.
Chemical	Borax increases the viscosity of the mucilage.
Uses	Used in the same way as tragacanth.

CAROTENE [7235-40-7]

CTFA name	Carotene
	Provitamin A; carotin; beta-carotene; CI 75130
	$C_{40}H_{56}$
Source	
NATURAL	Widely distributed in nature; found particularly in carrots and tomatoes.
Appearance	A yellow colouring matter.

70 Carrageen

Chemical Uses	Converted into vitamin A in the liver of animal organisms. As an active ingredient in rejuvenating cosmetics, especially in creams for the reduction or elimination of wrinkles: is more readily absorbed through the skin in oil solutions.
Comments	Beta- and alpha-carotene are widely distributed, the latter in small amounts.

CARRAGEEN [9000-07-1]

CTFA name	Carrageenan; carrageenin Irish moss; chondrus
Source	
GEOGRAPHIC	Seaweed found just below the low-water mark on the northern shores of the Atlantic, principally off the coasts of Brittany and north-west Ireland.
NATURAL	*Chondrus crispus*, Stackhouse, N.O. Gigartinaceae.
Isolation	Contains 90% mucilage, obtained by boiling the plant with water for about 0.5 h and straining: on cooling, the liquid sets to a jelly.
Physical	Completely soluble in hot water; dilute aqueous solutions are viscous and viscosity increases logarithmically with concentration.
Uses	A small percentage of the mucilage is used in vanishing creams, thus increasing the adhesive qualities of the product.

CARROT SEED OIL [801-58-8]

CTFA name	Carrot oil; daucus oil
Source	
GEOGRAPHIC	Carrots cultivated in the Maine and Loire districts of France, Hungary, Yugoslavia and the USSR.
NATURAL	Seeds of *Daucus carota*, L., N.O. Umbelliferae.
Isolation	The crop is collected during August: the seeds are separated by special beaters, giving two different products – the marketable seeds and a 'dust' containing the ground aggregate of the ripe seeds and 2–5% of seeds. The oil is obtained by steam distillation of the dried seeds.
Yield	0.8–1.6%.
Appearance	Limpid yellow to white liquid.
Odour	Earthy, woody, and root-like which varies slightly depending upon whether the oil comes from the seeds or the 'dust'.
Chemical	The chemistry of this oil is still obscure.
Uses	In natural-type fragrances, fougeres and chypres. On blending with cedarwood, a fairly good imitation of orris can be obtained. RIFM Monograph (1976) *FCT*, **14**, 705.

CARTHAMIN [36338-96-2]

Carmathic acid; safflor carmine; safflor red; CI 75140; CI Natural Red 26
$C_{43}H_{42}O_{22}$

History	This plant is said to have been introduced into Britain nearly 400 years ago. It was known to Pomet in his 'Histoire des Drogues' as German saffron, and was used among the feather sellers and for making Spanish red. In later years it was used for dyeing china, silks, crepes and Spanish wool.
Source	
GEOGRAPHIC	A plant cultivated particularly in India and China, but also in the Levant, Egypt and Southern Europe.
NATURAL	The safflower, *Carthamus tinctoria*, L., a prickly annual of N.O. Compositae, also known as cardoon and bastard saffron.
Isolation	Safflower contains two types of colouring matter: yellow, which is water soluble, and red. The yellow pigment is obtained by soaking safflower in water which has been slightly acidified with citric or acetic acid. The red pigment is obtained by drying the safflower, coarsely pulverizing it, and treating with a solution of Na_2CO_3 and precipitating the colouring matter with dilute acetic or citric acid.
Appearance	Dark red granular powder with a green lustre.
Physical	Slightly soluble in water, soluble in alcohol, practically insoluble in ether.
Uses	As rouge when mixed with talc or French chalk.
	RIFM Monograph (1976) *FCT*, **14**, 705.

CARVACROL [499–75–2]

2-Methyl-5-*iso*-propylphenol; 2-*p*-cymenol; isopropyl-orthocresol; isothymol
$C_{10}H_{14}O$

Source	
NATURAL	In origanum, thyme, marjoram and summer savory oils.
CHEMICAL	Chlorination of alpha-pinene with tert-butyl hypochlorite or from 2-bromo-*p*-cymol.
Appearance	Colourless liquid type.
Odour	Dry-medicinal, phenolic, herbaceous.
Physical	m.p. 1 °C, b.p. 238 °C. Volatile in steam: practically insoluble in water, freely soluble in alcohol or ether.
Uses	In dental preparations; a powerful antiseptic.
	RIFM Monograph (1979) *FCT*, **17**, 743.

CARVENE [5989–27–5]

Source	
NATURAL	A constituent of caraway oil.
Isolation	The residue after carvone has been extracted from caraway oil.

1-CARVEOL [2102–59–2]

Chemical Name	1-methyl-4-isopropenyl-6-cyclohexen-2-ol
	1-p-mentha-6,8-dien-2-ol; 6-cyclohexen-2-ol, 1-methyl-4-iso-propenyl-, 1-

Carvone

Source	
NATURAL	Present in spearmint oil in small amounts.
Isolation	Produced from l-carvone by selective reduction.
Appearance	A colourless liquid with a spearmint-like odour.
Physical	GC RIFM no 72–13; IRC RIFM no. 72–13.

CARVONE   [99-49-0] (mixed); [2244-16-8] (d-form); [6485-40-1] (l-form)

CTFA name	Carvone
	Carvol; 2-methyl-5-*iso*-propylcyclohexenone; 2-methyl-5-(1-methylethenyl)-2-cyclohexen-1-one
	$C_{10}H_{18}O$
Source	
NATURAL	In many essential oils: *l*-carvone is found in spearmint oil and *d*-carvone is found in dill and caraway seed oil.
Isolation	Distillation: can be synthesized from limonene; *l*-carvone is obtained from *d*-limonene and *d*-carvone from *l*-limonene.
Appearance	Colourless or light yellow liquid.
Odour	Attractive, spearmint, caraway, dill odour.
Physical	b.p. 231 °C.
Uses	As a flavouring agent and in fragrances to impart an aromatic note.
Comments	There are *d*-, *l*- and *dl*-forms.
	RIFM Monograph (1978) *FCT*, **16**, 673 (*d*-carvone)
	RIFM Monograph (1973) *FCT*, **11**, 1057 (*l*-carvone).

CARYOPHYLLENE   [87-44-5]

	beta-Caryophyllene
	$C_{15}H_{24}$
Source	
NATURAL	In several essential oils, notably in that of cloves.
Isolation	Distillation of clove leaf or clove stem oil.
Appearance	Colourless liquid.
Odour	Spicy, woody.
Uses	In soap perfumery as a 'light' clove oil.
Comments	The name used commercially as clove oil, from which the eugenol has been abstracted in the preparation of this phenol and also in the manufacture of vanillin.
	RIFM Monograph (1973) *FCT*, **11**, 1059.

CARYOPHYLLENE OXIDE   [1139-30-6]

Chemical name	4,11,11-trimethyl-8-methylene-5-oxatricyclo (8.2.0.0(4,6)) dodecane
	Caryophyllene epoxide; beta-caryphyllene oxide
Source	
NATURAL	Found in citrus fruits, mint, pepper and other plants and seeds.

Appearance	A white solid.
Physical	GC RIFM no. 78–24; IRC RIFM no. 78–24.

CASCARILLA OIL [8007-06-5]

History	In 1693, Stisser recorded that it was customary in this country to mix cascarilla bark with tobacco 'for the sake of correcting the smell of the latter when smoked'.
Source	
GEOGRAPHIC	A small tree indigenous to the Bahama Islands and Florida.
NATURAL	*Croton eluteria*, Bennett, and *Croton cascarilla*, Bennett.
Isolation	Steam distillation of the dried bark.
Yield	1.2–3%.
Odour	Aromatic and slightly peppery.
Chemical	Contains eugenol, amonene, etc., and cascarillin, vanillin.
Uses	In small quantities in chypre bases and masculine fragrances. RIFM Monograph (1976) *FCT*, **14**, 707.

CASEIN [9000-71-9]

CTFA name	Milk protein
Source	
NATURAL	Milk.
CHEMICAL	The solid, nitrogenous constituent of milk.
Isolation	Separated from milk by addition of a weak acid or by means of rennet.
Appearance	White amorphous powder or granules.
Odour	Without odour or taste.
Physical	Sparingly soluble in water and in non-polar organic solvents.
Chemical	Amphoteric.
Uses	Used in perfumery as the basis of many rolling massage creams.
Comments	Used industrially in the preparation of washable distemper, paints and artificial ivory.

CASSIA OIL [8007-80-5]

CTFA name	Cinnamon oil
	Chinese cinnamon oil; *Cinnamomum cassia* oil
History	Cassia oil and Chinese cinnamon oil were treated as two distinct, though closely allied, substances in the works of Theophrastus, Dioscorides, and Pliny. Cassia is also mentioned in the scriptures.
Source	
GEOGRAPHIC	A plant indigenous to Vietnam, and extensively cultivated in China.
NATURAL	*Cinnamomum cassia*, Blume, N.O. Lauraceae, which grows to a height of about 8 ft (2.5 m) after 6–7 years, with a trunk of 5 in. (127 mm) diameter.

Isolation	Steam distillation of the leaves and twigs.
Yield	0.5–1.9% depending upon raw material.
Appearance	Oily liquid.
Odour	Intense and cinnamon-like.
Chemical	Consists of cinnamic aldehyde, cinnamyl acetate, phenyl propyl acetate, methyl o-coumaric aldehyde, salicylic aldehyde, coumarin, benzoic acid, salicylic acid, benzaldehyde, methyl-salicylaldehyde.
Other	The cinnamic aldehyde content is 85–90%, which readily oxidizes to cinnamic acid.
Uses	At low levels as a blender in heavier-type perfumes.
Comments	Can contain lead as the cinnamic acid attacks the lead containers in which the oil may be stored. This should be rectified before use. RIFM Monograph (1975) *FCT*, **13**, 109 (Binder, p.194).

IFRA GUIDELINE

The Committee recommends that cassia oil should not be used as a fragrance ingredient at a level over 1% in a compound.

This recommendation is based on test results of RIFM showing weak sensitizing potential for the individual ingredient (*Food Cosmet. Toxicol.*, **13**, 109 (1975)) and absence of sensitizing reactions in a number of compounds containing cassia oil, tested on guinea pigs and humans (private communication to IFRA).

October 1974, amended October 1980

CASSIE CONCRETE/CASSIE ABSOLUTE [8023–82–3]

Source		
GEOGRAPHIC	A small shrub cultivated in southern France.	
NATURAL	*Acacia farnesiana*, N.O. Leguminoseae.	
Isolation	The concrete is obtained by solvent extraction of the flowers; further extraction produces the absolute.	
Yield	0.5% as concrete from which 30% is obtained as absolute.	
Odour	Exquisite, combining both spicy and floral notes.	
Chemical	Cassie oil contains farnesol and methyl salicylate among other substances.	
Uses	Widely in fine violet perfumes and blends with natural fragrances such as jasmine and rose.	

CASTILE SOAP

	Hard soap
Isolation	Made from lower grade olive oil and caustic soda: other oils such as coconut, sesame and tallow can be used.
Appearance	The better quality soap is white; poorer specimens are greenish.
Chemical	Consists mainly of sodium oleate.

CASTOR (CASTOREUM) [8023–83–4]

History	Castor has been used since the earliest times and is mentioned in Pomet's 'Histoire des Drogues'.

Source	
GEOGRAPHIC	Obtained from the wild beaver, *Castor fiber*, L., of Canada.
NATURAL	Extracted from the two pear-shaped sacs found in both male and female, situated between the anus and sexual organs.
Isolation	The dried preputial follicles and their secretions are extracted from the sacs, rubbed down with orris root powder or clean silver sand and macerated for several days with 20 times their weight of 90% alcohol. This is then filtered bright in the cold.
Yield	When benzene is used as the extractive agent, about 20% of the concrete is obtained. Ethyl alcohol yields the resinoid or absolute.
Appearance	Fresh castor is a creamy secretion; when dried in the sun it becomes a resinous reddish-brown substance. After filtration, the tincture is a dark liquid.
Odour	Strong and warm, resembling birch tar; becomes more pleasant on dilution.
Chemical	Contains a crystalline principle (castorin) together with resin, benzoic acid, and a volatile oil containing benzyl alcohol, acetophenone, p-ethyl-phenol, l-borneol and possibly a lactone.
Uses	In tincture form in fougeres, chypres and oriental bases; may be used in compounding artificial ambers.

RIFM Monograph (1973) *FCT*, **11**, 1061 (Castoreum).

CASTOR OIL [8001-79-4]

CTFA name	Castor oil
	Ricinus oil; oil of Palma Christi; tangantangan oil
Source	
GEOGRAPHIC	A plant indigenous to India, now cultivated in subtropical countries.
NATURAL	*Ricinus communis*, L., N.O. Euphorbiacae.
Isolation	Expressed from the seeds.
Appearance	Pale yellow viscous oil.
Odour	Characteristic, with a slightly acrid taste and nauseating after-taste.
Physical	Miscible with absolute alcohol, ether and chloroform.
Chemical	Consists of triglycerides of fatty acids, ricinoleic 87%, oleic 7%, linoleic 3%, palmitic 2% and stearic 1%: has excellent keeping qualities; does not turn rancid unless subjected to excessive heat.
Uses	In lipsticks, imparting a spreading quality and acting as a partial solvent for eosine: also an important constituent of transparent soaps. Oenanthic aldehyde, amyl cinnamic aldehyde, gamma-undecalactone, and methyl heptine carbonate are all prepared from castor oil.

CEDARWOOD OIL [85085-41-2]

History	Cedarwood was used by the King of Assyria to build his palace in Nineveh.

76   Cedrat oil

Source
GEOGRAPHIC   A tree native to North America.
NATURAL   *Juniperus virginiana*, L., N.O. Pinanceae.
Chemical   Composed chiefly of cedrene (a terpene) and cedral (cedar camphor).
Isolation   Steam distillation of the waste wood.
Yield   Approx. 3.5% depending upon quality of the raw material.
Appearance   Colourless or slightly yellow viscous liquid.
Odour   Soft and woody.
Physical   Insoluble in water, soluble in alcohol or ether.
Uses   Used extensively in the soap industry as it is cheap to manufacture: forms the basis of many perfumes, including violet, and cold cream: an adulterant of ionone and used occasionally in eau-de-botot.
Comments   An insect repellant.
    RIFM Monograph (1974) *FCT*, **12**, 845 (*Juniperus virginiana*, L.)
    (1976) *FCT*, **14**, 709 (*Cedrus atlantica*)
    (1976) *FCT*, **14**, 711 (*Juniperus mexicana*, Schiede).

CEDRAT OIL [85085-28-5]

Source
NATURAL   Usually a mixture of citrus oils. The citron of *Citrus medica*, Risso, N.O. Rutaceae yields an oil from the peel, which resembles lemon oil.

CEDRELA WOOD [93062-64-7]

Source
GEOGRAPHIC   Large trees indigenous to Central America and the West Indies.
NATURAL   The essential oil is distilled from the wood of different species of *Cedrela odorata*, L., N.O. Meliaceae.
Appearance   Oil.
Odour   Reminiscent of cedar.
Chemical   The oil from *Cedrela toona*, Roxh, known locally as red toon or Indian mahogany, was found by Pillai and Rao to contain 35% copaene, 45% cadinene and other dicyclic sesquiterpenes, and 13% cadinol.
Uses   The wood is used in the manufacture of cigar boxes.

alpha-CEDRENE EPOXIDE [13786-79-3]

    Cedroxyde
Chemical name   8,9-Epoxycedrane
    $C_{15}H_{24}O$
Appearance   Colourless to pale yellow liquid.
Odour   Woody, camphoraceous.

Chemical	The commercial product also contains beta-cedrene epoxide.
Uses	Used to give a tobacco note and blends well with patchouli and ionones to obtain herbal notes in lavender compounds: also used in soap compounds.
	RIFM Monograph (1980) FCT, **18**, 663.

CEDROL [77-53-2]

	$C_{15}H_{26}O$
Isolation	Isolated from cedarwood oil.
Appearance	White crystalline solid.
Odour	Dry, faint woody typical of cedarwood.
Physical	m.p. 86 °C.
	RIFM Monograph (1975) FCT, **13**, 745.

CEDRYL ACETATE [77-54-3]

	Cedran-8-yl acetate
	$C_{17}H_{28}O_2$
Isolation	Synthesized from cedrol and acetic anhydride.
Appearance	Colourless to pale yellow liquid or a white crystalline solid.
Odour	Warm and woody, typical of cedarwood oil.
Physical	m.p. 40–42 °C (ordinary commercial grades), up to 80 °C (refined grades).
Uses	As a low cost alternative to vetivert acetate.
	RIFM Monograph (1974) FCT, **12**, 847.

CELERY SEED OIL [8015-90-5]

	Oil of celery
Source	
GEOGRAPHIC	A celery plant cultivated in Central and Eastern Europe, India and California.
NATURAL	*Apium graveolens*, L., N.O. Umbelliferae.
Isolation	Steam distillation of the crushed seeds.
Yield	2.5–3% depending upon source of the raw material.
Appearance	Colourless to pale yellow liquid.
Odour	Persistent, spicy, celery-like and characteristic.
Physical	Slightly soluble in water, very soluble in alcohol.
Chemical	Consists of *d*-limonene, phenols, sedanolide and sedanoic acid.
Uses	Widely used in both natural-type and fantasy fragrances: used for sweet pea and tuberose.
	RIFM Monograph (1974) FCT, **12**, 849.

CERESIN [8001-75-0]

CTFA name	Ceresin
	Mineral wax; purified ozokerite; earth wax; cerosin; cerin

Ceryl cerotate

Source	
GEOGRAPHIC	A mineral wax found in Galacia and Tcheleken Island in the Caspian Sea.
NATURAL	Found in crude ozokerite, admixed with earthy impurities.
Isolation	Purified by repeated treatment with sulphuric acid: the yellow ceresin can be bleached by filtration through special earth and may then be used in several toilet preparations.
Appearance	Yellow in colour after treatment with sulphuric acid.
Odour	The white form is odourless, the yellow form has a slight odour.
Physical	m.p. of genuine samples is 60–80 °C.
Uses	As an alternative to beeswax or hard paraffin: used to raise the melting point of stick or non-aqueous cosmetics such as lipsticks.

CERYL CEROTATE [10210–18–1]

Hexacosyl hexacosanoate

Source	
NATURAL	The chief constituent of Chinese insect wax.
Appearance	A wax-like body resembling spermaceti.
Physical	m.p. 84 °C.
Uses	In eyebrow pencils and lipsticks because of its hardness. In other industries it is a useful substitute for carnauba wax.

CETACEUM [8023–73–2]

Spermaceti; spermwax

Source	
NATURAL	A concrete fatty substance obtained from the head of the sperm whale (Physeter macrocephalus).
Isolation	Separated from the sperm oil by freezing and hydraulic pressure: becomes almost neutral when refined with dilute caustic soda solution.
Appearance	Has a lustrous crystalline structure but becomes yellow and rancid on exposure to air.
Odour	Almost odourless and tasteless.
Physical	m.p. 45 °C.
Chemical	Consists mainly of cetyl palmitate (cetin): free cetyl alcohol is present in significant amounts; esters of lauric, stearic and myristic acid also are found.
Uses	In the manufacture of both liquid and solid creams and in lipsticks.

CETOMACROGOL 1000 [9004–95–9 (generic)]

CTFA name	Ceteth-20
	Polyethylene glycol 1000 monocetyl ether
Source	
CHEMICAL	A polyglycol ether containing 20 to 24 oxyethylene groups in the polyoxyethylene chain.
Appearance	A waxy, pale cream material.

CETYL ALCOHOL [36653-82-4]

CTFA name	Cetyl alcohol
	1-Hexadecanol; n-hexadecyl alcohol; palmityl alcohol; ethol
	$C_{16}H_{33}OH$
Source	
NATURAL	Can be obtained from cetyl palmitate found in spermaceti or indirectly from coconut oil.
Isolation	Saponification of cetyl palmitate found in spermaceti.
Appearance	Waxy, scaly powder.
Odour	Completely odourless and tasteless.
Physical	Melting range 48–49 °C; has an acid and ester value of 0: completely miscible with mineral and vegetable oils and fats: readily soluble in alcohol, ether, carbon disulphide, glycol and diglycol ethers.
Chemical	Completely stable in the presence of acids, alkalis, light and air.
Uses	A useful auxiliary emulsifying agent in both oil-in-water and water-in-oil emulsions. Addition of 1–5% in many cosmetics has a marked emollient effect on the skin: used in foundation creams, brushless shaving creams and lipsticks. Cosmetic grades contain approximately 70% cetyl alcohol, the remainder being a mixture of myristyl and stearyl alcohol with smaller proportions of unsaturated alcohols.
Comments	The material known as cetostearyl alcohol has a similar effect to cetyl alcohol and is used in a similar manner. It consists of approx. 45% cetyl alcohol and about 40% stearyl alcohol. The remainder is largely myristyl alcohol.
	RIFM Monograph (1978) *FCT*, **16**, 683.

CHALK (PRECIPITATED) [13397-25-6]

CTFA name	Chalk
	Calcium carbonate
	$CaCO_3$
Isolation	Obtained by adding a solution of sodium carbonate to a solution of calcium chloride.
Appearance	A fine white powder, composed of microscopic crystals of calcium carbonate.
Odour	Odourless and tasteless.
Uses	In the preparation of face powders and dentifrices.

CHALK (PREPARED) [13397-25-6]

CTFA name	Chalk
	Native calcium carbonate

Chamomile oil blue (German)

Isolation	Native calcium carbonate is purified by elutriation.
Appearance	Small conical pieces.
Uses	Preparation of compact powders.

CHAMOMILE OIL BLUE (GERMAN) See CAMOMILE OIL, GERMAN

CHAMOMILE OIL, ROMAN See CAMOMILE OIL, ROMAN

CHAMPACA OIL [94333-99-0]

Champaca concrete, champaca absolute

Source	
GEOGRAPHIC	A tree of the Magnolia family and native to the Phillipines and Indonesia.
NATURAL	Either *Michelia champaca*, L., or *M. longifolia*, Bl.
Isolation	Extracted from the flowers; the concrete is first obtained; further extraction yields the absolute.
Yield	Very low, usually less than 0.1% of concrete of which about 25% is absolute.
Appearance	The light yellow or reddish yellow oil obtained from the yellow flowers of *Michelia champaca* is slightly fluorescent.
Odour	The odour of the oil from *M. champaca* is fragrant and that from *M. longifolia* is more fruity.
Physical	Slightly soluble in water or alcohol, soluble in chloroform or ether.
Chemical	According to B. T. Brooks, the oil from *M. champaca* contains cineol, *p*-cresyl methyl ether (?), benzaldehyde, benzyl alcohol, benzoic acid, and phenylethyl alcohol, while that from *M. longifolia* contains linalol, methyleugenol, methylacetic acid and a phenol having an odour of thymol. Other chemists have demonstrated the presence of iso-eugenol, geraniol, and methyl anthranilate.
Uses	Almost all used in fine jasmine flower perfumes.
Comments	The majority of champaca perfumes are prepared artificially from these bodies as genuine champaca oil is very scarce, and limited to only a few kg per annum. (Champaca leaf oil is prepared on a large scale in Java and the odour recalls that of basil.) It should be kept well closed, cool and protected from light.

CHAMPACOL [489-86-1]

Guaiol; champaca camphor; guaiac alcohol
$C_{15}H_{26}O$

Source	
NATURAL	In the oil distilled from guaiac wood.
CHEMICAL	A sesquiterpene alcohol.
Appearance	Crystalline.

Odour	Reminiscent of tea roses.
Physical	m.p. 91 °C: soluble in alcohol and ether, insoluble in water.
Comments	The name 'guaiol' is also applied to 1,2-dimethyl-acrolein.

CHARCOAL (WOOD)

Source	
NATURAL	Generally willow.
Isolation	Heating different woods to redness in an enclosed space in the absence of oxygen.
Appearance	Black solid.
Uses	In some tooth powders and also in the manufacture of incense and fumigating pastilles.

CHAVICOL [501–92–8]

p-allyl phenol
$C_9H_{10}O$

Source	
NATURAL	In bay and other essential oils.
CHEMICAL	Aromatic phenol.
Appearance	Colourless liquid.
Odour	Powerful, dry, tar-like, medicinal note.
Physical	m.p. 16 °C, b.p. 236 °C. Miscible with alcohol, ether or chloroform.

CHERRY KERNEL OIL [8022–29–5]

CTFA name	Cherry pit oil
Isolation	Expressed from machine-separated cherry kernels (Prunus avium).
Appearance	A brownish-yellow colour, which becomes pale golden after purification.
Odour	Nutty.
Physical	Keeping qualities are said to be good.

CHLORAMINE [127–65–1]

Sodium salt of N-chloro-p-toluenesulphonamide; chloramine-T
$CH_3C_6H_4S(=O)(—ONa)(=NCl)$

Appearance	White crystalline powder.
Odour	Slight chlorine odour.
Physical	The anhydrous compound explodes at 175–180 mm Hg: incompatible with alcohol.
Other	Non-irritant and non-toxic.
Uses	An antiseptic.
Comments	Should be kept well closed.

Chloranthus inconspicuus

Source
GEOGRAPHIC   A Chinese shrub (Chu Lan Hwa).
Uses   The blossoms (known as 'pearl of flowers') are used for perfuming tea.

CHLORHEXIDINE DIACETATE [56-95-1]

CTFA name   Chlorhexidine diacetate
              1,1′-Hexamethylene-bis(5-(*p*-chlorophenyl) biguanide) diacetate
              $C_{22}H_{30}N_{10}Cl_2 \cdot 2CH_3COOH$
Appearance   White crystalline powder.
Odour   None.
Physical   Soluble 1 in 50 of water; soluble in alcohol, glycerol, propylene glycol and polyethylene glycols.
Uses   A bactericide against a wide range of Gram-negative and Gram-positive organisms, in shampoos and in aqueous or aqueous/alcoholic lotions: also used in cosmetic creams, and deodorant preparations, including vaginal sprays.
Comments   The related compound chlorhexidine dihydrochloride, $C_{22}H_{30}N_{10}Cl_2 \cdot 2HCl$, is an odourless white crystalline powder soluble 1 in 700 of water. It has similar bactericidal properties, but restricted use because of its lower solubility in water.

CHLOROFORM [67-66-3]

              Trichloromethane
              $CHCl_3$
Appearance   A highly refractive non-flammable, colourless, mobile liquid.
Odour   Diffusive and characteristic sweet-ethereal smell.
Physical   b.p. 61 °C, m.p. −63 °C.
Uses   Mainly as an anaesthetic: has had some limited use as a sweetening agent in dental preparations.
Comments   Pure chloroform is light-sensitive and should be kept cool and away from light. It is banned by the FDA from cosmetic use and has been listed as a carcinogen by the EPA.

CHLOROPHYLL [1406-65-1]

Source
NATURAL   The green colouring matter found in the leaves and stalks of plants.
CHEMICAL   Chlorophylls a, b, c and d exist.
Isolation   Extracted commercially from nettles and spinach by percolation with a volatile solvent, which is eventually removed by distillation at low temperatures or by treatment with dilute alkalis and precipitation by mineral acids.

Appearance	Varies according to the variety but commercial chlorophyll is an intensely dark-green aqueous, alcoholic or oil solution.
Other	Inclined to precipitate; long standing and decantation is desirable.
Uses	As a pigment for tinting oils and as a natural antiseptic and healing agent: also used as a deodorant and breath freshener and in toothpaste as an antiseptic and mouth freshener.
Comments	Dyestuffs have largely replaced chlorophyll in colouring aqueous and spirituous preparations and soaps.

CHLOROXYLENOL [88–04–0]

CTFA name	Chloroxylenol 4-Chloro-3,5-dimethylphenol; 4-chloro-3,5-xylenol; p-chloro-m-xylenol; 2-chloro-m-xylenol; $para$chloro-$meta$xylenol, PCMX C_8H_9OCl
Appearance	White or colourless crystals.
Odour	Characteristic, mild phenolic.
Physical	Soluble in alcohol, most organic solvents, and fixed oils; 0.03% in water; m.p. 115 °C, b.p. 246 °C.
Uses	As a bactericide for the preparation of antiseptic lotions; also an effective preservative.
Comments	Of limited use in cosmetic products because of its odour.

omega-CHLORSTYRENE [622–25–3]

	omega-Chlorstyrole; 1-chloro-2-phenylethylene $C_6H_5CH=CHCl$
Appearance	Colourless mobile liquid.
Odour	Penetrating harsh floral of the hyacinth type.
Physical	b.p. 199 °C.
Chemical	A mixture of the $trans$ and cis isomers (predominantly $trans$).
Comments	Not so fine as bromstyrol.

CHOLESTEROL [57–88–5]

CTFA name	Cholesterol Cholesterin; cholest-5-en-3-ol (3-beta-) $C_{27}H_{46}O$
Source	
NATURAL	The ester in lanolin or wool fat: occurs in glandular secretions.
Appearance	A white to yellowish substance which crystallizes in pearly plates containing water of crystallization.
Odour	None.
Physical	m.p. 148 °C: insoluble in water; soluble in ether, oils, fats, and hot alcohol.
Chemical	Anhydrous.

Other	Has water-absorbing properties.
Uses	A useful stabilizer for emulsion systems, particularly of the water-in-oil type: used in alcoholic hair lotions, creams and other cosmetic products for the treatment of greasy skin.

CHYPRE PERFUMES

Source	
NATURAL	Based upon oakmoss, patchouli, labdanum, and clary sage; the flowery note comes from rose, jasmine etc.: bergamot or lemon is a desirable addition.
Odour	The combination gives a perfume a sweet, appealing note.
Other	Remarkable strength and persistence.

CINEOLE [470-82-6]

CTFA name	Eucalyptol
	Cajeputol; 1,3,3-trimethyl-2-oxabicyclo-[2.2.2]-octane; 1,8-cineol; eucalyptol
	$C_{10}H_{18}O$
Source	
NATURAL	A constituent of many oils, including eucalyptus, cajuput and spike lavender: present in basil, champaca, and bois de rose oils.
Appearance	Colourless liquid.
Odour	Diffusive, camphor-like, spicy with a cooling taste.
Physical	b.p. 176–177 °C, m.p. 1.5 °C: 0.4% soluble in water at room temperature, miscible with alcohol, chloroform and ether.
Uses	As a flavour in dental preparations.
	Widely used in herbaceous, lavender, fougere types, etc.
	RIFM Monograph (1975) *FCT*, **13**, 105.

CINNAMIC ACID [140-10-3]

	trans-3-Phenylacrylic acid; 3-phenyl-2-propenoic acid; *trans*-cinnamic acid
	$C_6H_5CH{=}CHCOOH$
Source	
NATURAL	In several essential oils and gum resins, notably cassia, Tolu and Peru balsams, and storax.
Isolation	Synthesized by heating together benzylidene chloride and sodium acetate, or by mixing together benzylidene acetone, soda, and bleaching powder: the resulting sodium cinnamate is decomposed with sulphuric acid.
Appearance	White crystalline solid.
Odour	Pleasant balsamic and faintly honey-like.
Physical	b.p. 300 °C, m.p. 134 °C.

Uses	As a fixative; a good stabilizer in soaps. RIFM Monograph (1978) *FCT*, **16**, 687.

CINNAMIC ALCOHOL [104–54–1]

trans-Cinnamyl alcohol; 3-phenylallyl alcohol; styrone

Source	
NATURAL	In several essential oils: also in balsam of Peru and storax.
Isolation	May be synthesized, but is frequently extracted from storax after hydrolysis with an equal weight of 10% caustic soda solution.
Appearance	White or opaque crystalline solid.
Odour	Attractive warm balsamic, floral, sweet note; combines with almost any compound to add to its sweetness and persistence.
Physical	Almost insoluble in water, soluble in glycerol, freely soluble in organic solvents; m.p. 32–33 °C, b.p. 258 °C.
Chemical	Methyl and ethyl esters of the synthetic compound are excellent fixatives.
Other	Oxidized slowly on exposure to heat, light and air.
Uses	Useful in the preparation of hyacinth, lily and lilac perfumes.
Comments	The product obtained from storax is of a better and finer quality. RIFM Monograph cinnamic alcohol (1974) *FCT*, **12**, 855. Special Issue I (Binder, p. 214). IFRA GUIDELINE The Committee recommends that cinnamic alcohol should not be used as a fragrance ingredient at a level over 4% in fragrance compounds. This recommendation is made in order to promote Good Manufacturing Practice (GMP) for the use of cinnamic alcohol as a fragrance ingredient. It is based on test results showing a weak sensitizing potential for the individual ingredient (private communication to IFRA) and the absence of sensitizing reactions in tests of a considerable number of formulations containing cinnamic alcohol (R. J. Steltenkamp, K. A. Booman, J. Dorsky, T. O. King, A. S. Rothenstein, E. A. Schwoeppe, R. I. Sedlak, T. H. F. Smith and G. R. Thompson (1980), *Food Cosmet. Toxicol.*, **18**, 419–25).

October 1978, last amendment February 1983

CINNAMIC ALDEHYDE [104–55–2]

CTFA name	Cinnamal Cinnamaldehyde; 3-phenyl-2-propenal; *trans*-3-phenylprop-2-en-1-al, *trans*-Cinnamic Aldehyde, Cassia Aldehyde $C_6H_5CH{=}CHCHO$
Source	
NATURAL	In cinnamon or cassia oils and in the oils of patchouli, myrrh, and *Melaleuca bracteata*.

CHEMICAL	May be synthesized by condensation of benzaldehyde with acetaldehyde.
Isolation	Separated from cinnamon or cassia oil by sodium bisulphite.
Appearance	Yellow liquid.
Odour	Powerful warm spicy balsamic, suggestive of Cassia.
Physical	b.p. 252 °C (slight decomp.), m.p. -7.5 °C.
Uses	In perfumery as a powerful spicy ingredient in low cost household products and soaps; will discolour white soap and detergents. Used in conjunction with d-limonene or eugenol to remove its sensitizing potential. RIFM Monograph (1979) FCT, 17, 253. IFRA GUIDELINE The Committee recommends the use of cinnamic aldehyde as a fragrance ingredient in conjunction with substances preventing sensitization, as for example equal weights of eugenol or d-limonene. This recommendation is based on results showing sensitizing potential for the individual ingredient, but absence of sensitizing reaction in a number of compounds (private communication to IFRA), as well as in mixtures with equal weights of eugenol or d-limonene (D. L. J. Opdyke (1976), Food Cosmet. Toxicol., 14, 197, D. L. J. Opdyke (1979), Food Cosmet. Toxicol., 17, 253–8). October 1975, amended March 1978

CINNAMON BARK OIL [8015-91-6]

Source	Cinnamon oil
GEOGRAPHIC	A tree cultivated in Sri Lanka, but also grows wild here and in Madagascar, Southern India and the Comoro Islands.
NATURAL	*Cinnamomum ceylanicum*, Breyn, N. O. Lauraceae.
Isolation	Steam distillation of the dried bark: some trimmings and broken chips are distilled on the island and some are exported and distilled in Europe.
Yield	0.5–1.0% depending upon source of the raw material.
Appearance	A pale to dark yellow oil.
Odour	A warm, delicate odour of bark: English-distilled oils often possess a more delicate odour.
Chemical	Consists mainly of cinnamic aldehyde: the characteristic odour is probably due to the modifying influence of eugenol (4–10%), a higher aliphatic aldehyde and phellandrene.
Uses	Widely used in medicinal preparations, and as a flavour in dental preparations and aromatic cachous: used in small quantities in certain oriental-type perfumes.
Comments	Artificial cinnamon oil contains, in addition, amyl methyl ketone, cuminic aldehyde, phenyl propyl aldehyde, benzaldehyde, linalol and methyl eugenol.

RIFM Monograph (1975) *FCT*, **13**, 111 (Ceylon).
IFRA GUIDELINE
The Committee recommends that cinnamon bark oil, Ceylon, should not be used as a fragrance ingredient at a level over 1% in a compound.

This recommendation is based on test results of RIFM showing sensitizing potential for the individual ingredient (*Food Cosmet. Toxicol.*, **13**, 111 (1975)), and absence of sensitizing reactions in a number of compounds containing this material at low concentrations, tested on guinea pigs and humans (private communication to IFRA).

October 1974, amended October 1978

CINNAMON LEAF OIL [8015–96–1]

Source
GEOGRAPHIC A tree cultivated in Sri Lanka, South Kanara, and possibly other islands in the East Indies.
NATURAL *Cinnamomum ceylanicum*, Breyn, N. O. Lauraceae.
Isolation In Sri Lanka, the leaves are macerated in sea-water before steam distillation.
Yield 1.6–1.8%.
Appearance Bright yellow oil.
Odour Warm, recalling both cinnamon and clove.
Chemical Consists largely of eugenol with traces of safrole and benzaldehyde.
Uses A soap perfume.
RIFM Monograph (1975) *FCT*, **13**, 749.

CINNAMYL ACETATE [103–54–8]

3-Phenylallyl acetate, *trans*-cinnamyl acetate
$CH_3COOCH_2{=}CHC_6H_5$

Source
NATURAL In cassia oil.
Appearance A colourless oily liquid.
Odour Tenacious sweet, woody, rose–vetivert odour.
Physical b.p. 262 °C.
Uses As a base: blends well with vetivert, coumarin, santal and ambrette musk.
RIFM Monograph (1973) *FCT*, **11**, 1063.

CINNAMYL BENZOATE [5320–75–2]

3-Phenylallyl benzoate; *trans*-cinnamyl benzoate
$C_6H_5COOCH_2CH{=}CHC_6H_5$
Appearance White crystalline powder.
Odour Mild balsamic almost odourless.

Physical	b.p. 335 °C.
Uses	An excellent fixative for oriental and heavy floral fragrances. RIFM Monograph (1976) FCT, **14**, 717.

CINNAMYL iso-BUTYRATE [103–59–3]

Chemical name	3-Phenylallyl 2-methylpropanoate $(CH_3)_2CHCOOCH_2CH=CHC_6H_5$
Appearance	Colourless oily liquid.
Odour	Fruity, mild balsamic reminiscent of ripe banana.
Physical	b.p. 254 °C.
Uses	Similar to those of the normal butyrate. RIFM Monograph (1979) FCT, **17**, 523.

CINNAMYL n-BUTYRATE [103–61–7]

Chemical name	3-Phenylallyl butanoate $CH_3CH_2CH_2COOCH_2CH=CHC_6H_5$
Source	
NATURAL	None.
Isolation	Prepared synthetically.
Appearance	Colourless liquid.
Odour	Mild, and fruity balsamic.
Physical	b.p. 300 °C.
Chemical	Blends well with hydroxy-citronellal, cananga, and linaloe oils.
Uses	Useful in sweet pea fragrances and in traces to modify perfumes of the rose–lily type. RIFM Monograph (1978) FCT, **16**, 691.

CINNAMYL CINNAMATE [122–69–0]

	3-Phenylallyl cinnamate; styracin $C_6H_5CH=CHCH_2COOCHCHC_6H_5$
Source	
NATURAL	In storax, Peru balsam, and melaleuca oil.
Appearance	Crystalline solid.
Odour	Soft balsamic of great tenacity.
Physical	m.p. 44 °C, b.p. 370 °C.
Uses	As a fixative in heavy-type perfumes. RIFM Monograph (1975) FCT, **13**, 753.

CINNAMYL FORMATE [104–65–4]

	3-Phenylallyl formate $HCOOCH_2=CHC_6H_5$
Appearance	Liquid.
Odour	Green herbaceous, 'dry' balsamic.

Physical	b.p. 250 °C.
Uses	Oriental and chypre types and with woody or patchouli.

RIFM Monograph (1976) *FCT*, **14**, 719.

CINNAMYL NITRILE [1885-38-7]

Cinnamonitrile; 2-propenenitril, 3-phenyl-
$C_6H_5 \cdot CH:CH \cdot CN$

Isolation	From styryl bromide and potassium cyanide.
Appearance	Colourless oily liquid.
Physical	GC RIFM no. 74–174; IRC RIFM no. 74–174.

CINNAMYL PROPIONATE [103-56-0]

3-Phenylallyl propanoate
$C_2H_5COOCH_2CH=CHC_6H_5$

Appearance	Colourless liquid.
Odour	Fruity balsamic.
Physical	b.p. 289 °C.
Uses	In the manufacture of artificial fruit essences; in perfumery in narcissus compounds.

RIFM Monograph (1974) *FCT*, **12**, 859.

CINNAMYL *iso*-VALERATE [140-27-2]

3-Phenylallyl 3-methylbutanoate; cinnamyl valerianate
$(CH_3)_2CHCH_2COOCH_2CH=CHC_6H_5$

Appearance	Colourless liquid.
Odour	Fresh, floral, rosy with sweet tobacco undertones.
Physical	b.p. 313 °C.
Uses	In tea rose compounds as a modifier.

RIFM Monograph (1974) *FCT*, **12**, 857.

CINNAMYLIDENE METHYL CARBINOL

5-Phenylpenta-2,4-dien-2-ol; Homocinnamyl alcohol
$C_6H_5CHCHCHC(OH)CH_3$

Appearance	Colourless or pale yellow oily liquid.
Odour	Tenacious warm balsamic-green note.
Physical	b.p. 240 °C.
Uses	In perfumery in lilac and narcissus compounds.

CITRAL [5392-40-5]

CTFA name	Citral

3,7-Dimethylocta-2,6-dien-1-al; 3,7-dimethyl-2,6-octadienal; Neral
$C_{10}H_{16}O$

Citronellal

Source	
NATURAL	In lemon, lime, mandarin, verbena, lemon-grass, and *Backhousia citriodora* oils.
Isolation	Synthesized by the oxidation of geraniol: may be separated from lemon-grass and *Litsea cubeba*.
Appearance	Colourless oily liquid.
Odour	A fresh lemon odour: the difference in the odour of citral from lemon oils and lemon-grass oils has been attributed to a difference in the ratio of citral alpha to citral beta, which in lemon-grass oil occurs as 85–90% of the former and 10–15% of the latter.
Physical	Practically insoluble in water, miscible with alcohol, ether and other organic solvents; b.p. 228 °C.
Chemical	An aldehyde; consists of the two isomers, geranial and neral.
Uses	In the manufacture of ionone and in conjunction with *d*-limonene or other citrus terpenes to impart a fresh lemon odour to perfumes and in particular a peculiar note to odours of the rose type.
Comments	Will cause discoloration of white soaps and alkaline cosmetics. RIFM Monograph (1979) *FCT*, **17**, 259.

IFRA GUIDELINE

The Committee recommends the use of citral as a fragrance ingredient in conjunction with substances preventing sensitization, as for example 25% *d*-limonene, or mixed citrus terpenes, or α-pinene.

This recommendation is based on results showing sensitizing potential for the individual ingredient, but absence of sensitizing reactions in a number of compounds (private communication to IFRA), with lemon-grass oil, as well as in mixtures of 80 parts of citral to 20 parts of *d*-limonene, or mixed citrus terpenes, or α-pinene (D. L. J. Opdyke (1976), *Food Cosmet. Toxicol.*, **14**, 197, D. L. J. Opdyke (1979), *Food Cosmet. Toxicol.*, **17**, 259–66).

October 1975

CITRONELLAL [106–23–0]

Rhodinal; 3,7-dimethyloct-6-en-1-al
$C_{10}H_{18}O$

Source	
NATURAL	Isolated from citronella and *Eucalyptus citriodora* oils.
Isolation	Distilled from these oils using sodium bisulphite, or synthesized from pinene via myrcene and geraniol as intermediates.
Appearance	Colourless liquid.
Odour	Powerful fresh green-citrusy, herbaceous, woody.
Physical	Soluble in alcohols, very slightly soluble in water; b.p. 206 °C.
Uses	In cheap soap perfumes such as citronella, honey or cologne types. RIFM Monograph (1975) *FCT*, **13**, 755.

CITRONELLA OIL [8000-29-1]

Source	Citronella oil Java, citronella oil Ceylon
GEOGRAPHIC	Two varieties of grass which are cultivated on large plantations in China, Taiwan, Indonesia, Brazil, Colombia and Guatemala.
NATURAL	Citronella-grass, *Cymbopogon winterianus*, Jowitt, and *Cymbopogon nardus*.
Isolation	Steam distillation of the dried grass. The grass is dried for 3–4 days, which reduces bulk and moisture content, thus saving fuel and facilitating the separation of the oil from the water of distillation.
Yield	Approx. 1%: estimated yield per acre per annum (four crops) is about 8 tons (8.1 t) (about 68 lb (30 kg) oil).
Appearance	Oily liquid.
Odour	The oil from Java has a fresh, grassy odour; the Sri Lankan oil has a cruder, camphoraceous odour.
Physical	Slightly soluble in water.
Chemical	Identifiable constituents of citronella oil are: citronellal, camphene, dipentene, methyl heptenone, borneol, geraniol, methyl eugenol, esters of valeric acid, *l*-limonene, a body related to linalol, an alcohol resembling thujyl alcohol, esters of *d*-citronellol, nerol geranyl acetate and farnesol. The geraniol content of the oil varies: 80–90% for the Java oil, 60–65% for the Sri Lankan oil.
Uses	Largely used as the raw material for the working up of citronellal and second quality geraniol: used as the basis of cheap perfumes, of the curd and castile type, in the soap industry: may also be employed in fragrances for household and cleansing products.
Comments	An insect repellant: should be kept well closed, cool and away from light. RIFM Monograph (1973) *FCT*, **11**, 1067 (*Cymbopogon nardus*, Rendle).

CITRONELLOL [106-22-9]

dl-citronellol; 3,7-dimethyl-6-octen-1-ol; 6-octen-1-ol, 3,7-dimethyl-

CITRONELLYL ACETATE [150-84-5]

Chemical name	3,7-Dimethyloct-6-en-1-yl acetate $C_{12}H_{22}O_2$
Source	
NATURAL	In citronella oil and probably in the leaves of *Piper volkensii*.
Isolation	Synthesized from citronellol and acetic anhydride.
Appearance	Colourless, mobile liquid.

Odour	Fruity, resembling lime and bergamot oils.
Physical	b.p. 229 °C.
Uses	Extensively used in almost any type of perfume, particularly in rose, carnation, muguet and lavender compounds.
Comments	Citronellyl monochloracetate has a sharp odour of a similar type and is used in lilac and lily compounds. RIFM Monograph (1973) *FCT*, **11**, 1069.

CITRONELLYL BENZOATE [10482-77-6]

$C_6H_5COOCH_2CH_2CH(OH_3)CH_2CH\!=\!C(CH_3)_2$

Appearance	Colourless oily liquid.
Odour	Reminiscent of dried rose petals.
Physical	b.p. 340 °C.
Uses	A fixative in all rose compounds.

CITRONELLYL iso-BUTYRATE

3,7-Dimethyloct-6-en-1-yl 2-methylpropanoate
$(CH_3)CHCOOCH_2CH_2CH(CH_3)CH_2CH_2CH\!=\!C(CH_3)_2$

Properties	Similar to the *n*-butyrate. RIFM Monograph (1978) *FCT*, **16**, 693.

CITRONELLYL n-BUTYRATE [141-16-2]

3,7-Dimethyloct-6-en-1-yl *n*-butanoate
$CH_3CH_2CH_2COOCH_2CH_2CH(CH_3)CH_2CH_2CH\!=\!C(CH_3)_2$

Source	
NATURAL	In citronella oil Ceylon.
Appearance	Colourless liquid.
Odour	Fruity, citrus-rosy character.
Physical	b.p. 249 °C.
Uses	In small quantities with oakmoss resin to impart an odour of moss roses to a basic rose oil. RIFM Monograph (1973) *FCT*, **11**, 1071.

CITRONELLYL CAPROATE [10580-25-3]

3,7-Dimethyloct-6-en-1-yl *n*-hexanoate; citronellyl capronate; citronellyl hexylate
$CH_3(CH_2)_4COOCH_2CH_2CH(CH_3)CH_2CH_2CH\!=\!C(CH_3)_2$

Appearance	Colourless liquid.
Odour	Delicate floral with fruity rosaceous note.
Physical	b.p. 240 °C.

CITRONELLYL FORMATE [105–85–1]

3,7-Dimethyloct-6-en-1-yl formate
HCOOCH$_2$CH$_2$CH(CH$_3$)CH$_2$CH$_2$CH=C(CH$_3$)$_2$
Appearance — Colourless mobile liquid.
Odour — Powerful leafy green reminiscent of geranium leaves.
Physical — b.p. 235 °C.
Uses — In special types of rose perfumes: forms an excellent base for artificial muguet when suitably backed with ylang-ylang and hydroxyl citronellal.
RIFM Monograph (1973) *FCT*, **11**, 1073.

CITRONELLYL NITRILE [52671–32–6]

Source
NATURAL — Not found in nature.
Isolation — Treatment of citronellyl halide with potassium cyanide; or reacting citronellyl oxime with acetic anhydride.

CITRONELLYL OXYACETALDEHYDE [7492–67–3]

Chemical name — 6,10-Dimethyl-3-oxa-9-undecenal
Muguet aldehyde
C$_{12}$H$_{22}$O$_2$
Isolation — From citronellol acetal.
Appearance — Colourless viscous liquid.
Odour — Powerful and diffusive green muguet-lily-like character.
Physical — b.p. 239 °C.
Uses — Mainly as a floralizer in muguet-lily compositions but also useful in rose, lilac and other floral fragrance types.
Comments — Commercial product usually contains a significant amount of the geranyl isomer.
RIFM Monograph (1974) *FCT*, **12**, 861, Special Issue I (Binder, p. 241).

CITRONELLYL PROPIONATE [141–14–0]

3,7-Dimethyloct-6-en-1-yl propanoate
C$_2$H$_5$COOCH$_2$CH$_2$CH(CH$_3$)CH$_2$CH$_2$CH=C(CH$_3$)$_2$
Appearance — Colourless liquid.
Odour — Fruity rose odour.
Physical — b.p. 242 °C.
Uses — As a modifier in rose perfumes: valuable in honeysuckle compounds.
RIFM Monograph (1975) *FCT*, **13**, 759.

CITRONELLYL n-VALERATE

	3,7-Dimethyloct-6-en-1-yl pentanoate; citronellyl valerianate
	$CH_3(CH_2)_2COOCH_2CH_2CH(CH_3)CH_2CH_2CH=C(CH_3)_2$
Appearance	Liquid.
Odour	Warm-rosy, herbaceous with tobacco-like undertones.
Physical	b.p. 253 °C.
Uses	Frequently used as a flavour for tobaccos; may also be used as a blender in perfumery.

CIVET [68916-26-7]

History	In earliest times civet was used to perfume the gloves of Italians and Spaniards.
Source	
GEOGRAPHIC	A species of cat found principally in Ethiopia.
NATURAL	A glandular secretion of *Viverra civetta*.
Isolation	The cats (about the size of a fox), with grey fur and black spots, are kept in heated cages and fed on fresh meat. The quantity and quality of the secretion largely depends upon their food, although that of the male is superior. The secretion is collected from the pouch with a small ivory or bamboo spoon and packed in zebu horns for export. Extraction, using a volatile solvent, yields civet absolute.
Yield	The procedure is repeated several times a week and the average yield per animal is 4 g.
Appearance	Soft, fatty, and paste-like: after extraction it appears as a black or deep brown viscous liquid.
Odour	A powerful, characteristic, and animalic odour due to a ketone called zibetone.
Chemical	Liable to adulteration with yellow paraffin jelly, mucilage of gum acacia, honey, beeswax, lanolin, banana pulp, coconut fat, and butter.
Uses	Arab women are said to use civet for beautifying the hair and eyebrows: used in perfumery almost exclusively as a fixative.

RIFM Monograph (1974) *FCT*, **12**, 863 (absolute).

CIVET SYNTHETIC

Source	
CHEMICAL	A mixture of indole, skatole, phenylacetic acid etc.
Isolation	These chemicals and others are mixed, melted together, crystallized and powdered.
Odour	Lacks the characteristic softness found on dilution.
Comments	Unlikely to replace the natural product.

CIVETTONE [542-46-1]

Cycloheptadecen-9-one-l.

Source	
NATURAL	In natural civet.
Isolation	This ketone may be attributed to the transformation of oleic acid.
Appearance	White or colourless crystalline mass.
Odour	Delicate sweet animalic and musky.
Physical	Readily soluble in alcohol; m.p. 34 °C, b.p. 344 °C.
Uses	Employed in fine perfumery for chypres, oriental, citrus colognes and many other types.
	RIFM Monograph (1976) *FCT*, **14**, 727 (civetone).

CLARY SAGE ABSOLUTE [8016–63–5]

Source	
NATURAL	The clary sage plant cultivated in France, Spain and the USSR.
Isolation	Extraction with volatile solvents.
Appearance	Greyish or greenish-yellow with the consistency of hard honey.
Odour	A lasting subtle fragrance.
Physical	Soluble in the principal organic solvents and essential oils.
Chemical	The main constituents are free acetic acid and traces of free unsaturated acid, linalol and linalyl acetate, a substance similar to cedrene, sclareol (crystalline), and a sesquiterpene alcohol.
Uses	In chypres and semi-heavy toilet waters.

CLARY SAGE CONCRETE [8016–63–5]

Source	
GEOGRAPHIC	The clary sage plant, cultivated in France, Spain and the USSR.
NATURAL	*Salvia sclarea*.
Isolation	Extraction of the flowering tops of the plants with volatile solvents.
Yield	Approx. 0.5%.
Appearance	Green.
Odour	Resembles ambergris.
Chemical	Insoluble in alcohol due to the presence of waxes etc.
Uses	Incorporated in soap during the milling process.

CLARY SAGE OIL [8016–63–5]

History	The plant was introduced to England in 1562 and was for some time a popular medicine in the treatment of digestive disorders. It was first employed commercially by the German wine merchants and in 1908 was cultivated in the Piedmont district of Italy where the powdered flowers were used in the manufacture of various brands of vermouth.
Source	
GEOGRAPHIC	A xerophilous plant cultivated in France, Italy, and Northern Africa.
NATURAL	*Salvia sclarea*, N.O. Labiatae.

Isolation	Distillation of the inflorescences, or more frequently the whole herb. It is thought to be advantageous to distil the stalks and leaves separately from the tips. The amount of essential oil obtained increases with the fall of the blossoms until the seeds have reached maturity.
Yield	Approximately 1%.
Appearance	A bright olive-green oil.
Odour	The inflorescence and the herb have a musky-amber odour: the oil has a fine fragrance of ambergris, musk, neroli and lavender.
Chemical	Contains linalol and linalyl acetate among other constituents: has low solubility in alcohol.
Other	Subject to adulteration with linalol, linalyl acetate, and propionate or mixtures, which alters the odour of the oil, particularly its strength.
Uses	In eau-de-cologne to impart a sweetness, mellowness and persistence: invaluable in ambers, chypre, carnation, trefle, coin coupe, and orchidee, and a small quantity greatly improves most artificial perfumes. The water which distills over with the oil is used commercially in wines, and as a diluent for liquid dentifrices.

RIFM Monograph (1974) *FCT*, **12**, 865
(1982) *FCT*, **20**, 823.

CLERODENDRONS

C. trichotomum

Source	
NATURAL	A Japanese plant which bears large loose clusters of white flowers: grown to a limited extent in England and attains a height of about 5 ft (1.5 m).
Odour	The flowers have a sweet honeysuckle–verbena fragrance, with a distinct odour recalling jasmine, rich in indole, in the after-smell.
Comments	An artificial imitation could be based on hydroxyl citronellal, ionone, and ylang-ylang, with the floral note being imparted with jasmine and tuberose.

CLOVE OIL [8000–34–8]

CTFA name	Clove oil
	Clove bud oil, clove leaf oil
History	Cloves were known in China by 266 BC. In 1808 the Arabs brought the first cloves to Pemba and Tanzania.
Source	
GEOGRAPHIC	A tree cultivated in Tanzania, Indonesia, Madagascar and Sri Lanka.
NATURAL	*Eugenia caryophyllata*, Thunberg, N. O. Myrtaceae.

Isolation	Steam distilled from the dried unexpanded flower buds, or from the leaves and twigs.
Yield	Approx. 15% from the buds and 2–3% from the leaves and twigs.
Appearance	A colourless to pale yellow liquid; darkens and thickens with age.
Odour	Clove bud oil has a warm, sweet clove odour; that from the leaves and twigs has a typical clove odour, but is dry and transparent.
Physical	Insoluble in water.
Chemical	Consists chiefly of eugenol.
Uses	In Java, the natives mix cloves with tobacco for smoking. Madagascar clove oil, containing upwards of 85% of eugenol, is used as the main raw material in the synthesis of vanillin. In perfumery, it is used in floral perfumes, and blends well with geranium, bergamot, caraway and cassia oils: also widely used in dental preparations, such as toothpastes and mouth washes.
Comments	Has a local anaesthetic effect.

RIFM Monograph (1978) *FCT*, **16**, 695 (Leaf: Madagascar)
                        (1975) *FCT*, **13**, 761 (Bud)
                        (1975) *FCT*, **13**, 765 (Stem).

CLOVE ROOT OIL

Source	
NATURAL	*Eugenia caryophyllata*.
Isolation	Distillation of the roots.
Appearance	Bright yellow oil.
Odour	Comparable quality and composition with that obtained from the buds.

COCHINEAL [1343-78-8]

Source	
GEOGRAPHIC	Mexico, the West Indies, and Canary Islands.
NATURAL	Cochineal is the dried fecundated female insect, *Coccus cacti*, Linne.
Isolation	The silvery insects are killed by sulphur fumes, and the black ones by boiling water.
Yield	A good quality specimen produces about 3% ash.
Appearance	There are two important varieties; silver and dark brown to black, which produce either silver or black grains.
Chemical	Carminic acid is the colouring matter and is contained up to about 10%: isolated as tiny red prismatic crystals which are soluble in alkalis, alcohol and water, insoluble in fixed and volatile oils.
Other	May be adulterated (the silver grain with barium salts, and the black grain with magnetic iron sand, lead sulphide, etc.).
Uses	As a colouring agent. Carmine may be prepared from cochineal by the precipitation of decoctions with alum, etc.
Comments	Granilla is an inferior cochineal, probably the siftings.

COCONUT FATTY ACID MONOISOPROPANOLAMIDE [68333-82-4]

CTFA name	Cocamide MIPA
	Amides, coco, N-(2-hydroxypropyl)-; coconut isopropanolamide; monoisopropanolamine coconut acid amide
Source	
CHEMICAL	A mixture of isopropanolamides of coconut acid.
Appearance	A cream-coloured material.
Odour	Slight.
Chemical	Soluble in alcohol and most surface-active agents.
Uses	As a foam-regulating additive for detergents and shampoos in concentrations of 2–5%.

COCONUT MONOETHANOLAMIDE [68140-00-1]

CTFA name	Cocamide MEA
	Amides, coco, N-(2-hydroxyethyl)-; coco monoethanolamide; coconut fatty acid monoethanolamide; cocoyl monoethanolamine
Source	
CHEMICAL	A mixture of ethanolamides of coconut acid.
Appearance	A cream-coloured waxy material.
Physical	Dispersible in hot water.
Uses	As a foam booster and stabilizer in the formulation of hair shampoos and other liquid detergent compositions: ethylene oxide concentrates of coconut monoethanolamide are also used to increase detergency and foam stability: in liquid detergents such as washing-up liquids, and carpet and upholstery cleaners, they have solubilizing properties.

COCONUT OIL [8001-31-8]

CTFA name	Coconut oil
	Copra oil
Source	
GEOGRAPHIC	A tree indigenous to all tropical countries.
NATURAL	The coconut palm, *Cocos nucifera* and *C. butyracea*, L., N. O. Palmaceae.
Isolation	Either by expression or decoction from the fruit.
Appearance	A firm, white fat at normal temperatures.
Odour	Readily becomes rancid.
Physical	Practically insoluble in water, very soluble in chloroform, ether or carbon disulphide.
Chemical	Includes trimyristin, trilaurin, tripalmitin, tristearin and various other glycerides.
Uses	Has an abundant lather and is used in the manufacture of toilet and shaving soaps.

COFFIN-WOOD OIL [92347-03-0]

	Bois de Siam
Source	
GEOGRAPHIC	Three conifers found in the higher regions of Tonkin and Annam.
NATURAL	*Fokienia hodginsii*, A. Henry, *Cunninghamia senensis*, R. Br., and *Dacrydium elatum*, Wall.
Isolation	Distilled from the root stocks.
Odour	The wood has an odour of cedar and sandalwood.
Comments	Jewel boxes and rich coffins are manufactured from the wood.

COGNAC OIL [8016-21-5]

Source	
NATURAL	Produced during the fermentation of grape juice.
Isolation	Separated by fractionating the residues from the manufacture of brandy.
Uses	In the preparation of artificial flavourings and essences: in perfumery, ethyl heptoate is generally substituted.
	RIFM Monograph (1975) *FCT*, **13**, 769 (Green).

COLOBOT OIL [91771-50-5]

Source	
GEOGRAPHIC	A spiny tree which grows in the Philippines.
NATURAL	*Citrus hystrix*, D.C. var. *torosa* (Blanco), Webster, N. O. Rutaceae.
Isolation	Steam distillation from the peel of the fruit.
Yield	Approx. 2%.
Odour	Reminiscent of limes.
Chemical	Contains 26% citronellol, some terpenes and other unidentified aldehydes.
Uses	In the Philippines, the crushed fruits are used as a shampoo, and a fragrant brilliantine for the hair may be produced by mixing the crushed rinds with coconut oil.

COPAIBA BALSAM [8001-61-4]

	Balsam capivi; Jesuit's balsam
Source	
GEOGRAPHIC	A tree found in the valley of the Amazon and other parts of South America.
NATURAL	A natural oleoresin, which occurs in schizogenous ducts and lysigenous cavities of various species of Copaifera.
Isolation	Collected by cutting a large square hole, about 2 ft (60 cm) from the ground, well into the wood, and attaching a spout made from the bark. The oleoresin is discharged into the cavity.
Yield	Approximately 15 kg per tree.
Appearance	A natural oleoresinous exudation.

Coriander oil

Odour	A mild, but fresh peppery odour.
Physical	Vary according to the source.
Chemical	A large percentage of volatile oil is always present, containing the terpene caryophyllene.
Uses	Mainly as a fixative for soaps, blending well with geranium, cinnamon, clove and cassia.

RIFM Monograph (1976) FCT, **14**, 687.

CORIANDER OIL [8008-52-4]

CTFA name	Coriander oil
History	The spice has been known and used since the earliest times.
Source	
GEOGRAPHIC	A herb native to South-Eastern Europe, but largely cultivated in Russia, Germany, and Northern Africa.
NATURAL	*Coriandrum sativum*, L., N. O. Umbelliferae, a small herbaceous annual.
Isolation	Steam distilled from the dried, crushed fully ripened fruits.
Yield	0.8–1.0%.
Appearance	Colourless or pale yellow oil.
Odour	Spicy and aromatic.
Physical	Very soluble in chloroform or ether.
Chemical	Contains geraniol and *l*-borneol with their acetic acid esters, *n*-decylic aldehyde, dipentene, terpinene, cymene, and pinene: phellandrene may also be present.
Uses	Some use in small amounts in colognes and may be used as a substitute for linalol in fine perfumery.

RIFM Monograph (1973) FCT, **11**, 1077.

CORIANDROL [126-90-9]

The name given to impure linalol.
$C_{10}H_{18}O$

Source	
NATURAL	A constituent of linaloe oil and occurs in oils of Ceylon cinnamon, sassafras, orange flower.
Odour	Similar to that of bergamot or French lavender.
Chemical	The *d*-form of linalol.
Uses	Used in perfumery instead of bergamot or French lavender oil.

RIFM Monograph (1975) FCT, **13**, 827.

CORONATION OIL

History	The name given to the anointing oil used in Westminster Abbey at the Coronation: the formula dates back to the 17th century.

Source
NATURAL   A mixture of the essential oils of rose, orange blossom, jasmine and cinnamon, together with benzoin, musk, civet and ambergris in a base of sesame oil.

Cortinarius suaveolens

History   An aromatic fungus discovered by F. Bataille and L. Joachim in the woods of Fontainebleau.
Odour   A strong odour of orange blossom, which remains until the fungus has dried.

CORYLOPSIS PERFUMES

Source
NATURAL   Must resemble the odour of *Corylopsis spicata*, a shrub of Japanese origin, closely allied to witch hazel: it bears pendulous racemes of yellow cowslip-scented flowers.
Odour   Usually a composition of the lily, ylang-ylang type, with the addition of small quantities of patchouli oil.

COSTUS OIL [8023–88–7]

                  Koot, koost, kosht, kost, kashmirja, kushtha, kuth, kusta, kastam, chob-i-qut, kur, putchuk, goshtam, changala, sepuddy, upalet, ouplate, pachak
History   Costus roots have been used as incense for many years by the Chinese: also thought to be the cassia mentioned in the Bible.
Source
GEOGRAPHIC   A large plant which grows in North-Western India, the Himalayas, and China.
NATURAL   *Aplotaxis lappa*, Decaisne, N. O. Compositae: the flowers are bluish-black.
Isolation   Steam distillation of the comminuted dried roots.
Yield   Approx. 1%.
Appearance   Volatile oil.
Odour   Persistent, violet-like, and rooty.
Chemical   The following constituents have been identified: camphene, phellandrene, a terpene alcohol, alpha-costene, beta-costene, aplotaxene, costol, dihydrocostolactone, costolactone and costic acid.
Other   No longer used as it has been successfully replaced by synthetic substitutes.
Comments   Principally produced in the mountains of Kashmir: elecampane roots may be used with the costus during distillation as an adulterant.

RIFM Monograph (1974) FCT, **12**, 867 (Root, Absolute and Concrete).

IFRA GUIDELINE

The Committee recommends that costus root oil, absolute and concrete obtained from *Saussurea lappa*, Clarke, should not be used as a fragrance ingredient. Only preparations having this botanical origin which do not show a potential for sensitization should be used.

This recommendation is based on test results of RIFM on the sensitizing potential of some samples (D. L. Opdyke (1974), *Food Cosmet. Toxicol.*, **12**, 867) and the absence of sensitizing potential in other samples (private communication to IFRA). It is known that certain sesquiterpene lactones with α-methylene butyrolactone structure, present in costus root, have a potential for sensitization (J. Foussereau, J. C. Muller and C. Benezra (1975), *Contact Dermatitis*, **1**, 223–30; J. C. Mitchell and W. L. Epstein (1974), *Arch. Dermatol.*, **110**, 871–2; W. L. Epstein, G. W. Reynolds and E. Rodriguez (1980), *Arch. Dermatol.*, **116**, 59–60; A. Cheminat, C. Benezra, M. J. Farral and J. M. J. Fréchet (1981), *Can. J. Chem.*, **59**, 1405–14).

COUMARIN [91-64-5]

Chemical name	1,2-benzpyrone
CTFA name	Coumarin
	Tonka bean camphor
	$C_9H_6O_2$
Source	
NATURAL	In many plants including woodruff, sweet-scented vernal grass, liatris, and melilot: may also be found in French lavender, and tonka beans.
CHEMICAL	Synthesized from salicylic aldehyde, acetic anhydride and sodium acetate.
Appearance	White crystals or crystalline powder.
Odour	Sweet, warm, herbaceous, reminiscent of new-mown hay.
Physical	b.p. 291 °C, m.p. 68 °C: soluble in almost all the liquids used in perfumery.
Chemical	May undergo decomposition in strong light and in the presence of oxygen, forming salicylic aldehyde, dihydrodicoumarin and salicylic acid.
Uses	Widely used to impart strength to compounds such as tobacco and fern; blends well with vanillin and heliotropin: the methyl and ethyl esters of coumaric acid do not appear to be extensively used.
Comments	Should not be consumed orally and should, therefore, be avoided in the perfumes of lipsticks.
	RIFM Monograph (1974) *FCT*, **12**, 385.

CRAB APPLE

Source
GEOGRAPHIC These trees are widely distributed in the Northern hemisphere.
NATURAL The most odorous flowers are the large pale pink blossoms of *Pyrus coronaria* and *P. augustifolia*, indigenous to North America.
Isolation The perfume is not extracted from the flowers, but may be compounded.

CRAVO OIL

Known as laranja cravo in Brazil.

Source
GEOGRAPHIC Brazil.
NATURAL A type of hybrid tangerine of *Citrus aurantium*, L., and *Citrus reticulata*.
Isolation The machine-pressed peel of the ripened fruit.
Yield 0.5–1.2%.
Odour Fruity, between orange and mandarin.
Uses In perfumes and colognes.

p-CRESOL [106-44-5]

CTFA name p-Cresol
4-Methylphenol; p-methylphenol; 4-hydroxytoluene
C_7H_8O

Source
NATURAL In traces in many essential oils.
CHEMICAL From p-toluene sulphonic acid by alkali fusion or may be isolated from coal tar distillate.
Appearance White crystalline solid.
Odour Phenolic and tar-like, of the narcissus type.
Physical b.p. 202 °C, m.p. 35 °C.
Uses Used in traces in compounds of the narcissus and lily type.
RIFM Monograph (1974) *FCT*, **12**, 389.

p-CRESYL ACETATE [140-39-6]

p-Tolyl acetate; Acetyl *para*-cresol; Narceol
$C_9H_{10}O_2$

Source
NATURAL Occurs in traces in the oils of ylang-ylang and wallflower.
Isolation Synthesized by the acetylation of p-cresol.
Appearance Colourless liquid.
Odour Penetrating, suggestive of horse urine; in extreme dilution has a fragrance of narcissus.
Physical b.p. 212 °C.

Uses		Was indispensable for accurate representation of the odour of narcissus, but is now replaced by phenylacetate. RIFM Monograph (1974) FCT, **12**, 391.

p-CRESYL iso-BUTYRATE [103-93-5]

p-Tolyl 2-methylpropanoate
$(CH_3)_2CHCOOC_6H_4CH_3$

Appearance	Colourless liquid.
Odour	Narcissus-type.
Physical	b.p. 237 °C.

RIFM Monograph (1975) FCT, **13**, 773.

m-CRESYL CINNAMATE [65081-24-5]

m-Tolyl 3-phenylacrylate
$C_6H_5CH{=}CHCOOC_6H_4CH_3$

Appearance	White crystalline powder.
Physical	m.p. 65 °C.
Comments	Employed medicinally under the name hetocresol.

RIFM Monograph (1975) FCT, **13**, 773.

o-CRESYL ETHER [4731-34-4]

di-o-Tolyl ether; o-cresyl oxide
$CH_3C_6H_4OC_6H_4CH_3$

Odour	Metallic geranium–rose, similar to diphenyl oxide.
Physical	b.p. 274 °C.
Uses	Rarely used in perfumery.

p-CRESYL ETHYL ETHER [622-60-6]

Ethyl p-tolyl ether
$C_2H_5OC_6H_4CH_3$

Appearance	Colourless liquid.
Odour	Floral, similar to fine quality cananga.
Physical	b.p. 189 °C.
Uses	In perfumery in floral compositions and in soaps.

p-CRESYL FORMATE [1864-97-7]

p-Tolyl formate
$HCOOC_6H_4CH_3$

Appearance	Colourless liquid.
Odour	Powerful narcissus-type.
Uses	In perfumes of the hyacinth and lilac type and also in floral bouquets.

p-CRESYL METHYL ETHER [104-93-8]

Methyl p-tolyl ether; p-cresol methyl ether; methyl *para*-cresol
$C_8H_{10}O$

Source
NATURAL | In the oils of ylang-ylang, wallflower and probably champaca oil.
Isolation | Synthesized from p-cresol and dimethyl sulphate.
Appearance | Colourless mobile liquid.
Odour | Characteristic of narcissus and ylang-ylang on extreme dilution.
Physical | b.p. 176 °C.
Uses | In preparing synthetic wallflower, hyacinth, narcissus, jonquil, and ylang-ylang oils. Useful in floral notes, for jasmine, lilac, broom etc.
Comments | Most commonly used as cresol ether.
RIFM Monograph (1974) FCT, **12**, 393.

m-CRESYL PHENYLACETATE [122-27-0]

$C_6H_5CH_2COOC_6H_4CH_3$

Appearance | White granular crystals.
Odour | Ylang-ylang, narcissus-type odour.
Physical | m.p. 51 °C.
Uses | Imparts a greater persistence and tones down the harshness of benzyl acetate. Useful in honey, jasmine, narcissus bases etc.
RIFM Monograph (1975) FCT, **13**, 775.

p-CRESYL PHENYLACETATE [101-94-0]

p-Tolyl phenylacetate
$C_6H_5CH_2COOC_6H_4CH_3$

Isolation | Heating a mixture of freshly distilled p-cresol and phenylacetyl chloride until the production of hydrogen chloride ceases. After cooling, the mixture is stirred rapidly and poured into 6 N sodium hydroxide solution.
Appearance | White granular crystals.
Odour | Tenacious narcissus-jonquil type.
Physical | m.p. 74 °C.
Uses | Invaluable in flowery soap compounds, particularly jasmine and narcissus.
Comments | A better base for narcissus perfumes than the acetate as it does not possess the disagreeable odour of horse urine.
RIFM Monograph (1975) FCT, **13**, 775.

CUBEB OIL [8007-87-2]

Cubebs are also known as 'tailed pepper'.

Cucumber juice

Source	
GEOGRAPHIC	A shrub which grows in the East Indies, particularly in Borneo, Java and Sumatra.
NATURAL	*Piper cubeba*, Lf., N. O. Piperaceae, a climbing shrub.
Isolation	The green, berry-like fruits are dried in the sun to a greyish-black; the oil is extracted and an oleo-resin may also be obtained.
Yield	10–16% essential oil, 20% oleo-resin.
Appearance	Colourless to pale green essential oil.
Odour	Characteristic and peppery.
Physical	Insoluble in water; miscible with absolute alcohol or chloroform.
Chemical	Includes diterpene and cadinene.
Uses	In traces in soap perfumery in the same way as oil of pepper. RIFM Monograph (1976) *FCT*, **14**, 729.

CUCUMBER JUICE [8024–36–0]

CTFA name	Cucumber juice
Source	
NATURAL	*Cucumis sativus*, a trailing annual of N. O. Cucumbitaceae.
Isolation	Expression of the fresh cucumbers: alcohol, benzoic or salicyclic acids are added as preservatives. Concentrated cucumber perfume may be made by repeated extraction of the freshly sliced fruit with strong alcohol and concentration by distillation *in vacuo*.
Odour	Peculiar.
Uses	In the preparation of glycerine and cucumber creams.

CUDBEAR [93028–69–4]

Source	
NATURAL	*Rocella tinctoria*, D.C., and other lichens of N. O. Discomycetes.
Isolation	An ammoniacal infusion of the lichen is evaporated to dryness and then powdered.
Appearance	Purplish-red powder.
Chemical	Tinctures, prepared by maceration or percolation with alcohol, produce a colouring matter unaffected by acids.

CUIR DE RUSSE

	Cuir de Russie
Source	
GEOGRAPHIC	A Russian oil.
Isolation	A special formulation is prepared and added to birch tar oil.
Uses	Added to materials used in tanning leather to cover the objectionable smell.

CUMINALCOHOL [536-60-7]

p-iso-Propyl-benzyl alcohol; p-cymen-7-ol; cuminol; cuminyl alcohol; cuminic alcohol
$C_{10}H_{14}O$

Source
NATURAL In cumin oil and in the oils of cassie, myrrh, cinnamon, and certain types of eucalyptus.
Odour Powerful oily–spicy suggestive of dill seed and caraway.
Physical b.p. 248 °C; insoluble in water, miscible with alcohol or ether.
Uses In many different perfume compounds as a modifier for citrus notes.
Comments The commercial product is likely to contain a minor proportion of o-iso-propylbenzyl alcohol (o-cymen-7-ol).
RIFM Monograph (1974) FCT, **12**, 871.

CUMINALDEHYDE [122-03-2]

p-iso-Propylbenzaldehyde; cumic aldehyde; cuminal
$C_{10}H_{12}O$

Source
NATURAL Present at about 30% in cumin seed oil: has been found in traces in the oils from cassie flowers, cinnamon, myrrh, and probably boldo leaf.
Isolation Synthesized from cumene and carbon monoxide.
Appearance Colourless to yellowish liquid.
Odour Pungent, green herbaceous.
Physical 236 °C. Practically insoluble in water, soluble in alcohol or ether.
Uses Used in traces to prepare synthetic cassie, orris, foin coupe, lilac, lily, mimosa, and violet perfumes.
RIFM Monograph (1974) FCT, **12**, 395.

CUMIN OIL [8014-13-9]

Source
GEOGRAPHIC A plant cultivated in Malta, North-Eastern Africa, and India.
NATURAL A small herb, *Cuminum cyminum*, L., N.O. Umbelliferae.
Isolation Steam distillation of the dried and crushed seeds.
Yield 3–5% depending upon source of the raw material and production procedures.
Appearance Limpid pale yellow liquid.
Odour Powerful and characteristic: reminiscent of the spice itself.
Physical Almost insoluble in water, very soluble in chloroform or ether.
Chemical Constituents include cuminaldehyde (30–40%), p-cymene, beta-pinene and diterpene.
Uses In traces it is useful in compounding synthetic cassie: may be used to replace cuminic aldehyde.

RIFM Monograph (1974) *FCT*, **12**, 869, Special issue I (Binder, p. 274).
IFRA GUIDELINE
The Committee recommends for applications on areas of skin exposed to sunshine, excluding bath preparations, soaps and other products which are washed off the skin, to limit cumin oil to 2% in the compound (see remark on phototoxic ingredients in the introduction).

This recommendation is based on test results of RIFM on the phototoxicity of cumin oil (*Food Cosmet. Toxicol.*, **12**, 869 (1974)), on the observed no-effect level of 25% on the skin of the hairless mouse (private communication to IFRA) and on the no-effect level of 5% in tests with humans (private communication).

CURCUMA OIL [8024-37-1]

Turmeric oil

Source
GEOGRAPHIC A plant native to Southern Asia: cultivated mainly in India and Korea.
NATURAL *Curcuma longa*, L., N.O. Scitaminae, related to the ginger plant.
Isolation After the rhizomes have been boiled, they are dried in the sun, or by artificial heat; the oil is then obtained by steam distillation.
Yield 1–5% depending upon source of the plant material and techniques of production.
Appearance The colouring present in the yellow volatile oil is called curcumin: on addition of alkali the oil turns reddish-brown.
Odour Fresh and spicy, has a characteristic odour of the rhizomes.
Physical Soluble in alcohol.
Chemical The volatile oil contains phellandrene and a ketone called tumerone.
Uses Used by the Egyptians to perfume the red slippers sold in their bazaars: if the rhizomes are macerated in alcohol, a tincture of turmeric may be prepared, sometimes used for colouring face powders.

RIFM Monograph (1983) *FCT*, **21**, 839 (*Curcuma longa*).

CYCLAMEN [89998-12-9]

Source
GEOGRAPHIC This plant grows wild in Alpine regions, but is cultivated in greenhouses in England.
NATURAL The name given to a pretty genus of flowering plant with reddish-purple flowers, belonging to N. O. Primulaceae, and Sowbread *C. europaeum*.
Isolation The flowers are extracted with petroleum ether.
Yield 0.18%.

Appearance	A greyish-yellow hard waxy substance is extracted from the flowers.
Odour	The extract has a strong odour similar to jasmine; a smell like humus under forest trees predominates. Cyclamen recalls an odour of lily and violets, with a hint of hyacinth.
Other	Synthetic cyclamen contains hydroxyl citronellal, ionone, linalol, terpineol, and either cinnamic alcohol, or phenylacetic aldehyde.

CYCLAMEN ALDEHYDE [103–95–7]

Chemical name	2-Methyl-3-(p-iso-propylphenyl) propanal iso-Propyl alpha-methyl hydrocinnamic aldehyde; cyclamal; cyclosal $C_{13}H_{18}O$
Isolation	Synthesized from cuminaldehyde and propionaldehyde with subsequent hydrogenation.
Appearance	Colourless liquid.
Odour	Powerful floral–green with a characteristic cucumber–melon note used in floral types.
Uses	Extensively used in floral types especially in muguet and cyclamen-type perfumes. RIFM Monograph (1974) FCT, **12**, 389 RIFM Monograph (1979) FCT, **17**, 267 (Alcohol). IFRA GUIDELINE 3-(4-iso-propylphenyl)-2-methylpropanol The Committee recommends that cyclamen alcohol should not be used as a fragrance ingredient as such, but to accept a level of up to 1.5% in cyclamen aldehyde. This recommendation is based on the sensitizing potential of cyclamen alcohol and the absence of sensitization reaction in a sample of cyclamen aldehyde containing 1.5% cyclamen alcohol (D. L. J. Opdyke (1979), *Food Cosmet. Toxicol.*, **17**, 267).

iso-CYCLOCITRAL

Chemical name	Mixture of 3,5,6-trimethyl-3-cyclohexen-1-carboxalaldehyde and 2,4,6-trimethyl-4-cyclohexen-1-carboxalaldehyde *meta*-Cyclocitral $C_{10}H_{16}O$
Isolation	By Diels–Alder condensation of 2-methyl-2-4-pentadiene and crotonaldehyde.
Appearance	Colourless liquid.
Odour	Powerful and diffusive foliage green note with a distinctive 'flower shop' character.
Uses	In floral compositions at low concentrations to add natural green character. Also blends well with moss, ionones, geranium and galbanum notes. RIFM Monograph (1976) FCT, **14**, 313 (Binder, p. 459).

CYCLOHEXANOL [108-93-0]

Hexalin; hexahydrophenol
$C_6H_{11}OH$

Isolation	Reduction of phenol with hydrogen.
Appearance	Colourless hygroscopic crystals or a colourless viscous liquid.
Odour	Warm, camphoraceous.
Physical	b.p. 161 °C, m.p. 25 °C: and a good solvent.
Uses	May be used in extracting odours from flowers. Increases the cleansing value of soaps when added in small quantities, especially to detergents and scouring agents for woollen fabrics: used as a solvent for nitrocellulose.

RIFM Monograph (1975) *FCT*, **13**, 777.

CYCLOHEXYL ACETATE [622-45-7]

Cyclohexanyl acetate
$CH_3COOC_6H_{11}$

Appearance	Colourless oily liquid.
Odour	Sweet–fruity, chemical.
Physical	b.p. 174 °C. A good solvent which dissolves celluloids.
Other	Evaporates slower than amyl acetate.
Uses	In the preparation of nail varnishes.

RIFM Monograph (1975) *FCT*, **13**, 875.

CYCLOHEXYL n-BUTYRATE [1551-44-6]

Cyclohexyl *n*-butanoate; cyclohexanyl butyrate
$CH_3CH_2CH_2COOC_6H_{11}$

Appearance	Colourless oily liquid.
Odour	Fruity, slightly floral.
Physical	b.p. 212 °C.
Uses	In sweet pea and apple blossom bouquets: the iso-butyrate is more suited to opoponax and amber-type bouquets.

RIFM Monograph (1979) *FCT*, **17**, 751.

CYCLOHEXYL ETHYL ALCOHOL [4442-79-9]

2-Cyclohexylethanol; cyclohexanethanol

Appearance	Colourless liquid.
Odour	Dry, camphoraceous with suggestion of rose and patchouli.
Physical	b.p. 204 °C.
Uses	Sometimes used in floral compositions instead of phenyl ethyl alcohol.

RIFM Monograph (1975) *FCT*, **13**, 785.

CYCLOHEXYL OXYACETIC ACID ALLYL ESTER

$C_{11}H_{18}O_3$

Isolation — From methyl phenoxy acetate ester by hydrogenation and subsequent transesterification with allyl alcohol.
Appearance — Colourless liquid.
Odour — Strong, green, and rather fruity.

CYCLOPENTADECANOLIDE [106–02–5]

Exaltolide; Pentadecanolide; Muscolide; Thisbetolide
$C_{15}H_{28}O_2$

Source
NATURAL — In angelica root oil.
CHEMICAL — Synthesized from hydroxypentadecanoic acid.
Appearance — White crystalline mass.
RIFM Monograph (1975) *FCT*, **13**, 787.

CYCLOPENTADECANONE [502–72–7]

Exaltone; Muscosyn

Source
CHEMICAL — A synthetic product similar to muscone.
Physical — b.p. 306 °C, m.p. 63 °C.
RIFM Monograph (1976) *FCT*, **14**, 735.

CYMENE [99–87–6]

CTFA name — *p*-Cymene
1-Methyl-4-(1-methylethyl)-benzene
p-$CH_3C_6H_4CH(CH_3)_2$

Source
NATURAL — An aromatic hydrocarbon found in the oils of cumin, aniseed, coriander, cinnamon, thyme, etc.
Appearance — Colourless mobile liquid.
Odour — Citrusy, reminiscent of bergamot when pure.
Physical — b.p. 179 °C.
RIFM Monograph (1974) *FCT*, **12**, 401.

CYMYLACETIC ALDEHYDE [4395–92–0]

para-iso-propyl phenylacetaldehyde; cortexal; cuminic aldehyde
$(CH_3)_2CHC_6H_4CH_2CHO$

Isolation — Condensing magnesium cymyl chloride with ethyl orthoformate, then hydrolysing the product with dilute sulphuric acid and separating the aldehyde by means of its bisulphite compound.
Appearance — Colourless liquid.

Odour	Powerful, green–woody note.
Physical	b.p. 243 °C.
Uses	As a topnote modifier for many fragrance types including citrus, green floral and mossy notes.

CYPERUS

Source
GEOGRAPHIC	Distributed throughout tropical and subtropical regions of both hemispheres.
NATURAL	The name given to several species of the sedge family, Cyperaceae.
Isolation	The roots are dried and powdered and may be mixed with different resins. An essential oil may be extracted from the rhizomes.
Yield	Up to 0.5% of volatile oil.
Odour	The volatile oil has a camphoraceous odour.
Uses	The dried and powdered roots are used by Indian ladies to perfume their clothes and hair. The rhizomes are used medicinally or as a perfume for dyed fabrics. The roots of the species *C. longus*, L., known as 'English galingale', are said to be used in the preparation of lavender water: said to have an odour resembling violets. It is also used by the natives of Togo where the roots are ground and mixed with different resins, and used to perfume garments suspended above glowing coals, on to which the powder has been thrown.

CYPRESS OIL [8013–86–3]

Source
GEOGRAPHIC	A tree cultivated throughout the whole of the Mediterranean area, especially in Algeria, and Southern France.
NATURAL	*Cupressus sempervirens*, L., and other species of this evergreen tree.
Isolation	Steam distillation from the leaves and twigs.
Yield	1.3–1.5% depending upon source of the plant material.
Appearance	Yellowish viscous liquid.
Odour	Refreshing and spicy; distinctly resembles ambergris on evaporation.
Physical	Slightly soluble in water.
Chemical	Main constituents include furfural, d-pinene, cymene, d-camphene, d-terpineol and l-cadinene.
Uses	Useful in artificial ambers.
Comments	Should be kept well closed, cool and away from light. RIFM Monograph (*Cupressus sempervirens*, L.) (1978) *FCT*, **16**, 699.

DAMASCENONE [23726–93–4]

Chemical name	Trimethyl cyclohexadienyl butenone $C_{13}H_{18}O$

Source	
NATURAL	Rose oil.
Isolation	Synthesized from cyclocitral.
Appearance	Colourless liquid.
Odour	Fruity odour of plums and raisins.
Uses	In perfumery in a wide range of modern floral and fruity floral compositions.

alpha-DAMASCONE [43052–87–5]

Chemical name	1-(2,6,6-Trimethylcyclohex-2-en-1-yl)but-2-en-1-one $C_{13}H_{20}O$
Source	
NATURAL	Tea aroma.
Isolation	Synthesized from the methyl ester of geranic acid.
Appearance	Almost colourless liquid.
Odour	Intense fruity, reminiscent of plums and roses: on dilution it has an intense odour of roses.

DARWINIA OILS

History	The plants from which these oils are obtained were first studied in 1899 by Baker and Smith: they were found to contain about 60% geranyl acetate and 13% geraniol.
Source	
GEOGRAPHIC	Plants grown in New South Wales.
NATURAL	*Darwinia fascicularis*, Rudge, *D. taxifolia* var. *grandiflora*, Bentham, and other species of N. O. Myrtaceae.
Isolation	Distillation from the fresh leaves and twigs.
Yield	0.5%.
Appearance	Oil.
Odour	Very aromatic.
Uses	Of great interest to perfumers since they may become a source for the manufacture of geranyl esters.

DAVANA OIL [8016–03–3]

Source	
GEOGRAPHIC	A plant grown in Bangalore.
NATURAL	The flowering herb, *Artemisia pallens*, Wall.
Isolation	Steam distillation of all the overground parts of the herb.
Yield	0.2–0.5%.
Appearance	Dark oil.
Odour	Strong, herbaceous, and slightly tea-like.
Uses	Limited use in fine fragrances.
	RIFM Monograph (*Artemisia pallens*, Wall) (1976) FCT, **14**, 737.

DECAHYDRO-*beta*-NAPHTHYL ACETATE [10519-11-6]

Decahydro-2-naphthyl acetate; jasmalol; decalinyl acetate
$CH_3COOC_{10}H_{17}$
Odour　Somewhat crude, oily, jasmine-type.
Physical　b.p. 200 °C.
Uses　In combination with benzyl acetate, linalyl acetate, and amyl cinnamic aldehyde in cheap soap compounds.
RIFM Monograph (1979) *FCT*, **17**, 755.

DECAHYDRO-*beta*-NAPHTHYL FORMATE [10519-12-7]

Decahydro-2-naphthyl formate; santalozone; santiat; beta-decalinyl formate
$HCOOC_{10}H_{17}$
Odour　Sweet-woody, musty, sandalwood-type.
Uses　In the preparation of cheap soaps in combination with santal, vetivert, patchouli, and musk ambrette.
RIFM Monograph (1979) *FCT*, **17**, 757.

delta- AND *gamma-*DECALACTONE

δ[706-14-9]
γ[705-86-2]

$C_{10}H_{18}O_2$
Isolation　Many procedures for synthesis.
Appearance　Colourless liquid.
Odour　Gamma-decalactone has a powerful, peachy odour; delta-decalactone has a creamy–nutty odour.
RIFM Monograph (1976) *FCT*, **14**, 739 (δ).
FCT, **14**, 741 (γ).

9-DECENAL [39770-05-3]

Chemical name　9-Decen-1-al
$C_{10}H_{18}O$
Appearance　Colourless liquid.
Odour　Fresh aldehydic with fruity, rose character.
Physical　b.p. 230 °C.
Uses　Useful at low concentrations in floral compositions particularly rose, tuberose, and muguet notes. Also combines well with green notes to give natural tonality.

9-DECENOL [13019-22-2]

Rosalva; omega-decenol
$CH_2\!\!=\!\!CH(CH_2)_7CH_2OH$
Appearance　Colourless liquid.
Odour　Powerful and tenacious waxy-rosy note.

Uses   In soap and detergent fragrances: in small proportions it gives a waxy aldehydic character to rose compounds and floral bouquets.

n-DECYL ACETATE [112-17-4]

Acetate C-10; Decanyl acetate
$C_{12}H_{24}O_2$

Isolation	Synthesized from n-decanol and either acetic acid or acetic anhydride.
Appearance	Colourless liquid.
Odour	Sharp orange–rose.
Physical	b.p. 244 °C.
Uses	May be used in jasmine and fancy bouquets as a modifier.

RIFM Monograph (1975) FCT, 13, 689.

n-DECYL ALCOHOL [112-30-1]

1-Decanol; n-decanol; nonylcarbinol; alcohol C_{10}
$CH_3(CH_2)_8CH_2OH$

Source	
NATURAL	Probably in traces in neroli oil.
Appearance	Colourless, moderately viscous, strongly refractive liquid.
Odour	Sweet, fatty-oily-rose character.
Physical	b.p. 233 °C, m.p. 7 °C: insoluble in water, soluble in alcohol or ether.
Uses	Used up to 5% in some types of rose perfumes and may also give an odour of distinction to ordinary perfumes when only used in traces: used in synthetic rose, violet and neroli.
Comments	This and the other aliphatic alcohols in the series are well established in perfumery.

RIFM Monograph (1973) FCT, 11, 105, (1973) FCT, 11, 1079 and (1975) FCT, 13, 701.

DECYL ALDEHYDE [112-31-2]

n-Decanal; Aldehyde C_{10}; Capric aldehyde
$CH_3(CH_2)_8CHO$

Source	
NATURAL	In certain essential oils such as coriander, cassie, orange, orris, lemon-grass, and rose.
Appearance	Colourless oily liquid.
Odour	Powerful waxy–orange-peel-like character.
Physical	b.p. 209 °C.
Chemical	Oxidation yields capric acid.
Uses	In conjunction with other aliphatic aldehydes to perfect synthetic floral compounds of cassie, jasmine, violet, neroli, and rose.

RIFM Monograph (1973) FCT, 11, 477 and (1973) FCT, 11, 1080.

n-DECYL FORMATE [5451–52–5]

$HCOOCH_2(CH_2)_8CH_3$

Odour	Sweeter and sharper than that of the acetate.
Physical	b.p. 240 °C.
Other	Properties are similar to those of the acetate.
Uses	In neroli and orange blossom compounds.

DEXTRIN [9004–53–9]

CTFA name	Dextrin
	Dextrine; pyrodextrin; torrefaction dextrin
	$(C_6H_{10}O_5)_n \cdot H_2O$
Source	
NATURAL	Made from the flours of corn, tapioca and potato.
Isolation	The starches in the flour are incompletely hydrolysed to dextrins by moistening with dilute acid or alkali and roasting in a rotary kiln. Further heating produces a sugar-dextrose.
Appearance	Amorphous and colourless.
Odour	Odourless and tasteless.
Physical	Soluble in water; not coloured by iodine.
Uses	Used commercially in adhesives, mud packs and some hair fixers.

DIACETONE ALCOHOL [123–42–2]

4-Hydroxy-4-methylpentan-2-one; 2-hydroxy-2-methyl-4-pentanone; diketone alcohol
$CH_3COCH_2C(CH_3)_2OH$

Appearance	Colourless liquid.
Odour	Almost none.
Physical	b.p. 168 °C, m.p. -47 °C: flammable.
Uses	In the lacquer industry as a solvent and to a small extent in the production of nail enamels.

DIACETYL [431–03–8]

2,3-Butanedione; biacetyl; dimethyl diketone; dimethyl glyoxal; 2,3-diketobutane
$C_4H_6O_2$

Source	
NATURAL	In many essential oils, including angelica root, clove and caraway oils: probably present due to the decomposition of plant products in the course of distillaton.
Isolation	Oxidation of methyl ethyl ketone.
Appearance	Yellowish-green, oily liquid.
Odour	Powerful butter odour, said to be similar to quinone.
Physical	b.p. 88 °C: miscible with alcohol or ether.

Di-*n*-AMYL PHTHALATE [131–18–0]

	Di-*n*-pentyl *o*-phthalate; amyl phthalate C₆H₄(COOC₅H₁₁)₂
Physical	b.p. 204–206 °C at 11 mm Hg.
Uses	As a fixer and plasticizer in nail enamels.

DIANTHINE

A fragrance name.

Di-*n*-BUTYL PHTHALATE [84–74–2]

CTFA name	Dibutyl phthalate
	Butyl phthalate; di-*n*-butyl *o*-phthalate; dibutyl 1,2-benzene-dicarboxylate; DBP $C_{16}H_{22}O_4$
Appearance	Colourless oily liquid.
Odour	Very faint.
Physical	b.p. 340 °C: miscible with most organic solvents, practically insoluble in water (about 1 in 2500).
Uses	As an insect repellant and, although not considered as effective as dimethyl phthalate, is sometimes preferred, particularly for impregnating clothing: also used as a solvent and fixative.

DIBUTYL SULPHIDE [544–40–1]

	Di-*n*-butyl sulphide; *n*-butyl sulphide; 1,1'-thiobisbutane $(CH_3CH_2CH_2CH_2)_2S$
Appearance	Colourless liquid.
Odour	Powerful sulphuraceous rose–geranium character: on extreme dilution, reminiscent of methyl heptin carbonate.
Physical	b.p. 182 °C.
Uses	In small proportions, particularly in rose compounds, to impart a floral topnote: also used in soap compounds. RIFM Monograph (1979) *FCT*, **17**, 769.

DICHLOROXYLENOL [133–53–9]

CTFA name	Dichloro-*m*-xylenol
	2,4-Dichloro-3,5-dimethylphenol; dichlorometaxylenol; DCMX $C_8H_8OCl_2$

Appearance	White to creamy-white crystalline powder.
Odour	Characteristic, but less than parachlorometaxylenol.
Physical	Soluble in water (1 in 5000), alcohol and fixed oils.
Uses	A bactericide, preservative and deodorant: often used in the preparation of antiseptic and deodorant lotions and bath preparations.

DIETHYL MALONATE [105-53-3]

Malonic ester; diethyl propandioate
$C_2H_5OCOCH_2COOC_2H_5$

Appearance	Colourless liquid.
Odour	Sweet, fruity green reminiscent of apples.
Physical	b.p. 199 °C, miscible with alcohol or ether.
Uses	In the preparation of apple and other fruity or citrus cologne notes. RIFM Monograph (1976) FCT, **14**, 745.

DIETHYL PHTHALATE [84-66-2]

CTFA name	Diethyl phthalate
	Diethyl 1,2-benzenedicarboxylate; solveol; anozol; DEP
	$C_6H_4(COOC_2H_5)_2$
Appearance	Colourless oily liquid.
Odour	Virtually odourless.
Physical	b.p. 298 °C.
Uses	Widely used as a stable synthetic fixative and solvent in the perfume industry: in certain countries is permitted as a denaturant for ethyl alcohol: used as a plasticizer for materials such as cellulose acetate and cellulose nitrate in nail lacquers.

DIETHYL SEBACATE [110-40-7]

CTFA name	Diethyl sebacate
	Diethyl decandioate
	$C_{14}H_{26}O_4$
Appearance	Colourless liquid.
Odour	Faint, winy–fruity reminiscent of melon.
Physical	b.p. 306 °C, m.p. 1.3 °C.
Uses	The manufacture of fruit essences. RIFM Monograph (1978) FCT, **16**, 705.

DIETHYL SUCCINATE [123-25-1]

	Diethyl butandioate
Appearance	Colourless liquid.
Odour	Faint, winy-ethereal.
Physical	b.p. 218 °C, m.p. −21 °C.

Uses	In a similar way to the ethyl esters of citric and oleic acids. RIFM Monograph (1978) *FCT*, **16**, 707.

DIETHYLENE GLYCOL [111–46–6]

CTFA name	Diethylene glycol Diglycol; 2,2'-oxybisethanol; di-(2-hydroxyethyl) ether; 2,2'-oxydiethanol $HOCH_2CH_2OCH_2CH_2OH$
Appearance	Colourless, hygroscopic liquid.
Odour	Almost odourless but has a sharply sweetish taste.
Physical	b.p. 245 °C, m.p. − 10.5 °C: will dissolve oxygenated products to the exclusion of terpene derivatives, the former being subsequently washed out with water: miscible with water, ether, acetone, alcohol and ethylene glycol: insoluble in benzene and carbon tetrachloride.

DIETHYLENE GLYCOL MONOETHYL ETHER [111–90–0]

CTFA name	Ethoxydiglycol Carbitol; 2-hydroxyethyl 2-ethoxyethyl ether $C_2H_5OCH_2CH_2OCH_2CH_2OH$
Appearance	Colourless, slightly hygroscopic liquid.
Odour	Mild, ethereal pleasant.
Physical	b.p. 202 °C: miscible with water.
Other	Has a smoothing action on the skin: does not cause stickiness and increases the spreading power and the intensity of dyestuffs in cosmetics.
Uses	In cold and cleansing creams, shaving creams, and as a solvent for eosin in lipsticks. RIFM Monograph (1974) *FCT*, **12**, 517.

DIETHYLENE GLYCOL MONOLAURATE [141–20–8]

CTFA name	PEG-2 laurate Diglycol laurate; diglycol monolaurate $C_{11}H_{23}COOCH_2CH_2OCH_2CH_2OH$
Appearance	Pale-yellow, oily liquid.
Physical	Insoluble in water, soluble in methanol, ethanol, benzene, toluene and acetone.
Other	An emulsifying agent.
Uses	In the preparation of oil-in-water emulsions, particularly lotions and beauty milks.

DIETHYLENE GLYCOL MONOOLEATE [106–12–7]

CTFA name	PEG-2 oleate Diglycol oleate; glycol monooleate $C_{17}H_{33}COOCH_2CH_2OCH_2CH_2OH$

Appearance	Dark-brown oily liquid.
Physical	Dispersible in water and miscible with alcohol.
Uses	In conjunction with an alkali as an oil-in-water emulsifying agent for creams and lotions.

DIETHYLENE GLYCOL MONOSTEARATE [106-11-6]

CTFA name	PEG-2 stearate
	Diglycol stearate; glycol monostearate
	$C_{17}H_{35}COOCH_2CH_2OCH_2OH$
Appearance	Cream-coloured waxy solid.
Physical	m.p. 42 °C: dispersible in hot water and hot oils.
Other	Excellent emulsifying agent.
Uses	In the preparation of oil-in-water emulsions and also with surface-active agents for the preparation of pearly lotion and cream shampoos.

DIHEXYL KETONE [462-18-0]

	Tridecan-7-one; oenanthone
	$CH_3(CH_2)_5CO(CH_2)_5CH_3$
Appearance	Crystalline solid.
Odour	Warm herbaceous with woody-floral undertones.
Physical	b.p. 264 °C, m.p. 33 °C.
Uses	In new-mown hay, fougere and other fragrance types.

DIHYDROCARVEOL [619-01-2]

	6-methyl-3-isopropenylcyclohexanol; cyclohexanol, 2-methyl-5-(1-methylethenyl)-
Chemical name	8-p-menthen-2-ol
Source	
NATURAL	Some essential oils.
Isolation	Reduction of carvone followed by separation.
Appearance	Colourless or pale straw-coloured liquid.
Physical	GC RIFM no. 77-120; IRC RIFM no. 77-120.

DIHYDROCOUMARIN [119-84-6]

	1,2-Benzdihydropyrone
	$C_9H_8O_2$
Appearance	Colourless liquid.
Odour	A fine Tonka, sweet nutty, hay character.
Physical	b.p. 272 °C, m.p. 23 °C.
Uses	No longer used as a fragrance ingredient.
	RIFM Monograph (1974) *FCT*, **12**, 521.
	IFRA GUIDELINE
	The Committee recommends that dihydrocoumarin should not be used as a fragrance ingredient.

This recommendation is based on the findings of RIFM on the sensitizing potential of this material (*Food Cosmet. Toxicol.*, **12**, 521 (1974)).

DIHYDRO-*nor*-DICYCLOPENTADIENYL ACETATE [54830-99-8]

	Verdyl acetate
	$C_{12}H_{16}O_2$
Isolation	Synthesized by adding acetic acid to dicyclopentadiene.
Appearance	Colourless liquid.
Odour	Powerful, tenacious, herbaceous-green woody.
Uses	Widely used in fragrances for soaps and detergents.

DIHYDROJASMONE [1128-08-1]

	2-Methyl-2-pentyl-cyclopent-2-en-1-one;
	2-pentyl-3-methyl-2-cyclopenten-2-one-1
	$C_{11}H_{18}O$
Isolation	May be synthesized from heptanal and butenone.
Appearance	Colourless liquid.
Odour	Intense floral very similar to *cis*-jasmone.
Physical	b.p. 230 °C.
Comments	Should not be mistaken for iso-jasmone or dihydro iso-jasmone, which are two related isomeric ketones.
Uses	Used widely in jasmine compositions and as a floralizer for many other types.
	RIFM Monograph (1974) *FCT*, **12**, 523.

DIHYDROMYRCENOL [18479-58-8]

	2,6-Dimethyloct-7-en-2-ol
	$C_{10}H_{20}O$
Isolation	Synthesized by catalytic hydrogenation of myrcenol, or from pinene via dimethyl octadiene as an intermediate.
Appearance	Slightly viscous, colourless liquid.
Odour	Fresh, lime–citrus.
Uses	Extensively used in soap and detergent perfumes. May successfully replace natural lime oil and is used in concentrations of up to 20% in lime colognes to impart freshness.
	RIFM Monograph (1974) *FCT*, **12**, 525.

DIHYDRO TERPINYLACETATE [80-25-1]

Chemical name	Dihydro-*alpha*-terpinylacetate
	Menthanyl acetate
	$C_{12}H_{22}O_2$

Isolation	Synthesized by the hydrogenation of terpineol and subsequent acetylation, or by acetylating, followed by hydrogenation.
Appearance	Colourless liquid.
Odour	Fresh-piney, slightly woody, similar to terpinyl acetate.
Chemical	A mixture of isomers, more stable than terpinyl acetate.
Uses	Widely used in low cost lavenders, citrus and fougere types for detergents and household products.

DIHYDROXYACETONE [96–26–4]

CTFA name	Dihydroxyacetone
	1,3-Dihydroxymethyl ketone; 1,3-dihydroxy-2-propanone; oxantin
	$HOCH_2COCH_2OH$
Appearance	Crystalline powder.
Odour	Characteristic, faint–sweet.
Physical	Fairly hygroscopic: soluble in one part of water and in 15 parts of alcohol.
Chemical	The common form is dimeric, oxidizes readily in air.
Other	Sweet cooling taste.
Uses	In the preparation of artificial suntan lotions and creams.

DILL SEED OIL [8006–75–5]

Oil of Dill; Dill fruit oil, Aneth oil

Source	
GEOGRAPHIC	A plant cultivated in the USA and the Balkan countries.
NATURAL	The fruits of the herb *Anethum graveolens*, N. O. Umbelliferae.
CHEMICAL	Contains some 50% carvone, with *d*-limonene, phellandrene and other terpenes.
Isolation	Steam distillation of the whole herb.
Yield	0.3–1.5%.
Appearance	Colourless or pale yellow liquid.
Odour	Sweet, spicy, and somewhat minty.
Physical	Insoluble in water, soluble in alcohol.
Uses	As a replacement for caraway oil in soap perfumery.
Comments	Should be kept well closed, cool and away from light.
	RIFM Monograph (Dill Seed Oil (Indian)) (1982) *FCT*, 20, 673.

2,4-DIMETHYLACETOPHENONE [89–74–7]

	Acetyl-*meta*-xylene
	$CH_3COC_6H_3(CH_3)_2$
Appearance	Colourless liquid.
Odour	Intense sweet floral. Mimosa-like on dilution.
Physical	b.p. 228 °C.
Uses	In much the same way as methyl methophenone, but has a finer fragrance.

DIMETHYL BENZYL CARBINOL [611–69–8]

	2-Methyl-1-phenylpropan-2-ol; alpha, alpha-dimethyl-phenylethyl alcohol; DMBC
	$C_6H_5CH_2C(OH)(CH_3)_2$
Appearance	Translucent crystalline mass.
Odour	Warm herbaceous floral reminiscent of lilac and elderflowers.
Physical	b.p. 215 °C, m.p. 24 °C: readily liquefied.
Uses	In lilac, apple blossom, and other floral bouquets.
	RIFM Monograph (1974) *FCT*, **12**, 531.

DIMETHYL BENZYL CARBINYL ACETATE

	2-Methyl-1-phenylprop-2-yl acetate; alpha, alpha-dimethyl-phenethyl acetate; DMBCA
	$C_{12}H_{16}O_2$
Isolation	Synthesized by the acetylation of dimethylbenzyl carbinol.
Appearance	Translucent crystals, which melt at ordinary temperatures, to a colourless liquid.
Odour	Fresh and powerful, floral-fruity reminiscent of the, jasmine–lily type.
Physical	b.p. 250 °C, m.p. 30 °C.
Uses	Widely used in compositions for soap, detergent and cosmetic fragrances to give a fresh note to floral bouquets: also useful in lilac and hyacinth-type compounds.
	RIFM Monograph (1974) *FCT*, **12**, 533.

DIMETHYL BENZYL CARBINYL BUTYRATE [10094–34–5]

Chemical name	1,1-Dimethyl-2-phenyl ethyl butyrate
	DMBC butyrate
	$C_{14}H_{20}O_2$
Isolation	By direct esterification of dimethylbenzylcarbinol with *n*-butyric acid.
Appearance	Colourless slightly oily liquid.
Odour	Mild herbaceous, fruity note in the direction of plum or prune.
Uses	Finds useful application in compositions of the Oriental-rose or chypre type and as a modifier in many other floral types.
	RIFM Monograph (1980) *FCT*, **18**, 667.

DIMETHYLCYCLOHEXENE ALDEHYDE [68039–49–6]

	2,4-Dimethyl butadienacrolein
	$C_9H_{14}O$
Isolation	Composed of two isomers and may be obtained by the Diels–Alder reaction from acrolein and dimethyl butadiene.

2,6-Dimethyl-2-heptanol

Appearance   Colourless liquid.
Odour   Powerful, sweet–green, leafy character.

2,6-DIMETHYL-2-HEPTANOL [13254-34-7]

$C_9H_{20}O$

Source
NATURAL   Not found in nature.
CHEMICAL   Synthesized by a Grignard reaction of 2-methyl-2-hepten-6-one and methyl magnesium chloride, followed by hydrogenation.
Appearance   Colourless liquid.
Odour   Delicate flowery, with a discrete herbaceous character suggestive of lavender and clary sage.
Physical   b.p. 170–172 °C.
Uses   Used for flowery compositions.

DIMETHYL HYDANTOIN FORMALDEHYDE RESIN (DMHF) [9065-13-8]

CTFA name   DMHF
Source
CHEMICAL   Obtained when MDMH (monomethyloldimethylhydantoin) is heated in the dry state.
Appearance   White to amber, brittle resin.
Odour   Faint.
Physical   Soluble in water and alcohol; compatible with gelatin, polyvinyl acetate and ethyl cellulose.
Uses   In concentrations of 5–7% in hair sprays and at lower concentrations in the preparation of hair setting lotions; produces a rigid film with a good gloss depending upon the proportion of plasticizer used in the formulation.

DIMETHYL MALONATE [108-59-8]

             Dimethyl propandioate; methyl malonate
Appearance   Colourless oily liquid.
Odour   Powerful ethereal fruity.
Physical   b.p. 181 °C.
Uses   As a modifier for apple, gooseberry, apricot etc.
             RIFM Monograph (1979) *FCT*, **17**, 363.

DIMETHYL OCTANOL [106-21-8]

             3,7-Dimethyloctan-1-ol; dihydro citronellol tetrahydrogeraniol; pelargol
             $C_{10}H_{22}O$
Isolation   Hydrogenation of geraniol, nerol or citronellol.
Appearance   Colourless liquid.

Odour	Distinctive waxy-rose odour.
Physical	b.p. 213 °C.
Uses	Extensively used in conjunction with other rose alcohols and is employed in both cosmetic and soap compounds.
Comments	Isomeric with decyl alcohol.
	RIFM Monograph (3,7-dimethyl-1-octanol) (1974) *FCT*, **12**, 535.

DIMETHYL PHENYL CARBINOL [617-94-7]

2-Phenylpropan-2-ol; phenyl dimethyl carbinol; DMPC
$C_6H_5C(OH)(CH_3)_2$

Appearance	Translucent crystalline mass.
Odour	Woody–rose with slightly green undertones.
Physical	b.p. 199 °C (slight decomp.), m.p. 37 °C.
Uses	In rose, muguet or lilac compounds.
	RIFM Monograph (1982) *FCT*, **20**, 675.

DIMETHYL PHTHALATE [131-11-3]

CTFA name	Dimethyl phthalate
	DMP; methyl phthalate; dimethyl-1,2-benzenedicarboxylate; phthalic acid dimethyl ester
	$C_6H_4(COOCH_3)_2$
Appearance	Colourless oily liquid.
Odour	Faint ethereal.
Physical	Soluble in most organic solvents, and slightly soluble in water (about 1 in 250); b.p. 284 °C.
Uses	An excellent plasticizer and solvent: used in creams and lotions as an insect repellant, but may cause irritation if in contact with the sensitive areas around the eyes, neck, lips, or mucous membranes.

DIMYRCETOL [18479-58-8]

Is a mixture of dihydromyrcenol and dihydromyrcenyl formate;
2-6-dimethyl-7-octen-2-ol; 3,7-dimethyl-1-octen-7-ol
$CH_3 \cdot CH_2 \cdot C(:CH_2) \cdot [CH_2]_3 \cdot C(CH_3)_2 \cdot OH$ and
$CH_3 \cdot CH_2 \cdot C(:CH_2) \cdot [CH_2]_3 \cdot C(CH_3)_2 \cdot OOC \cdot H$

Source	
NATURAL	Does not occur in nature.
Appearance	Colourless liquid.

DIOCTYL ADIPATE [103-23-1]

CTFA name	Dioctyl adipate
	Bis(2-ethylhexyl)hexanedioate; di-(2-ethylhexyl) adipate; hexanedioic acid, bis(2-ethylhexyl) ester
	$C_{22}H_{42}O_4$

Dipentene

Source
CHEMICAL   The diester of a 2-ethylhexyl alcohol and adipic acid.

DIPENTENE [138–86–3]

4-iso-Propenyl-1-methyl-cyclohex-1-ene, *dl*-limonene
$C_{10}H_{16}$

Source
NATURAL   In many essential oils: the optically inactive (*dl*-) form of limonene.
Isolation   A by-product in the chemical processing of pinene or turpentine oil.
Appearance   Colourless liquid.
Odour   Pleasant, citrusy and almost lemon-like when pure.
Physical   b.p. 178 °C. Practically insoluble in water, miscible with alcohol.
Uses   In low cost industrial and household fragrances.
                  RIFM Monograph (1974) *FCT*, **12**, 703.

DIPHENYL METHANE [101–81–5]

CTFA name   Diphenyl methane
                  1,1'-Methylenebisbenzene; benzylbenzene; ditan
                  $C_{13}H_{12}$
Isolation   Obtained by the Friedel–Crafts reaction from benzyl chloride and benzene in the presence of aluminium chloride.
Appearance   Colourless liquid, or a crystalline mass (orthorhombic needles).
Odour   Harsh geranium leaf, orange blossom character.
Physical   b.p. 262 °C, m.p. 27 °C.
Uses   Principally in soap perfumery, and as it is not affected by traces of alkali, is useful in cheap soaps: used in the preparation of synthetic pelargonium oil and in traces in some floral ottos: it blends well with coumarin, dimethyl hydroquinone and hydroxycitronellal.
                  RIFM Monograph (1974) *FCT*, **12**, 705.

DIPHENYL OXIDE [101–84–8]

Geranium crystals; phenyl ether; 1,1'-oxybisbenzene; diphenyl ether
$C_{12}H_{10}O$
Isolation   A by-product from chlorobenzene in the manufacture of phenol.
Appearance   Light yellow liquid or crystalline solid.
Odour   Harsh green floral geranium type.
Physical   b.p. 252 °C, m.p. 28 °C: very stable.
Uses   Has excellent solvent properties: used in the preparation of geranium oils, and in perfumes for soaps, detergents and household products.
                  RIFM Monograph (Diphenyl oxide) (1974) *FCT*, **12**, 707.

DI-*iso*-PROPYL ADIPATE [6938-94-9]

Appearance	Colourless, mobile liquid.
Odour	Almost none.
Physical	b.p. 242–246 °C: miscible with alcohol, and aqueous–alcoholic solutions.
Other	A non-greasy emollient.
Uses	In cosmetic formulations as a blending agent or as a replacement for vegetable and mineral oils: a useful ingredient in hand preparations.

DIPROPYLENE GLYCOL [110-98-5]

CTFA name	Dipropylene glycol
	Di-1,2-propylene glycol; 1,1′-oxybis-2-propanol
	$C_6H_{14}O_3$
Source	
CHEMICAL	A mixture of glycols.
Uses	Widely used as a fragrant solvent, diluent or fixative, replacing the role of diethyl phthalate.
	RIFM Monograph (1978) *FCT*, **16**, 729.

DI-*p*-TOLYLMETHANE [4957-14-6]

	Ditolyl methane
	$CH_3C_6H_4CH_2C_6H_4CH_3$
Appearance	Colourless liquid or crystalline mass.
Odour	Faint, woody-rosy resembling geraniol.
Physical	b.p. 291 °C, m.p. 23 °C.
Uses	In rose, sweet pea and sweet–woody fragrances.

DODECAHYDRO-3a,6,6,9a-TETRAMETHYL-(2,1-b)NAPHTHO-FURAN [3738-00-9]

	Ambrox
	$C_{16}H_{28}O$
Isolation	Synthesized from sclareol (obtained from clary sage).
Appearance	Crystalline mass.
Odour	Intense and extremely tenacious dry–woody, and ambergris-like.

delta-DODECALACTONE [713-95-1]

	Dodecanolide-1,5; delta-n-heptyl-delta-valerolactone; n-heptyl-delta-valerolactone; 5-hydroxydodecanoic acid delta-lactone; 2H-pyran-2-one, 6-heptyltetrahydro
Source	
NATURAL	Found in peach and coconut and some animal products.
Isolation	Lactonization of 5-hydroxydodecanoic acid.

gamma-DODECALACTONE [2305–05–7]

Chemical name Dodecanolide-1,4
gamma-*n*-octyl-gamma-*n*-butyrolactone; 2(3 h)-furanone, dihydro-5-octyl-; 4-hydroxydodecanoic acid, gamma-lactone; 4-*n*-octyl-4-hydroxybutanoic acid lactone

Source
NATURAL   *Prunus persica.*
Isolation   Lactonization of 4-hydroxydodecanoic acid.
Appearance   Colourless oily liquid.
Physical   GC RIFM no. 74–189; IRC RIFM no. 74–189.

n-DODECANAL [112–54–9]

Duodecyl aldehyde; dodecyl aldehyde; lauric aldehyde; aldehyde C-12 lauric
$CH_3(CH_2)_{10}CHO$

Source
NATURAL   Has been identified in pine needle oil.
Appearance   Solid at room temperature.
Odour   Sweet, waxy-herbaceous: when diluted it recalls the earthy odour associated with wood violets.
Physical   b.p. 249 °C, m.p. 44 °C.
Uses   Extensively used to impart a persistence to violet, jasmine, muguet, lilac, pine needle, and many other modern floral types.
             RIFM Monograph (1973) *FCT*, **11**, 482, 1080.

DODECYL ALCOHOL [112–53–8]

n-Dodecanol; lauryl alcohol; duodecyl alcohol; alcohol C-12 lauric
$CH_3(CH_2)_{10}CH_2OH$

Appearance   Colourless microcrystalline mass.
Odour   Mild, waxy–green floral.
Physical   b.p. 259 °C; m.p. 24 °C: insoluble in water, soluble in alcohol or ether.
Chemical   There is a full range of the esters of this alcohol.
Uses   Mainly in pine, pine needle, modern floral or aldehyde compositions where it is often used in conjunction with the corresponding aldehyde.
             RIFM Monograph (Alcohol C-12) (1973) *FCT*, **11**, 109 and (1973) *FCT*, **11**, 1080.

DOUGLAS FIR OIL [8050–89–3]

CTFA name   Balsam Oregon
             Oregon Balsam
Source
GEOGRAPHIC   A species of tree widely distributed in North America.
NATURAL   Oregon Douglas Fir (*Pseudotsuga taxifolia*).

Isolation	Distilled from the leaves of the trees.
Appearance	Oil.
Odour	Fragrant and reminiscent of pineapples.
Chemical	Contains 31.5% geraniol, bornyl acetate and traces of citral.

DWARF PINE OIL [97676-05-6]

Oil of dwarf pine needles; oil of mountain pine; *Pinus montana* oil; *Pinus pumilio* oil

Source	
GEOGRAPHIC	A species of tree native to the European alps.
NATURAL	*Pinus montana, Pinus mugho* and *Pinus pumilio*.
Isolation	Steam distillation of the young shoots and needles.
Yield	Approximately 1%.
Appearance	Colourless or faintly yellow liquid.
Odour	Typical of pine needles.
Physical	Insoluble in water, soluble in alcohol.
Chemical	Constituents include *l*-pinene, *l*-phellandrine, sylvestrine, dipentene and cadinene.
Uses	In conifer-type perfumes.
Comments	Should be kept well closed, cool and away from light.

EAGLE-WOOD

Source	
GEOGRAPHIC	A tree found in parts of Cambodia and southern Annam.
NATURAL	*Aquilaria crassna*, Pierre, N.O. Thymeleaceae.
Isolation	The tree is subject to attacks of an unknown disease resulting in the appearance of secretions of resin.
Odour	The tree has a fine penetrating odour of incense in this condition.

EAU DE BROUTS ABSOLUTE

Source	
GEOGRAPHIC	A tree which grows throughout the Mediterranean region.
NATURAL	The bitter orange tree, *Citrus bigaradia*, Risso.
Isolation	The distillation waters from the essential oils of neroli, petitgrain and others are extracted.
Yield	Approximately 0.1%.
Odour	Harsh bread-like odour of orange blossom.
Uses	In colognes and some perfumes.
	RIFM Monograph (1976) *FCT*, **14**, 753.

ELDER FLOWERS [68916-55-2]

Sureau oil

Source	
GEOGRAPHIC	Shrubby tree common to the British countryside.

NATURAL	*Sambucus nigra*, L., N.O. Caprifoliaceae.
Isolation	The oil is obtained by distillation of the blossoms, leaves and bark.
Appearance	Oily liquid.
Odour	Possesses the characteristic honey-like odour of the flowers; most fragrant on extreme dilution.
Chemical	The hot alcoholic extract of the blossoms, leaves, and bark contains a glucose yielding benzaldehyde (similar to amygdalin).
Uses	The water obtained by distillation of the flowers is used in some toilet preparations and in medicine. The black berries may be used to make wine or for colouring port wines.

ELECAMPANE OIL [92128-75-1]

Allantroot oil

Source	
GEOGRAPHIC	A perennial plant found in the USA, which originated from Europe and Central Asia.
NATURAL	*Inula helenium*, L., N.O. Compositae. The source plant is also known as scabwort, elfwort, and horse-heal.
Isolation	Distilled from the roots and rhizomes.
Appearance	Dark yellow, viscous liquid.
Odour	Resembles costus, cistus, and orris.
Chemical	The main constituent is alantol.
Uses	No longer used as a fragrance ingredient due to its sensitizing potential.
Comments	Comparatively rare.
	RIFM Monograph (1976) *FCT*, **14**, 307 (Binder, p. 34).
	IFRA GUIDELINE
	The Committee recommends that allantroot oil should not be used as a fragrance ingredient.
	This recommendation is based on the findings of RIFM on the sensitizing potential of this material (*Food Cosmet. Toxicol.*, **14**, 307 (1976)).

June 1975

ELEMI RESINOID/ELEMI OIL [8023-89-0]

Source	
GEOGRAPHIC	A tree cultivated in the Philippines.
NATURAL	*Canarium luzonicum*, Gray, N.O. Burseraceae.
Isolation	The resin is exuded from the trees when the leaves are developing; incisions are made in the trunks to yield the resin, which is collected before it has darkened. The resinoid is obtained by solvent extraction and the oil by steam distillation.
Yield	A mature tree yields 4–5 kg resin per annum: yield of resinoid is 80–90% of the crude gum and some 20–30% is obtained in the form of oil.

Appearance	Fresh elemi is soft, white, and granular, but hardens with age and becomes yellow.
Odour	Fresh, spicy-lemon.
Chemical	The essential oil contains a crystalline sesquiterpene alcohol, elemol, with an odour recalling fennel.
Uses	As a fixative in heather and verbena bouquets: an elemi tincture may be employed in soap perfumery.
Comments	Most of the Grasse houses prepare an elemi resinoid of brilliant yellow clarity.
	RIFM Monograph (Elemi oil) (1976) FCT, **14**, 755.

ENFLEURAGE

History	The process by which 'pomades' are made.
Isolation	A thin layer of lard or fat is uniformly distributed on the surface of glass trays covered in fresh flowers, which are periodically renewed. The fat thus becomes fully charged with the perfume of the flowers. This process gives the 'pomade' from which the first, second, and third infusions or 'washings' are made.

EOSIN [548–26–5]

Bromo-acid (water-insoluble acid form); eosine; D & C Red 22; CI Acid Red 87; CI 45380

$C_{20}H_6Br_4Na_2O_5$

Source	
CHEMICAL	Disodium salt of tetrabromofluorescein.
Appearance	Gives yellow–red fluorescence in dilute aqueous solution, and yellow–green fluorescence in dilute alcoholic solutions.
Other	The acid water-insoluble forms of halogenated fluoresceins stain the mucous membrane of the lip and are used in lipsticks to give permanence to the colour.
Uses	In red ink and as a colouring in solutions, powders, and tablets: dibromo and di-iodo fluoresceins give a yellowish-red stain and the effect may be increased with tetrabromofluorescein to an intense bluish red with tetrachlorotetrabromofluorescein. Maximum staining may be obtained in lipsticks by using a solvent together with a coupling agent such as propylene glycol monomyristate.
Comments	By using solvents in lipsticks to ensure that there are no minute particles of undissolved material, the incidence of allergy to halogenated fluoresceins may be reduced.

Eriocephalus africanus

Source	
GEOGRAPHIC	A cold-resistant plant indigenous to eastern Africa: cultivated on the Riviera.

Erythrosine

NATURAL	Belongs to the Compositae which comprises 17 species of shrubby plants.
Isolation	Preliminary extraction of the plant, with volatile solvents, by Trabaud and Sabetay yields a concrete oil.
Yield	0.3% concrete oil which produces 10–15% of an essential oil on steam distillation.
Appearance	The plant has white flowers with a balsamic odour: the concrete oil is a green liquid: the essential oil is a yellow, viscous liquid.
Odour	The essential oil has a herbaceous, balsamic odour.

ERYTHROSINE [16423–68–0]

Tetra-iodo-fluorescein; FD & C Red 3; CI Acid Red 51; CI 45430
$C_{20}H_6I_4Na_2O_5$

Appearance	Brilliant red dye.
Physical	Water-soluble: similar to eosin and phloxine.
Uses	Liquid rouges.

ESCULIN [531–75–9]

6,7-Dihydroxycoumarin 6-glucoside; esculoside; bicolorin
$C_{15}H_{16}O_9$

Source	
NATURAL	Obtained from the inner bark of the horse-chestnut, *Aesculus hippocastanum*, L., N.O. Hippocastaneae.
Appearance	White crystalline substance.
Physical	m.p. about 200 °C: very soluble in hot water, especially in the presence of alkalis: in solution has a faint blue fluorescence.
Comments	A bitter taste.

ESTRAGON OIL [8016–88–4]

Tarragon oil

Source	
GEOGRAPHIC	A culinary herb cultivated in France, Yugoslavia, Spain and Italy.
NATURAL	*Artemisia dracunculus*, L., N.O. Compositae.
Isolation	Steam distillation of the whole plant.
Yield	0.8–1.0%.
Appearance	Yellowish-green oil.
Odour	Powerful, aromatic, and anise-like.
Chemical	Contains methyl chavicol, phellandrene, an aliphatic hydrocarbon resembling ocimene, and probably also *p*-methoxy-cinnamic aldehyde.

| Uses | Up to 1.5% in fancy bouquets, colognes, chypres, and ferns. RIFM Monograph (1974) *FCT*, **12**, 709. |

ETHANOLAMINE

1. Ethanolamine, 2-aminoethanol (monoethanolamine);
2. Diethanolamine, di-(2-hydroxyethyl)amine;
3. Triethanolamine, tri-(2-hydroxyethyl)amine.
1. $NH_2CH_2CH_2OH$;   [141–43–5]
2. $NH(CH_2CH_2OH)_2$;   [111–42–2]
3. $N(CH_2CH_2OH)_3$.   [102–71–6]

Source	
CHEMICAL	These three compounds differ slightly in physical and chemical properties, but are all substituted ammonia compounds, in which one or more of the hydrogen atoms of NH_3 has been replaced by the ethanol group —CH_2CH_2OH.
Isolation	All synthesized from ammonia.
Appearance	1. A slightly viscous, colourless liquid; 2. A colourless liquid, of similar viscosity to glycerine; 3. A white crystalline solid in pure form.
Odour	1. Faint and ammoniacal; 2. None.
Physical	All miscible with water, alcohol, acetone, and glycerine: 1. b.p. 171 °C at 757 mm Hg, m.p. 10.5 °C: extremely hygroscopic; 2. b.p. 270 °C at 748 mm Hg, m.p. 28 °C; 3. b.p. 277–9 °C at 150 mm Hg, m.p. 21 °C. Aqueous solutions of all are strongly alkaline to phenolphthalein.
Uses	Combination of any of these with a fatty acid gives a neutral compound: triethanolamine, which contains the mono- and di-compounds, is invariably used commercially.

ETHYL ACETATE [141–78–6]

Acetic ether; acetic acid ethyl ester; vinegar naphtha
$C_4H_8O_2$

Isolation	Synthesized from ethyl alcohol and acetic acid: two types obtainable; one from duty-free alcohol, one from industrial spirit.
Appearance	Colourless, volatile, flammable liquid.
Odour	Refreshing, fruity–ethereal with a pleasant taste when diluted.
Physical	b.p. 77 °C, miscible with alcohol, acetone, chloroform or ether: a good solvent for nitrocotton.
Uses	In artificial fruit essences such as cherry, peach, raspberry and strawberry: employed, in traces, in magnolia and ylang-ylang perfumes: as a fumigant in America where it is combined with carbon tetrachloride: also used in nail enamels. RIFM Monograph (1974) *FCT*, **12**, 711.

ETHYL ACETOACETATE [141-97-9]

Ethyl 3-ketobutanoate; 3-oxybutanoic acid ethyl ester; acetoacetic acid ethyl ester; ethyl 3-oxobutanoate; acetoacetic ester
$CH_3COCH_2COOC_2H_5$

Source	
CHEMICAL	Prepared from ethyl acetate.
Appearance	Colourless liquid.
Odour	Sweet ethereal–fruity and rum-like character.
Physical	b.p. 181 °C: soluble in 35 parts of water, miscible with the usual organic solvents.
Chemical	Mixture of keto and enol forms.
Uses	In fresh citrusy and light non-floral types as a topnote material.
Comments	Moderately irritating to the skin and mucous membranes.

RIFM Monograph (1974) *FCT*, **12**, 713.

ETHYL ANISATE [94-30-4]

Ethyl *p*-methoxybenzoate
$CH_3OC_6H_4COOC_2H_5$

Appearance	Colourless liquid.
Odour	Mild, sweet–anisic, fennel-like.
Physical	b.p. 270 °C, m.p. 8 °C.
Uses	Useful in lime blossom, lilac, carnation and fancy bouquets.

RIFM Monograph (1976) *FCT*, **14**, 757.

ETHYL ANTHRANILATE [87-25-2]

Ethyl *o*-aminobenzoate
$NH_2C_6H_4COOC_2H_5$

Appearance	Colourless liquid.
Odour	Sweet–fruity, grape-like character more delicate than the methyl ester.
Physical	b.p. 267 °C, m.p. 13 °C.
Uses	In neroli, jasmine and other floral fragrances.

RIFM Monograph (1976) *FCT*, **14**, 759.

ETHYL BENZOATE [93-89-0]

Benzoic acid ethyl ester
$C_6H_5COOC_2H_5$

Appearance	Colourless, clear, refractive liquid.
Odour	Warm fruity-floral, smoother than the methyl ester (Niobe oil); the vapour may cause coughing.
Physical	b.p. 212 °C, almost insoluble in water, miscible with alcohol, chloroform or ether.
Uses	In synthetic ylang-ylang and florals with a heavy character.

RIFM Monograph (1974) *FCT*, **12**, 717.

ETHYL n-BUTYRATE [105-54-4]

	Butyric ether; butanoic acid ethyl ester; butyric acid ethyl ester; ethyl-n-butanoate
	$CH_3CH_2CH_2COOC_2H_5$
Appearance	Colourless mobile liquid.
Odour	Powerful and diffusive ethereal–fruity, suggestive of pineapple and banana.
Physical	b.p. 122 °C, partially soluble in water, miscible with alcohol or ether.
Uses	In traces as a modifier in perfumes: one of the constituents of artificial pineapple, apricot, and peach essence.
	RIFM Monograph (Ethyl Butyrate) (1974) *FCT*, **12**, 719.

ETHYL n-CAPROATE [123-66-0]

	Ethyl n-hexanoate; ethyl hexanoate; hexanoic acid ethyl ester
	$CH_3(CH_2)_4COOC_2H_5$
Appearance	Colourless liquid.
Odour	Powerful and diffusive, fruity-winy odour suggestive of apple, pineapple and banana.
Physical	b.p. 165 °C, insoluble in water, miscible with alcohol or ether.
Uses	In artificial fruit essences and modern fruity–floral topnotes.
	RIFM Monograph (1976) *FCT*, **14**, 761.

ETHYL n-CAPRYLATE [106-32-1]

	Ethyl-n-octanoate; ethyl octoate; ethyl octylate; octanoic acid ethyl ester
	$CH_3(CH_2)_6COOC_2H_5$
Appearance	Colourless, clear, very mobile liquid.
Odour	Fruity, less fatty than that of ethyl caprylate; resembling pineapple.
Physical	b.p. 209 °C, insoluble in water, miscible with alcohol or ether.
Uses	In the preparation of fruit essences.
	RIFM Monograph (Ethyl Caprylate) (1976) *FCT*, **14**, 763.

ETHYL CINNAMATE [103-36-6]

	Ethyl 3-phenylacrylate
	$C_{11}H_{12}O_2$
Source	
NATURAL	Occurs in storax and in the volatile oil of *Koempferia galanga*.
Isolation	Synthesized for commercial use from either benzaldehyde and ethyl acetate by the Claisen reaction, or from cinnamic acid and ethyl alcohol.
Appearance	Colourless liquid.
Odour	Persistent honey-like, sweet balsamic, amber odour.
Physical	b.p. 271 °C, m.p. 12 °C.

Uses	In a wide range of fragrances including rose, chypres and oriental types.

RIFM Monograph (1974) FCT, **12**, 721.

ETHYL CYCLOHEXYL ACETATE [21722-83-8]

2-Cyclohexylethyl acetate; cyclohexanethyl acetate; cyclohexyl ethyl acetate

Appearance	Colourless liquid.
Odour	Powerful, sweet–fruity.
Uses	In floral compositions.

RIFM Monograph (1975) FCT, **13**, 783.

ETHYL DECYLATE [110-38-3]

Ethyl n-decanoate; ethyl caprate
$CH_3(CH_2)_8COOC_2H_5$

Appearance	Colourless liquid.
Odour	Sweet, oily with a slight nutty character.
Physical	b.p. 245 °C.
Uses	Seldomly used, but useful in fragrances which require a fatty–waxy note.

RIFM Monograph (Ethyl Caprate) (1978) FCT, **16**, 733.

ETHYL p-DIETHYLAMINOBENZOATE   diethyl [10287-54-4]
AND p-DIMETHYLAMINOBENZOATE   dimethyl [10287-53-3]

Physical	Neither is soluble in water: ethyl-p-diethylaminobenzoate is more soluble in aqueous/alcoholic solutions.
Uses	Effective sunscreening agents; they filter out ultraviolet wavelengths and therefore are used in concentrations of 2% in suntan preparations.

ETHYL FORMATE [109-94-4]

Formic acid ethyl ester; ethyl formate; ethyl methanoate
$HCOOC_2H_5$

Appearance	Colourless mobile liquid.
Odour	Diffusive, ethereal, rum-like odour.
Physical	b.p. 54 °C, soluble in about 10 parts of water, miscible with alcohol or ether, very flammable; its vapour forms explosive mixture with air.
Chemical	Aqueous solutions gradually decompose into free acid and alcohol.
Uses	Sometimes used as a modifier; its main use is in the manufacture of artificial fruit essences.
Comments	Keep tightly closed and in contact with $CaCl_2$.

RIFM Monograph (1978) FCT, **16**, 737.

ETHYL FUROATE (*alpha*) [614-99-3]

	Ethyl furan-2-carboxylate; ethyl pyromucate
	$C_4H_3OCOOC_2H_5$
Appearance	White crystalline substance.
Odour	Warm, fruity floral, resembles methyl benzoate.
Physical	b.p. 195 °C, m.p. 34 °C.
Uses	In compounds where methyl benzoate would be too coarse.

ETHYL FURYL *beta*-HYDROXYPROPIONATE [25408-95-1]

	Ethyl 3-(2-furyl)-3-hydroxypropanoate; ethyl furyl hydracrylate
	$C_4H_3OCHOHCH_2COOC_2H_5$
Appearance	Colourless or pale yellow liquid.
Odour	Warm, woody, somewhat tea-like in character.
Physical	b.p. 220 °C.
Uses	Its odour is useful in cassie, orris, oriental, tabac, and violet perfume compounds.

ETHYL HEPTIN CARBONATE [10519-20-7]

	Ethyl octynoate; ethyl *n*-oct-2-ynoate
Odour	Powerful foliage note, with waxy undertones reminiscent of violet leaves.
Physical	b.p. 220 °C.
Uses	Used in a similar way to methyl heptine carbonate, severely restricted due to skin sensitizing properties.

IFRA GUIDELINES
The Committee recommends that the following materials should not be used as fragrance ingredients:

- Allylisothiocyanate
- Chenopodium oil
- 3,7-Dimethyl-2-octen-1-ol (6,7-Dihydrogeraniol)
- Furfurylideneacetone
- Methyl methacrylate
- Phenylacetone (Methyl benzyl ketone)
- Esters of 2-octynoic acid, except those covered elsewhere in these Guidelines i.e. methyl and allyl heptine carbonate
- Esters of 2-nonynoic acid, except methyl octine carbonate
- Thea sinensis absolute

These recommendations are based on the absence of reports on the use of these materials as fragrance ingredients and inadequate evaluation of possible physiological effects resulting from their use in fragrances.

March 1988, last amendment July 1990

ETHYL n-HEPTOATE [106-30-9]

Ethyl oenanthylate; ethyl oenanthate; heptanoic acid ethyl ester; ethyl heptanoate; oenanthic ether; synthetic cognac oil; oil of grapes; oleum vitis viniferae
$CH_3(CH_2)_5COOC_2H_5$

Source	
CHEMICAL	Esterification of heptanoic acid or coconut oil (which yields a mixture of ethyl esters).
Odour	Powerful fruity wine-like odour and taste; burning after-taste.
Physical	b.p. 189 °C; the pure material is a liquid: insoluble in water, miscible with alcohol, ether or chloroform.
Other	Artificial ethyl oenanthate is a mixture of amyl and ethyl caprates with ethyl and iso-amyl butyrates and caprylates. In topnotes of fruity–floral and citrus compositions.
Uses	May be sold as a synthetic cognac: used in traces in the preparation of toilet waters.

RIFM Monograph (Ethyl Heptoate) (1981) *FCT*, **19**, 247.

ETHYL HEPTYLATE [106-30-9]

Ethyl heptoate; ethyl-*n*-heptanoate
$C_9H_{18}O_2$

Isolation	Synthesized from *n*-heptanoic acid and ethanol.
Appearance	Colourless liquid.
Odour	Powerful fruity–wine-like.
Physical	b.p. 189 °C.
Uses	In topnotes of modern fruity floral types and citrus odours.

ETHYL LACTATE [97-64-3]

Ethyl 2-hydroxypropanoate; 2-hydroxypropanoic acid ethyl ester; alpha-hydroxypropionate
$CH_3CHOHCOOC_2H_5$

Appearance	Syrupy, colourless liquid.
Odour	Mild, ethereal-buttery.
Physical	b.p. 154 °C: soluble in water, alcohol or ether.
Chemical	Has an acid reaction.
Uses	In the preparation of lacquers and nail enamels: a useful solvent in dyestuffs and perfumes.
Comments	Should be kept well closed.

RIFM Monograph (1982) *FCT*, **20**, 677.

ETHYL LAURATE [106-33-2]

Ethyl *n*-dodecanoate; ethyl laurinate; dodecanoic acid ethyl ester
$CH_3(CH_2)_{10}COOC_2H_5$

Appearance	Colourless oily liquid.

Odour	Mild, oily–fatty with a fruity–flowery character.
Physical	b.p. 269 °C, m.p. −10.7 °C: insoluble in water, very soluble in alcohol or ether.
Uses	A good fixative: used in perfumes of the tuberose type and in fancy bouquets.

RIFM Monograph (1975) *FCT*, **13**, 93.

ETHYL MALTOL [4940-11-8]

3-hydroxy-2-ethyl-4-pyrone; 2-ethyl-3-hydroxy-4H-pyran-4-one; 2-ethylpyronmeconic acid; 4H-pyran-4-one, 2-ethyl-3-hydroxy-; Veltol-plus.

Source	
NATURAL	Does not occur in nature.
CHEMICAL	Alkaline hydrolysis of streptomycin salts.
Physical	IRC RIFM no. 74–195.

ETHYL METHYL KETONE [78-93-3]

2-Butanone; methyl ethyl ketone; ketone C-4
$CH_3COC_2H_5$

Source	
CHEMICAL	Refluxing methyl acetoacetate and dilute H_2SO_4 or by oxidation of secondary butyl alcohol.
Appearance	Colourless mobile liquid.
Odour	Unpleasant, acetone-like.
Physical	b.p. 80 °C, m.p. −86 °C: partially soluble in water, miscible with alcohol, ether or benzene.
Uses	A solvent, rarely used in perfumes.

RIFM Monograph (Methyl Ethyl Ketone) (1977) *FCT*, **15**, 627.

ETHYL METHYL PHENYL GLYCIDATE [77-83-8]

Strawberry aldehyde; Aldehyde C_{16} (so-called); methylphenyl glycidic acid, ethyl ester; EMPG; Fraiseol
$C_{12}H_{14}O_3$

Isolation	Synthesized from acetophenone and ethyl monochloroacetate.
Appearance	Colourless liquid, slightly viscous.
Odour	Sweet–fruity, strawberry-like.
Comments	The name Aldehyde C_{16} is misleading as this is an ester.
Uses	Widely used to impart long-lasting sweet, fruity character to fragrance compositions.

RIFM Monograph (1975) *FCT*, **13**, 95.

ETHYL MYRISTATE [124-06-1]

Ethyl-*n*-tetradecanoate; ethyl myristinate; myristic acid ethyl ester
$CH_3(CH_2)_{12}COOC_2H_5$

140 Ethyl *beta*-naphthyl ether

Source	
NATURAL	Myristic acid occurs in nutmeg butter.
Appearance	Colourless oily liquid, solidifying when cold.
Odour	Mild violet-like.
Physical	b.p. 295 °C, m.p. 12 °C: insoluble in water, slightly soluble in ether, soluble in alcohol.
Other	Has marked fixative properties.
Uses	In artificial violet compounds.
	RIFM Monograph (1978) *FCT*, **16**, 745.

ETHYL *beta*-NAPHTHYL ETHER [93-18-5]

2-Ethoxynaphthalene; beta-naphthol ethyl ether; ethyl beta-naphtholate; nerolin
$C_{10}H_7OC_2H_5$

Source	
CHEMICAL	Prepared by heating beta-naphthol with potassium ethyl sulphate.
Appearance	Lustrous crystals.
Odour	Tenacious, soft–floral with some neroli character.
Physical	b.p. 282 °C, m.p. 37 °C: practically insoluble in water, soluble in alcohol, chloroform, ether and other organic solvents.
Other	Similar properties to the methyl ether: about equal in strength to neroli oil.
Uses	As a sweetener and floralizer in many floral cologne fragrance types.
	RIFM Monograph (beta-Naphthyl Ethyl Ether) (1975) *FCT*, **13**, 883.

ETHYL NONANOATE [123-29-5]

Ethyl pelargonate, Ethyl nonylate
$C_{11}H_{22}O_2$

Isolation	By esterification of ethanol with nonanoic acid.
Appearance	Colourless oily liquid.
Odour	Delicately fruity with a slight oily–nutty undertone.
Physical	b.p. approx. 228 °C.
Uses	In small amounts to introduce a natural fruity effect in rose and citrus types and to provide power and freshness to the topnote of some floral compositions.
	RIFM Monograph (1978) *FCT*, **16**, 747, Special Issue IV.

ETHYL OCTIN CARBONATE [10031-92-2]

Ethyl nonynoate; ethyl *n*-non-2-ynoate

Appearance	Colourless liquid.
Odour	Intense violet-leaf, with green-vegetable, cucumber-like character.

Physical	b.p. 221 °C.
Uses	In fancy bouquets at low dosages.
Other	It is a skin sensitizer; its use in perfumes is thus severely limited.

RIFM Monograph (Ethyl Octine Carbonate) (1975) *FCT*, **13**, 97.

IFRA GUIDELINE

The Committee recommends that the following materials should not be used as fragrance ingredients:

- Allylisothiocyanate
- Chenopodium oil
- 3,7-Dimethyl-2-octen-1-ol (6,7-Dihydrogeraniol)
- Furfurylideneacetone
- Methyl methacrylate
- Phenylacetone (Methyl benzyl ketone)
- Esters of 2-octynoic acid, except those covered elsewhere in these Guidelines i.e. methyl and allyl heptine carbonate
- Esters of 2-nonynoic acid, except methyl octine carbonate
- Thea sinensis absolute

These recommendations are based on the absence of reports on the use of these materials as fragrance ingredients and inadequate evaluation of possible physiological effects resulting from their use in fragrances.

March 1988, last amendment July 1990

ETHYL OLEATE [111-62-6]

	Ethyl *cis*-octadec-9-enoate
	$CH_3(CH_2)_7CH=CH(CH_2)_7COOC_2H_5$
Appearance	Pale yellowish oily liquid.
Odour	Very weak flowery note.
Physical	b.p. 320 °C, insoluble in water, miscible with alcohol or ether.
Uses	As an adulterant of essential oils, and as a solvent for some fragrances.

RIFM Monograph (1982) *FCT*, **20**, 683.

ETHYL PALMITATE [628-97-7]

	Ethyl hexadecanoate; ethyl cetylate
Isolation	Prepared from palmitic acid which also contains other fatty acids such as stearic, myristic, lauric and oleic acid.
Appearance	Colourless, needle-like crystals.
Odour	Faint, waxy-sweet.
Physical	m.p. 25 °C, b.p. 303 °C.
Uses	As an emollient in cosmetic creams and lotions: also used as a plasticizer in nail lacquers where it improves water resistance and gloss.

ETHYL PHENYLACETATE [101-97-3]

Benzeneacetic acid ethyl ester; alpha-toluic acid ethyl ester
$C_{10}H_{12}O_2$

Isolation: Synthesized from phenylacetic acid and ethanol.
Appearance: Colourless liquid.
Odour: Intense honey-like with slight animalic–fruity undertones.
Physical: b.p. 229 °C.
Uses: In small amounts in fragrance compositions for artificial rose otto, orange blossom, honey, sweet-pea, and oriental types. Widely used in flavour compositions for honey, tobacco, apricot etc.
RIFM Monograph (1975) *FCT*, **13**, 99.

ETHYL PROPIONATE [105-37-3]

Propanoic acid ethyl ester
$C_2H_5COOC_2H_5$

Appearance: Colourless, mobile liquid.
Odour: Sweet, ethereal, fruity–rum-like.
Physical: b.p. 99 °C, soluble in about 60 parts of water, miscible with alcohol or ether.
RIFM Monograph (1978) *FCT*, **16**, 749.

ETHYL PYRUVATE [617-35-6]

Ethyl 2-ketopropanoate; ethyl acetyl formate
$CH_3COCOOC_2H_5$

Appearance: Colourless liquid.
Odour: Sweet, fruity–floral.
Physical: b.p. 155 °C.
Uses: In fancy compositions as a topnote material.

ETHYL SALICYLATE [118-61-6]

Ethyl o-hydroxybenzoate; salicylic acid ethyl ester; salicylic ether; sal ethyl
$HOC_6H_4COOC_2H_5$

Appearance: Colourless, clear, very refractive liquid: becomes yellowish-brown on long exposure to air and sunlight.
Odour: Heavy, sweet, floral–fruity of the wintergreen type.
Physical: b.p. 231 °C, m.p. 1 °C: slightly soluble in water, miscible with alcohol or ether.
Uses: In small proportions in synthetic cassie and ylang bouquets and in other sweet floral types.
Comments: Should be protected from light.
RIFM Monograph (1978) *FCT*, **16**, 751.

ETHYL UNDECYLATE [627–90–7]

Ethyl *n*-hendecanoate; ethyl undecanoate
$CH_3(CH_2)_9COOC_2H_5$

Appearance	Colourless oily liquid.
Odour	Powerful, winy–fruity with nutty undertones.
Physical	b.p. 255 °C.
Uses	Occasionally in citrus colognes, chypres, and modern fruity–aldehydic types.

ETHYL UNDECYLENATE [692–86–4]

Ethyl *n*-hendec-10-enoate
$CH_2=CH(CH_2)_8COOC_2H_5$

Source	
NATURAL	Costus oil.
Appearance	Colourless oily liquid.
Odour	Powerful winy–fruity character.
Physical	b.p. 264 °C.

RIFM Monograph (1982) *FCT*, **20**, 687.

ETHYL UNDECYNOATE [78491–10–8]

Ethyl *n*-hendec-2-ynoate; ethyl decine carbonate: 'mignonette'

Appearance	Colourless liquid.
Odour	Leafy green, herbaceous resembling methyl heptine carbonate, but less pungent.
Uses	As a topnote ingredient for certain floral, herbaceous and citrus types.

ETHYL *iso*-VALERATE [108–64–5]

Ethyl valerianate
$(CH_3)_2CHCH_2COOC_2H_5$

Appearance	Colourless mobile liquid.
Odour	Strong, diffusive ethereal, vinous-fruity.
Physical	b.p. 135 °C.
Uses	In the preparation of artificial fruit essences, rarely used in perfume compositions.

RIFM Monograph (1978) *FCT*, **16**, 743.

ETHYL VANILLIN [121–32–4]

4-Hydroxy-3-ethoxy-benzaldehyde; ethylprotocatechuic aldehyde; bourbonal; ethylprotal; vanillal
$C_9H_{10}O_3$

Isolation	Synthesized from pyrocatechol via guethol.

Ethylene brassylate

Appearance	White to pale yellow crystalline solid.
Odour	Intense, vanilla-like odour and taste.
Physical	m.p. 78 °C: sparingly soluble in water, soluble in alcohol, ether and other organic solvents.
Uses	Widely used for its odour and fixative properties, in a similar way to vanillin, but in lower proportions: causes less discoloration. RIFM Monograph (1975) *FCT*, **13**, 103.

ETHYLENE BRASSYLATE [105–95–3]

Ethylene tridecane dioate; Astrotone; musk T
$C_{15}H_{26}O_4$

Isolation	Synthesized from ethylene glycol and brassylic acid.
Appearance	Colourless viscous liquid.
Odour	Warm, powdery, and musk-like of great tenacity.
Uses	Extensively as a fixative and intensifier of floral notes. RIFM Monograph (1975) *FCT*, **13**, 91.

ETHYLENE GLYCOL [107–21–1]

Ethan-1,2-diol; 1,2-ethanediol
$HOCH_2CH_2OH$

Source	
CHEMICAL	Prepared on a large scale by the hydration of ethylene oxide.
Appearance	Colourless, slightly viscous liquid.
Odour	Faint, musty and sweet taste.
Physical	b.p. 190 °C: considerably hygroscopic and absorbs twice its weight of water at 100% relative humidity: miscible with water and most organic solvents.
Uses	In the preparation of anti-freeze solutions and as a solvent for dyestuffs: also in toilet preparations, as a substitute for glycerine: should not be used in concentrations above 5%.
Comments	Toxic when taken orally: the symptoms include vomiting, sleepiness and convulsions.

ETHYLENE GLYCOL MONOSTEARATE [111–60–4]

2-Hydroxyethyl stearate

Appearance	Waxy, white or pale yellow solid.
Uses	In cosmetic creams and lotions as an auxiliary emulsifying agent: also as an opacifying and pearling agent for liquid and cream shampoos.

ETHYLENE OXIDE [75–21–8]

Oxirane
C_2H_4O

Eucalyptus globulus Oil   145

Source
CHEMICAL   Prepared from ethylene chlorhydrin and potassium hydroxide.
Appearance   Colourless, flammable gas at room temperature and pressure, which forms a mobile, colourless liquid on cooling.
Odour   Ethereal.
Physical   b.p. 14 °C: explosive: soluble in water, alcohol or ether.
Other   Its vapour, though highly toxic to pests, is relatively non-toxic to man. Used in conjunction with carbon dioxide, the toxic effect is accelerated, thus reducing the dosage and exposure time: no residual odour or taste in the treated material.
Uses   In the cosmetic industry as a fumigant for talc and other powder materials: also used to prepare surface-active agents which are good solubilizers and emulsifiers and have high foaming properties, for bubble baths and shampoos.

Eucalyptus citriodora OIL   [85203–50–1]

Source
GEOGRAPHIC   A tree native to Australia, but cultivated in Brazil, South Africa, China and India.
NATURAL   *Eucalyptus citriodora*.
Isolation   Steam distillation of leaves and twigs.
Yield   2% from the cultivated trees and bushes: only 1% from the wild trees.
Appearance   Oil.
Odour   Rose-like, similar to citronella oil.
Chemical   Contains about 60% citronellal and pinene: blends with benzaldehyde and terpineol.
Uses   In low-cost household products and in soap perfumery.
                 RIFM Monograph (1988) *FCT*, **26**, 323.

Eucalyptus dives OIL   [90028–48–1]

Source
GEOGRAPHIC   A tree cultivated in Australia.
NATURAL   *Eucalyptus dives*, Schau.
Isolation   Steam distillation of the leaves and twigs.
Yield   2–4%.
Appearance   Oil.
Odour   Distinctive and minty.
Chemical   Contains a high proportion of piperitone and phellandrene.
Uses   In the preparation of synthetic thymol from piperitone, and as a toner in the flavour industry.

Eucalyptus globulus OIL   [84625–32–1]

                 Dinkum oil

Eucalyptus staigeriana oil

Source
GEOGRAPHIC   A tree known as blue gum cultivated in Spain, Portugal and Australia.
NATURAL   *Eucalyptus globulus*, Lab.
Isolation   Steam distillation of the fresh leaves and twigs.
Yield   1.8–2% depending upon source of the plant material.
Appearance   Colourless to pale yellow oil.
Odour   Powerful and pungent.
Physical   Almost insoluble in water, miscible with absolute alcohol, oils or fats.
Chemical   Principal constituent is cineole, together with pinene.
Uses   In low-cost soaps and detergents, but mainly in dental preparations as a flavouring agent.
             RIFM Monograph (Eucalyptus oil) (1975) *FCT*, **13**, 107.

Eucalyptus staigeriana OIL [91771-69-6]

Source
GEOGRAPHIC   A tree native to Australia, but cultivated mainly in Brazil.
NATURAL   *Eucalyptus staigeriana*.
Isolation   Steam distillation of the leaves.
Yield   1.9–2.1%.
Appearance   Oily.
Odour   Distinctly herbaceous, reminiscent of verbena.
Chemical   Contains limonene, citral, geraniol and its acetic acid ester.
Uses   In citric-type perfumes; blends with lemon-grass, clove, rosemary, and spike lavender for a verbena soap perfume.

EUGENOL [97-53-0]

4-Allyl-2-methoxyphenol; allylguaiacol; eugenic acid; caryophyllic acid
$C_{10}H_{12}O_2$

Source
NATURAL   Up to 85% in clove oil, and also bay, pimento, cinnamon leaf, sassafras, patchouli and *Magnolia kobus*.
Isolation   Distillation or extraction of either clove leaf or clove stem oil.
Appearance   Colourless or pale yellow, viscous liquid.
Odour   Characteristic, medicinal, clove-like.
Physical   b.p. 253 °C, practically insoluble in water, miscible with alcohol, chloroform or ether.
Uses   As the basis of many carnation, oriental and modern spicy fragrances.
Comments   Large quantities in the commercial production of vanillin: can be used as an analgesic in dentistry.
             RIFM Monograph (1975) *FCT*, **13**, 545.

iso-EUGENOL [97-54-1]

4-Propenyl-2-methoxyphenol
$CH_3CH=CHC_6H_3(OCH_3)OH$

Source
NATURAL — Oils of ylang-ylang, nutmeg, and champaca flowers.
Isolation — Synthesized by treating eugenol with (amyl) alcoholic potash. The commercial product is a mixture of *cis*- and *trans*-isomers.
Appearance — Pale yellow viscous liquids.
Odour — Warm, tenacious, resembling carnation, sweet-williams or wallflower.
Physical — b.p. 266 °C.
Uses — Important in the manufacture of vanillin. Extensively used as a base for carnation perfumes: combined with musk and civet in the production of oriental and modern spicy fragrances. Now replaced to a large extent by other, less skin sensitizing material.
RIFM Monograph (Isoeugenol) (1975) *FCT*, **13**, 815.
IFRA GUIDELINE
The Committee recommends that isoeugenol should not be used as a fragrance ingredient at a level over 1% in fragrance compounds.

This recommendation is based on an evaluation of the testing results of RIFM at 8 and 10% showing the sensitizing potential of the material, and many other RIFM and other data on consumer products, fragrances and the material itself at lower dosages (private communication to IFRA and G. R. Thompson, K. A. Booman, J. Dorsky *et al.* (1983), *Food Cosmet. Toxicol.*, **21**, No. 6, 735–40).

EUGENYL ACETATE [93-28-7]

4-Allyl-2-methoxyphenyl acetate; eugenol acetate; aceteugenol
$C_{12}H_{14}O_3$

Source
NATURAL — Traces in clove oil.
Isolation — Synthesized by acetylating eugenol.
Appearance — Pale yellow liquid, or white crystalline solid.
Odour — Mild, sweet-balsamic and slightly spicy, reminiscent of clove pinks.
Physical — b.p. 282 °C, m.p. 29 °C.
Uses — In carnation-type perfume compounds.
RIFM Monograph (1974) *FCT*, **12**, 877.

iso-EUGENYL ACETATE [93-29-8]

2-Methoxy-4-propenylphenyl acetate; iso-eugenol acetate; acetisoeugenol
$C_{12}H_{14}O_3$

Eugenyl benzoate

Source	
NATURAL	Traces in clove bud oil.
Isolation	Synthesized by reacting acetic anhydride with eugenol: usually consists of a mixture of *trans-* and *cis-*isomers.
Appearance	White granular crystals.
Odour	Balsamic, warm and faintly spicy. Reminiscent of clove.
Physical	*trans-*Isomer: b.p. 282–283 °C, m.p. 80 °C, *cis-*isomer.
Uses	In carnation, floral and herbaceous perfumes.

RIFM Monograph (1975) FCT, **13**, 815.

EUGENYL BENZOATE [531-26-0]

4-Allyl-2-methoxyphenyl benzoate; benzoyl eugenol
$C_{17}H_{16}O_3$

Source	
NATURAL	In cascarilla oil.
Appearance	Colourless needles.
Odour	Very mild spicy-balsamic.
Physical	b.p. 360 °C, m.p. 70 °C.
Uses	Finds little use as a fragrance material.

*iso-*EUGENOL BENZYL ETHER [120-11-6]

Benzyl Iso eugenol
$C_{17}H_{18}O_2$

Isolation	Synthesised from benzyl chloride and eugenol, with subsequent isomerization.
Appearance	Fine lamellae or a light, flocculent powder.
Odour	Sweet, weak, clove-like.
Other	Possesses the property of providing tenacity and 'body' to violet compositions.
Physical	m.p. 57 °C.
Chemical	Isomeric with eugenol benzyl ether.
Uses	In synthetic rose and as a fixative for carnation compounds up to about 10%.

*iso-*EUGENYL *n-*HEXYL ETHER [92666-20-1]

1-*n*-Hexoxy-2-methoxy-4-propenylbenzene; hexyl iso-eugenol
$C_6H_3(OC_6H_{13})(OCH_3)CH=CHCH_3$

Appearance	Viscous yellow liquid.
Odour	Tenacious, balsamic–vanilla-like with a cocao-like undertone.
Uses	In carnation, heavy floral and oriental fragrances as a modifier and fixative.

EUGENYL METHYL ETHER [93-15-2]

Methyl eugenol; 4-allyl-1,2-dimethoxybenzene
$CH_2=CHCH_2C_6H_3(OCH_3)_2$

Source	
NATURAL	Essential oils of bay, citronella, pimento, Canadian snake root, laurel, enodia, betel, and 'Brisbane sassafras': is frequently present in the same oils as eugenol.
Appearance	Almost colourless oily liquid.
Odour	Warm, slightly spicy with a musty-tea-like character.
Physical	b.p. 249 °C.
Chemical	When boiled with alcoholic potash, yields methyl iso-eugenol.
Uses	As a modifier for carnation, lilac and oriental-spicy fragrance types.

RIFM Monograph (Methyl Eugenol) (1975) *FCT*, **13**, 857.

iso-EUGENYL METHYL ETHER [93–16–3]

iso-Eugenol methyl ether; methyl iso-eugenol; 1,2-dimethoxy-4-propenylbenzene
$C_{11}H_{14}O_2$

Source	
NATURAL	Laurel leaf, citronella and pimento oils.
Isolation	Synthesized from eugenol and dimethyl sulphate.
Appearance	Colourless liquid, solidifying in the cold.
Odour	Tenacious, warm, sweet–spicy.
Physical	b.p. 270 °C, m.p. 6 °C.
Chemical	Commercial samples are a mixture of *cis*- and *trans*-isomers.

EVERLASTING FLOWER OIL [8023–95–8] Immortelle oil
[90045–56–0] Helichrysum augustifolium extract

An absolute prepared by the Grasse houses is known as Immortel.

Source	
GEOGRAPHIC	A plant which grows wild on the Dalmatian Islands: also found in Provence and parts of Italy.
NATURAL	*Helichrysum augustifolium*, D.C., N.O. Compositae.
Isolation	Distillation of the fresh brilliant yellow flowers and leaves. If the flowers are left for 24 h, fermentation increases the yield. A concrete may be obtained by extracting the flowers with petroleum ether or benzene. The resulting absolute is yellowish in colour.
Appearance	Intense olive-green liquid.
Odour	Reminiscent of a mixture of rose, orange flower and camomile.
Chemical	Principal constituent is nerol.
Uses	In perfumery to give a characteristic sweetness: blends well with clary sage, patchouli and vetivert.

RIFM Monograph (1979) *FCT*, **17**, 821 (Immortelle absolute).

EXALTOLIDE *See* CYCLOPENTADECAROLIDE

FARNESOL [4602–84–0]

Chemical name	*trans*:*trans*-3,7,11-Trimethyldodeca-2,6,10-trien-1-ol $C_{15}H_{26}O$

150 Fatty acids

Source	
NATURAL	Ambrette seed oil, cassie, rose, Ceylon citronella, neroli, cananga flower oils, cabreuva oil and musk grains.
Isolation	The allylic rearrangement of nerolidol.
Appearance	Colourless oily liquid.
Odour	Delicate, floral, reminiscent of muguet.
Chemical	Four possible stereoisomers but the *trans–trans* form is the only one present in many essential oils.
Physical	b.p. 263 °C.
Uses	Occasionally used in perfumery especially for blending in sweet floral perfumes and oriental bouquets.

IFRA GUIDELINE

The Committee recommends that farnesol should only be used as a fragrance ingredient if it contains a minimum of 96% of farnesol isomers as determinated by GLC.

This recommendation is based on the test results of RIFM on the sensitizing potential of farnesol samples containing variable amounts of impurities and the absence of sensitization reactions in tests with samples containing at least 96% farnesol isomers (private communication to IFRA).

FATTY ACIDS

EFA (essential fatty acids)

Source	
NATURAL	Oils of coconut, palm kernel, olive, peanut, linseed and sunflower, where they are combined with glycerine, as triesters or triglycerides.
CHEMICAL	A mixture of the unsaturated fatty acids, linoleic, linolenic, and arachidonic, is sometimes referred to as vitamin F; known as essential fatty acids (EFA).
Isolation	Saponification with caustic soda, or by heating with water under pressure. Glycerine is obtained as a by-product. The various constituents of the fatty acids may be separated out by fractional distillation under reduced pressure.
Other	Appear to have skin-healing properties.
Uses	In cosmetic creams or in treating dry and roughened skin.

FATTY ALCOHOLS

Source	
NATURAL	Natural fats and oils. The higher fatty alcohols (C_{16} (cetyl), C_{18} (stearyl)) are derived from sperm whale oil.
Isolation	Prepared by copper-catalysed hydrogenation of fatty esters from natural fats and oils: mixed alcohols are derived from palm kernel oil.
Uses	C_{14} (myristyl) and C_{16} (cetyl) alcohols have emollient properties and are used in creams and lotions: also as auxiliary emulsifiers or emulsion stabilizers and to increase viscosity.

FENCHYL ACETATE [1195-79-5]

Chemical name	1,3,3-trimethyl-2-norbornanyl acetate
	Bicyclo[2.2.1]heptan-2-ol, 1,3,3-trimethyl-, acetate
	$C_{10}H_{17} \cdot OCO \cdot CH_3$
Source	
NATURAL	Oil from leaves of *Juniperus rigida*; present in rosemary and fennel oils.
Isolation	Acetylation of fenchyl alcohol.
Appearance	Colourless liquid.

FENCHYL ALCOHOL [2217-02-9]

Chemical name	1,3,3-Trimethyl-2-norbornanol
	Fenchol, 2-Fenchanol
	$C_{10}H_{18}O$
Isolation	Prepared synthetically from pine terpenes.
Appearance	Colourless solid crystalline mass.
Odour	Powerful, diffusive camphor-like, with a sweetish, lime-like characteristic.
Physical	m.p. 48–49 °C.
Uses	Imparts a fresh camphoraceous or citrus-lime note to herbaceous, pine, citrus and even floral compositions for soaps, detergents and household products.
Comments	Exhibits good stability in strong alkaline and oxidizing situations. RIFM Monograph (α-Fenchyl alcohol) (1976) *FCT*, **14**, 775, Special Issue III (Binder, p. 384).

FENNEL OIL [8006-84-6]

CTFA name	Fennel
	Fennel oil bitter, fennel oil sweet, foeniculum
Source	
GEOGRAPHIC	Plants grow around the Mediterranean.
NATURAL	*Foeniculum dulce*, D.C., or *F. sativum*, N.O. Umbelliferae.
Isolation	Steam distillation of the crushed seeds.
Yield	2–4%.
Appearance	Oil.
Odour	Sweet odour of aniseed.
Chemical	Sweet fennel oil contains a higher proportion of anethole.
Uses	In herbaceous-type perfumes and in fern-type soaps.
	RIFM Monograph (1974) *FCT*, **12**, 879.

FENUGREEK RESINOID [68990-15-8]

CTFA name	Fenugreek extract
	Foenugreek resinoid/seeds
Source	
GEOGRAPHIC	A wild herb which is also cultivated in India, Egypt, Morocco, Arabia, Iran and Greece.

152 Fir needle oil siberian

NATURAL	*Trigonella foenum graecum*, L., (Greek hay), N.O. Leguminosae. The seeds contain the alkaloid trigonelline and a steroidal sapogenin diosgenin: used in making curry.
Isolation	The fixed oil is obtained by solvent extraction of the crushed seeds. Steam distillation yields about 0.014% of volatile oil.
Yield	Approx. 6%; varies depending upon raw material and isolation procedure.
Appearance	The volatile oil is brown.
Odour	Intense and woody, resembling that of beta-methyl umbelliferone ethyl ether.
Chemical	A synthetic substance of the hydroxy coumarin series.
Uses	In traces as a toner in many perfume compositions.
Comments	Has been used in making imitation maple syrup and for culinary spices other than curry.

RIFM Monograph (Fenugreek Absolute) (1978) *FCT*, **16**, 755.

FIR NEEDLE OIL SIBERIAN [91697-89-1]

Source	
GEOGRAPHIC	Siberian trees.
NATURAL	*Abies sibirica*, Ledeb.
Isolation	Steam distillation of the young shoots.
Yield	0.8–1.2%.
Odour	Powerful, fresh, pine-type.
Uses	In perfumery to produce a masculine note, and in pine-type creations.

RIFM Monograph (1975) *FCT*, **13**, 450.

FIXATIVES

Chemical	There are four classes: 1. Balsams, gums and oleo-resins; 2. Animal; 3. Essential oils of low volatility; 4. Synthetic aromatic chemicals.
Uses	In perfumes to give a bouquet of persistent and unchanging fragrance: when added to a perfume composition, evaporation is retarded.

FLOUVE OIL [68916-09-6]

Source	
GEOGRAPHIC	A wild grass cultivated in southern France.
NATURAL	*Flouve odoronte* (*Anthoxanthum odoratum*).
Isolation	Steam distillation of the dried grass.
Yield	Varies with source of the raw material.

Odour	Sweet and hay-like, reminiscent of mimosa.
Uses	At low levels in many different perfumes.
	RIFM Monograph (1978) *FCT*, **16**, 757.

FORMALDEHYDE [50–00–0]

Chemical name	Formaldehyde solution
CTFA name	Formaldehyde
	Formalin; as the gas – methanal, oxomethane, oxymethylene, methylene oxide, formic aldehyde, methyl aldehyde
	CH_2O
Source	
CHEMICAL	37–40% formaldehyde gas, in water.
Isolation	Obtained by dissolving formaldehyde gas in water.
Appearance	Colourless liquid.
Odour	Pungent, with an irritating vapour.
Physical	Miscible in water and alcohol.
Chemical	May convert to *para*-formaldehyde, becoming cloudy and forming a white deposit; this may be inhibited by the presence of 10–15% methanol.
Uses	An antiseptic, particularly in shampoos: is an excellent germicide and may be employed in the cleansing treatment of contaminated plant or packaging materials.
Comments	There are various restrictions on use.

FRANGIPANNI

History	This perfume was very popular to the middle of the nineteenth century. The name is derived from that of an ancient Roman family a descendant of which invented the method of perfuming gloves. More recently, each perfumer has a frangipanni which possesses a different bouquet – though jasmine is predominant.

FREESIA

History	Freesia flowers were introduced into Europe in 1883.
Source	
GEOGRAPHIC	Native to South Africa, but grown in Guernsey.
NATURAL	Three species (N.O. Iridaceae.)
Isolation	Early attempts by Messrs Roure-Bertrand Fils (1912) to extract a perfume from the flowers were unsuccessful – both with volatile solvents and with enfleurage.
Odour	The flowers are sometimes odourless, but the perfume is generally of the sweet, fruity type.
Other	Every freesia perfume is different; they are based upon vetivert and santal, and sometimes are of the heavy narcissus-type.

FURFURAL [98–01–1]

Chemical name	2-Furancarboxaldehyde
	2-Furaldehyde; furan-2-carboxaldehyde; pyromucic aldehyde; artificial oil of ants; furfurol
	$C_5H_4O_2$
Source	
NATURAL	Identified in the essential oils or aqueous distillates of ambrette, clove, cinnamon, cypress, caraway, lavender, orris, petitgrain, sandalwood, vetivert and ylang.
CHEMICAL	Prepared industrially from pentosans contained in cereal straws and brans.
Isolation	Believed to develop as a result of the action of heat and acid on the carbohydrates in the raw material.
Appearance	Colourless oily liquid.
Odour	Peculiar; somewhat resembling the odour of bread with an almond-cinnamon-like character.
Physical	b.p. 162 °C.
Chemical	Turns yellow to brown on exposure to air and light (a polymerization aided by hot alkali).

RIFM Monograph (1978) *FCT*, **16**, 759.

GALBANUM RESINOID/OIL [8023–91–4]

Source	
GEOGRAPHIC	This resinoid/oil comes from a plant indigenous to Iran.
NATURAL	The gum-resin is obtained from *Ferula galbaniflua* and *Ferula rubricaulis*, N.O. Umbelliferae.
Isolation	The sap from the plants dries on contact with the air; the resinoid is obtained by solvent extraction, the oil by steam distillation.
Yield	Resinoid 35–50%, oil 20%.
Appearance	Volatile oil.
Odour	The resinoid has a musky, balsamic odour: the oil has a sharp odour.
Uses	A fixative: a strong alcoholic extract is useful in combination with traces of ammoniacum R. and large quantities of bissabol: may be employed as a basis for opoponax perfumes.

RIFM Monograph (Galbanum oil) (1978) *FCT*, **16**, 765.

GALLATES

	These are the gallic acid esters of propyl, octyl and dodecyl alcohol.
Chemical	Frequently used together with citric or phosphoric acid, giving a synergistic effect.
Uses	May be used singly or as mixtures to prevent rancidity in oils, fats, cosmetics and perfumes.
Comments	Contact with metals should be avoided. Gallates should be protected from the light.

GALLS

Source	
GEOGRAPHIC	Galls (also known as nutgall, Aleppo-galls, Turkey-galls or Mecca-galls) are found on trees in Persia and Asia Minor.
NATURAL	The dyer's oak, *Quercus infectoria*, Olivier, N.O. Cupuliferae.
Isolation	Appear as excrescences on the branches as a result of deposition of eggs by an insect. During growth they feed on the plant tissues, setting up irritation, and causing formation of the galls.
Chemical	Constituents are 50–60% tannic acid, 2–4% gallic acid, ellagic acids and resin.
Uses	'Gallotannic acid' was widely used in the preparation of hair dyes.
Comments	Primarily used for the manufacture of tannin and ink, for dyeing or tanning; astringent.

GARDENIA [92475–01–7]

Source	
GEOGRAPHIC	A shrub native to tropical Asia and South Africa.
NATURAL	*Gardenia florida*, L., N.O. Rubiaceae.
Appearance	A beautiful, highly-scented flower from which can be derived a yellow coloured essential oil.
Odour	Gardenia oil has an odour reminiscent of jasmine.
Chemical	The following constituents have been identified: primarily benzyl acetate but also linalol, linalyl acetate, terpineol and methyl anthranilate, together with traces of benzoic acid as ester: the characteristic odour is due to styrolyl acetate.
Uses	Has been used in China for scenting teas.

GAULTHERIA OIL [90045–28–6]

Wintergreen oil

Source	
GEOGRAPHIC	A plant which grows in Canada and the United States.
NATURAL	*Gaultheria procumbens*, L., N.O. Ericaceae, or the 'tea berry'.
Isolation	The leaves, harvested from early April to September, are placed in a still and covered overnight with water at 49 °C, which induces hydrolysis of the glucoside gaultherin. The first quality oil (about 90%) passes into the condenser within the first 3 h, the distillation being complete after 5–6 h.
Yield	0.7%.
Uses	In medicine, and also in perfumery as a component of synthetic cassie and foin coupe: may be used in soap perfumes.
Comments	The genuine oil is almost impossible to obtain. An almost identical product is offered commercially, distilled from the bark of *Betula lenta*, L., N.O. Betulaceae, and known as wintergreen oil. The principal constituent is methyl salicylate.

GERANIOL [106–24–1]

Chemical name	3,7-Dimethyl-2,6-octadien-1-ol
	trans-3,7-Dimethylocta-2,6-dien-1-ol, lemonol
	$C_{10}H_{18}O$
Source	
NATURAL	In oils such as rose, champaca, ginger-grass, geranium, Mexican linaloe, palmarosa and citronella.
CHEMICAL	An olefinic terpene alcohol which constitutes the chief part of oil of rose and oil of palmarosa: isomeric with linalool.
Isolation	Synthesized from pinene or isolated from the natural oils of palmarosa and citronella by fractional distillation under reduced pressure.
Appearance	Colourless oily liquid.
Odour	Sweet, rose-type, varying with source.
Physical	b.p. 230 °C.
Other	Forms an acetate, butyrate and formate with the odour of roses or green rose leaves.
Uses	Extensively as the basis to artificial floral oils, such as rose: may also be employed, in combination with phenylethyl alcohol, as the basis of rose compounds for soaps. The butyrate is used for compounding an artificial attar of roses; the formate as a constituent of artificial neroli oil and artificial orange blossom oil. RIFM Monograph (1974) *FCT*, **12**, 881.

GERANIUM OIL [8000–46–2]

CTFA name	Geranium extract
Source	
GEOGRAPHIC	Various species of a perennial plant cultivated in Northern Africa, Southern France, Spain, Réunion, Kenya, Italy and Corsica.
NATURAL	*Pelargonium graveolens*, N.O. Geraniaceae and other species.
Isolation	The plants yield oil for about 7 years – harvesting takes place when in blossom. Yield can be increased by cutting and leaving to dry and fermenting for 24 h: steam distillation of the leaves and stalks.
Yield	0.1–0.2% depending upon plant material.
Appearance	Oil.
Odour	Varies depending upon species and where it was grown and produced. The oil from Réunion has a full leaf-like rose odour, while that from North Africa has a lighter, less minty odour, with a stronger rose note.
Chemical	Principal constituents of the oil are geraniol and geranyl tiglate.
Uses	Very popular and widely used in perfumery. RIFM Monograph (1975) *FCT*, **13**, 451 and (1976) *FCT*, **14**, 781.

GERANIUM SUR ROSE

Source	
GEOGRAPHIC	Prepared in the south of France.
NATURAL	French geranium oil and rose petals.
Isolation	10 kg of French geranium oil are distilled with 500 kg of rose petals.
Yield	8 kg of oil.
Appearance	Oil.
Odour	Fine.
Chemical	Principal constituents are geraniol and citronellol, together with small quantities of linalol, terpineol, phenylethyl alcohol, menthol and probably borneol and an amyl alcohol. Traces of citral, *l*-menthone, tiglic, valerianic, acetic and butyric acids, pinene, phellandrene and dimethyl sulphide have been found.
Uses	Widely used in all toilet preparations; invaluable in blending floral bouquets.
Comments	Fractional distillation of the Réunion oil yields commercial rhodinol.

GERANYL ACETATE [105–87–3]

Chemical name	*trans*-3,7-Dimethylocta-2,6-dien-1-yl acetate
	$C_{12}H_{20}O_2$
Source	
NATURAL	Many volatile oils, such as sassafras leaves, French lavender, lemon, neroli and Ceylon citronella.
Isolation	Synthesized from geraniol and acetic anhydride: commercial samples never exceed 95% purity.
Appearance	Colourless liquid.
Odour	Sweet, fragrant, reminiscent of rose and lavender.
Physical	b.p. 245 °C.
Uses	In rose, jasmine and lavender perfumes: in certain types of synthetic rose otto, and as a sweetening agent in all types of perfumes. RIFM Monograph (1974) *FCT*, **12**, 885.

GERANYL BENZOATE [94–48–4]

Chemical name	*trans*-3,7-Dimethylocta-2,6-dien-1-yl benzoate
	$C_6H_5COOC_{10}H_{17}$
Appearance	Viscous almost colourless liquid.
Odour	Faint, rosy and mildly floral.
Uses	As a modifier and fixative in rose, violet and jasmine perfume compounds.
Physical	b.p. 305 °C.
	RIFM Monograph (1974) *FCT*, **12**, 887.

GERANYL iso-BUTYRATE [2345-26-8]

Chemical name trans-3,7-Dimethylocta-2,6-dien-1-yl 2-methyl-propanoate
$(CH_3)_2CHCOOC_{10}H_{17}$
Appearance Colourless oily liquid.
Odour Has a richer floral fragrance that the n-butyrate.
Physical b.p. 245 °C.
Uses In perfume compounds of rose, chypre, fougere, lavender types etc.
RIFM Monograph (1975) FCT, 13, 451.

GERANYL n-BUTYRATE [106-29-6]

Chemical name trans-3,7-Dimethylocta-2,6-dien-1-yl n-butanoate
$CH_3CH_2CH_2COOC_{10}H_{17}$
Source
NATURAL Essential oil of *Darwinia grandifolia*.
Isolation Obtained commercially from geraniol and butyric anhydride.
Odour Sweet, fruity with a note of red roses and apple.
Physical b.p. 253 °C.
Uses May be used to cheapen Algerian geranium oil: employed, in rose muguet, lavender floral and citrus fragrances.
RIFM Monograph (Geranyl Butyrate) (1974) FCT, 12, 889.

GERANYL CAPROATE [10032-02-7]

Chemical name trans-3,7-Dimethylocta-2,6-dien-1-yl n-hexanoate
$CH_3(CH_2)_4COOC_{10}H_{17}$
Appearance Colourless oily liquid.
Odour Fruity rose–geranium with green undertones.
Physical b.p. 290 °C.
Uses In synthetic geranium oils and for heavy floral-type bouquets.
RIFM Monograph (1976) FCT, 14, 783.

GERANYL CAPRYLATE [51532-26-4]

Chemical name trans-3,7-Dimethylocta-2,6-dien-1-yl n-octanoate
$CH_3(CH_2)_6COOC_{10}H_{17}$
Appearance Colourless liquid.
Odour Sweet, oily-herbaceous and rosy.
Physical s.g. 0.880, b.p. 304 °C.
Uses To enhance the odour of geranium oils and in floral-type bouquets.

GERANYL FORMATE [105-86-2]

Chemical name trans-3,7-Dimethylocta-2,6-dien-1-yl formate
$C_{11}H_{18}O_2$
Source
NATURAL Geranium oil.

Isolation	Synthesized from geraniol and formic acid.
Appearance	Colourless liquid.
Odour	Dry, rose-leaf odour, which has been compared to that of guaiacwood.
Physical	b.p. 216 °C.
Uses	May be used in synthetic rose oils, but should not exceed 5%: also necessary in the composition of artificial neroli oil: orange blossom oil may be reproduced when 2.5% is added to terpeneless petitgrain oil.

RIFM Monograph (1974) *FCT*, **12**, 893.

GERANYL METHYL ETHER [2565-82-4]

	Methyl geranyl ether
	$C_{11}H_{20}O$
Isolation	Synthesized by the methylation of geraniol.
Appearance	Colourless liquid.
Odour	Green-leafy, slightly waxy and vegetable-like.
Physical	b.p. 210 °C.

GERANYL NITRILE [5585-39-7]

Chemical name	3,7-Dimethyl-2,6-octadiene-1-nitrile
	Citralva
	$C_{10}H_{15}N$
Isolation	From geranyl chloride and sodium cyanide.
Appearance	Pale yellow oily liquid.
Odour	Powerful, penetrating, oily-green, citrus-green character.
Uses	Commonly used as a more stable alternative to citral in lemon and other citrus fragrances for soap, detergents, etc.
Comments	Not stable in acid conditions.
	RIFM Monograph (1976) *FCT*, **14**, 787, Special Issue III (Binder, p. 407).

GERANYL PELARGONATE

Chemical name	*trans*-3,7-Dimethylocta-2,6-dien-1-yl *n*-nonanoate
	Geranyl nonanoate
	$CH_3(CH_2)_7COOC_{10}H_{17}$
Appearance	Colourless liquid.
Odour	Fresh, leafy, rose–geranium character.
Uses	In geranium and floral bouquet type compositions.

GERANYL PHENYLACETATE [102-22-7]

Chemical name	*trans*-3,7-Dimethylocta-2,6-dien-1-yl phenylacetate
	$C_6H_5CH_2COOC_{10}H_{17}$

Appearance	Colourless or pale-straw-coloured liquid.
Odour	Soft, sweet, mildly floral and honey-like.
Physical	s.g. 0.903.
Uses	Useful in floral types where a honey note is required.

RIFM Monograph (1974) *FCT*, **12**, 895.

GERANYL PROPIONATE [105–90–8]

Chemical name	*trans*-3,7-Dimethylocta-2,6-dien-1-yl *n*-propanoate
	$C_2H_5COOC_{10}H_{17}$
Appearance	Colourless liquid.
Odour	Sweet fruity rose.
Physical	b.p. 253 °C.
Uses	As a fruity topnote in perfume compositions of the bergamot, geranium, and rose type.

RIFM Monograph (1974) *FCT*, **12**, 897.

GERANYL TIGLATE [7785–33–3]

Chemical name	*trans*-3,7-Dimethylocta-2,6-dien-1-yl *trans*-2,3-dimethyl acrylate
	$CH_3CH{=}C(CH_3)COOC_{10}H_{17}$
Appearance	Colourless liquid.
Odour	Sweet-herbaceous geranium-like.
Physical	b.p. 269 °C.
Uses	Preparation of artificial rose and geranium notes.

RIFM Monograph (1974) *FCT*, **12**, 899.

GERANYL *n*-VALERATE [10402–47–8]

Chemical name	*trans*-3,7-Dimethylocta-2,6-dien-1-yl *n*-pentanoate
	Geranyl valerianate
	$CH_3(CH_2)_3COOC_{10}H_{17}$
Source	
NATURAL	Oils of sassafras leaves and some angophora species.
Isolation	May be synthesized.
Appearance	Colourless liquid.
Odour	Fruity, herbaceous-rosy character.
Physical	b.p. 291 °C.
Uses	In rose, apple blossom, sweet pea and other floral compositions.

GIANT HYSSOP OIL

Source	
GEOGRAPHIC	A plant which grows on the Pacific Coast of North America.
NATURAL	*Agastache pallidiflora*, Heller, Rydbo., N.O. Labiatae.
Isolation	Distillation of the flowers.
Appearance	Oil.

Odour	Before the buds expand, the flowers possess a fragrant odour recalling thyme. After blooming, the odour is a mixture of thyme and peppermint.
Comments	Not produced commercially: should not be confused with hyssop oil.

GINGER-GRASS OIL [8023–92–5]

Source	
GEOGRAPHIC	A grass cultivated from the Afghan frontier to the bend in the Ganges, and from the subtropical zone of the Himalayas to Madras and Bangalore.
NATURAL	*Cymbopogon martini*, Stapf., N.O. Graminaceae, related to palmarosa grass.
Isolation	Steam distillation of the fresh grass: primitive stills are erected by streams and the condenser is submerged near the river bank.
Yield	Approx. 0.2%.
Appearance	Brownish-yellow oil, resembling palmarosa oil.
Odour	Spicy, herbaceous.
Chemical	The chief constituent is geraniol, together with dihydrocuminic alcohol, phellandrene, limonene, dipentene, carvone, and an aldehyde with an odour reminiscent of heptaldehyde and citronellal.
Uses	Little used, but may be employed in soap perfumery.
Comments	May be adulterated with mineral oil, turpentine, and gurgun balsam oil. A pale yellow oil may be obtained from *Cymbopogon nervatus*, or 'naal'. This comes from the inflorescences and consists mainly of perilla alcohol and limonene; closely resembles ginger-grass oil.

GINGER LILY [94334–08–4]

Source	
GEOGRAPHIC	A plant cultivated in the United States.
NATURAL	*Hedychium coronarium*, L., N.O. Scitaminae.
Isolation	Has been cultivated experimentally: the flowers possess a fragrant odour reminiscent of tuberose and gardenia.

GINGER OIL [8002–60–6]

Source	
GEOGRAPHIC	A plant native to India, but cultivated in Jamaica, Africa, Japan and China.
NATURAL	*Zingiber officinale*, Roscoe, N.O. Scitaminae.
Isolation	The plants are propagated by cutting the rhizomes into 'fingers', each containing a bud. The plant flowers in the autumn: when the aerial stems wither, the rhizomes are dug up, dried and

	exported. The oil is obtained by steam distillation of the dried, freshly ground rhizomes.
Yield	Approx. 4.4% depending upon source of the plant material and isolation procedure.
Appearance	Pale yellow volatile oil.
Odour	Typical of ginger.
Chemical	The following constituents have been identified in the essential oil: citral, methylheptenone, nonylaldehyde, linalol, d-borneol, esters, a phenol and a sesquiterpene alcohol.
Uses	A toner may be added to oriental perfumes to obtain a peculiar effect which is extremely difficult to imitate.

RIFM Monograph (1974) *FCT*, **12**, 901.

GLUCOSE (LIQUID)

[50–99–7] (anhydrous)
[5996–10–1] (hydrated)

Chemical name	d-Glucose
CTFA name	Glucose
	Dextrose; blood sugar; grape sugar; corn sugar
	$C_6H_{12}O_6$
Isolation	Hydrolysis of starch.
Uses	An excipient in the manufacture of toothpastes.

GLYCEROL

[56–81–5]

Chemical name	1,2,3-Propanetriol
CTFA name	Glycerin
	Propan-1,2,3-triol; glycerin; trihydroxypropane
	$CH_2OHCHOHCH_2OH$
Source	
CHEMICAL	A by-product in the manufacture of soap.
Isolation	Formed during the saponification of fixed oils and fats by alkali: may be prepared by passing superheated steam into these substances, and purified by re-distillation.
Appearance	Colourless, hygroscopic, syrupy mixture.
Odour	Odourless but has a sweet, warm taste.
Uses	In vanishing creams, and hair creams: as an excipient in dental preparations and in hand creams.

GLYCERYL ACETATE

[26446–35–5]

Chemical name	1,2,3-Propanetriol monoacetate
	Monacetin; also glyceryl monoacetate (mono-acetin), glyceryl diacetate (di-acetin), glyceryl triacetate (tri-acetin); acetin (generic name for a commercial material consisting of mixed isomers)
Appearance	Colourless hygroscopic liquid.
Odour	Characteristic.

Physical	Monoacetate b.p. 158 °C at 165 mm Hg; diacetate: b.p. 175–176 °C at 40 mm Hg; triacetate: b.p. 172 °C at 40 mm Hg: miscible with 45% alcohol.
Uses	A diluent and plasticizer for cellulose lacquers.
Comments	An adulterant of citrus and other essential oils.

GLYCERYL MONOSTEARATE [31566-31-1]

Chemical name	1,2,3-Propanetriol octadecanoic acid monoester Monostearin $C_{17}H_{35}COOCH_2CHOHCH_2OH$
Appearance	Either a white waxy solid, or a free-running white powder.
Physical	Manufactured as both self-emulsifying and non-self-emulsifying material: the self-emulsifying grade is incompatible with electrolytes.
Chemical	Has a monostearate content of 40%, and a maximum free glycerine content of 7%: the self-emulsifying grade has an acid value of 16–18 by the presence of about 5% sodium or potassium stearate: the non-self-emulsifying grade has an acid value of 4–6: m.p. 54–57 °C.
Other	Emulsions prepared with glyceryl monostearate may be reheated without separation.
Uses	The self-emulsifying grade is used in the preparation of oil-in-water emulsions to be used as bases for cosmetic creams and lotions. The non-self-emulsifying grade is used in oil-in-water emulsion systems which contain electrolytes, e.g. in the preparation of antiperspirant creams and lotions: also used to increase viscosity.

GLYCINE [56-40-6]

CTFA name	Glycine Aminoacetic acid; glycocoll; aminoethanoic acid $C_2H_5NO_2$
Appearance	Sweet monoclinic prisms; the hydrochloride is hygroscopic.
Comments	A non-essential amino acid.

GORSE [90131-76-3]

Source	
GEOGRAPHIC	A prickly shrub common to the British countryside.
NATURAL	*Ulex europaeus*, N.O. Leguminosae.
Appearance	Yellow flowers.
Odour	The inflorescence exhales a honey-like fragrance.
Comments	Perfumes of the same name are usually a product of the perfumer's imagination.

GRAINS OF PARADISE [90320–48–2]

Source
GEOGRAPHIC   A plant indigenous to West Africa.
NATURAL   *Amomum melegueta*, Roscoe, N.O. Scitaminae.
Isolation   The essential oil is extracted from the seeds.
Appearance   The essential oil is yellowish.
Odour   Pepper-like and aromatic.
Uses   Grains of paradise are used in the manufacture of some types of incense.

GRAPEFRUIT OIL [90045–43–5]

History   First produced commercially in Florida in 1933.
Source
GEOGRAPHIC   A tree cultivated in Israel and the USA.
NATURAL   Grapefruit, *Citrus decumana*, *C. paradisi*, *C. maxima*, L., N.O. Rutaceae.
Isolation   Mechanized pressing of the fresh peel.
Yield   0.05–0.1% depending upon source of the plant material.
Appearance   Amber-coloured liquid.
Odour   Bright, fresh and bitter, suggestive of lemon.
Physical   On standing, the oil deposits a flocculent precipitate, which disappears on warming.
Chemical   Shown by Nelson and Mottern to contain approx. 90% limonene, 2–3% oxygenated, volatile constituents and sesquiterpenes, and 7–8% non-volatile, waxy materials. In the terpene-free oil the following constituents were identified: octyl alcohol (both free and as acetates), cadinene and small quantities of citral and methyl anthranilate.
Uses   In colognes; mainly in the flavouring industry.
              RIFM Monograph (1974) *FCT*, **12**, 723.

GRAPE SEED OIL [8024–22–4]

CTFA name   Grape seed oil
History   First produced in Southern France in about 1900 as a substitute for olive oil.
Source
GEOGRAPHIC   Produced almost exclusively in California.
NATURAL   Raisins.
Isolation   The seeds, a waste product of the raisin industry, are mechanically removed, washed to remove any adherent pulp, dried, ground, pre-heated and hydraulically pressed. The resultant oil is settled, refined, bleached and then winterized.
Yield   About 17%.
Appearance   Colourless oil.

Odour	Odourless and tasteless.
Chemical	Contains over 50% linoleic acid: a useful source of 'vitamin F'.
Other	Has no known allergic effect upon the skin.
Uses	Useful in the cosmetic industry; to keep well it requires some anti-oxidant.

GUAIACOL [90–05–1]

Chemical name	alpha-Methoxyphenol
	Methylcatechol; o-hydroxyanisole; 1-hydroxy-2-methyoxybenzene
	$C_7H_8O_2$
Source	
NATURAL	Neroli oil.
Isolation	Synthesized from o-nitrophenol or pyrocatechol.
Appearance	Colourless prismatic crystals.
Odour	Powerful, medicinal, somewhat phenolic.
Physical	b.p. 204–206 °C, m.p. 28 °C: miscible with alcohol, chloroform, ether and oils, 1.7% soluble in water.
Comments	Protect from light: medicinally, used as an expectorant.
	RIFM Monograph (1982) FCT, 20, 697.

GUAIACWOOD OIL [89958–10–1]

Champaca-wood oil

Source	
GEOGRAPHIC	A wild tree which grows in the Argentine, and in the jungles of Paraguay.
NATURAL	*Bulnesia sarmienti*, Lor., N.O. Zygophyllaceae.
Isolation	Steam distillation of the waste wood, branches, sawdust etc.
Yield	Up to 5% depending upon isolation procedure.
Appearance	The wood develops a greenish-blue colour on exposure to the air, due to the presence of guaiac resin. The distilled oil melts at about 45 °C.
Odour	Balsamic and sweet, recalling tea roses.
Uses	May be blended with other oils in the production of rose and violet compounds: an effective adulterant of otto of roses and of sandalwood oil: used in delicate soap perfumes, but almost all sweet odours may benefit by its addition.
	RIFM Monograph (1974) FCT, 12, 905.

GUAIYL ACETATE [17431–48–0]

Guaiacwood acetate
$C_{17}H_{28}O_2$

Isolation	Acetylation of guaiacwood oil.
Appearance	Yellow, viscous oil.
Odour	Sweet, delicate-woody, reminiscent of tea rose.

Physical	b.p. approx. 269 °C.
Uses	Useful in perfumery, in the preparation of tea rose, lily and cyclamen compounds.
	RIFM Monograph (Guaiacwood Acetate) (1974) *FCT*, **12**, 903.

GURJUN BALSAM/GURJUN BALSAM OIL [8030-55-5]

Source	
GEOGRAPHIC	A tree which grows wild in South East Asia, especially in India.
NATURAL	A natural oleo-resin exuded from Dipterocarpus trees, especially *D. turbinatus*, and *D. tuberculatus*.
Isolation	The oleo-resinous secretion is similar to copaiba balsam and is obtained by drilling holes into the trunk of the tree: the oil is then produced by steam distillation of the balsam.
Yield	60–70% of oil from the balsam.
Odour	Rather weak odour of warm wood.
Uses	A fixative.
	RIFM Monograph (Gurjun Balsam) (1976) *FCT*, **14**, 789.

HAMAMELIS [84696-19-5]

CTFA name	Witch hazel; Witch hazel distillate; Witch hazel extract; Liquor Hamamelidis Dest
Source	
NATURAL	*Hamamelis virginiana*, L., N.O. Hamamelidaceae.
Isolation	Distillation of the leaves: about 17% alcohol is added as preservative.
Appearance	Colourless liquid.
Odour	Characteristic.
Uses	In cosmetics as an astringent, in skin lotions, and in stearic creams of the witch hazel snow type.

HAWTHORN [84603-61-2]

CTFA name	Crataegus
Source	
GEOGRAPHIC	A hedge tree native to Northern Europe and Asia.
NATURAL	*Crataegus oxycantha*, N.O. Rosaceae.
Odour	The blossoms have an exquisite fragrance, recalled by anisic aldehyde.
Uses	A base for all perfumes of this type with the addition of acetophenone and rose.

HAYFIELDS [8031-00-3]

	Foin coupé
Source	
NATURAL	Two plants contribute largely to the odour of hayfields: *Alyssum*

	compactum (sweet alyssum) and *Anthoxanthum odoratum* (sweet-scented vernal grass).
Isolation	These perfumes are based upon coumarin in combination with terpeneless oils of bergamot and lavender: the odour is rounded off with methyl salicylate, clary sage oil, and oakmoss resin.

HELIOTROPE CONCRETE

Source	
NATURAL	A mixture of aromatic substances.
Isolation	Heliotropin, vanillin, cinnamic alcohol, iso-eugenyl acetate, anisic alcohol, benzyl salicylate are mixed together, warmed, and allowed to solidify: stearic acid may be added if necessary.
Appearance	Waxy solid mass.

HELIOTROPIN [120–57–0]

Chemical name	1,3-Benzodioxole-5-carboxaldehyde
	Piperonal; piperonylaldehyde; 3:4-methylenedioxy-benzaldehyde; dioxymethyleneprotocatechuic aldehyde
	$C_8H_6O_3$
Source	
NATURAL	Pepper oil, spirea ulmaria, *Robinia pseudo-acacia*, and probably also in several vanillas.
Isolation	Oxidation of iso-safrole: was originally obtained from pepper by the oxidation of piperic acid with potassium permanganate in alkaline solution.
Appearance	White or whitish-yellow crystalline solid.
Odour	Very sweet floral and slightly spicy. Reminiscent of cherry pie.
Physical	b.p. 263 °C, m.p. 36 °C: turns reddish-brown on exposure to light for any period.
Chemical	Methylene ether of protocatechuic aldehyde.
Other	Blends well with anisic aldehyde, vanillin, musks, coumarin, benzyl iso-eugenol, and the essential oils of bergamot, neroli and rose.
Uses	An indispensable base for heliotrope, lilac and sweet pea perfumes, and is used in carnation, muguet, honeysuckle and mimosa types: one of the most popular modifiers: may be used in soap perfumery, but the soap should be coloured, as it develops brown patches.
Comments	Should be kept in a cool, dark place.
	RIFM Monograph (1974) *FCT*, **12**, 907.

Hemerocallis flava

	Day lily, yellow tuberose
Source	
NATURAL	Bulbous flowering plant.
Appearance	Has yellow lily-like blossoms which seldom last longer than one day.

Odour	Pleasant.
Uses	Hemmerocalle ottos may be prepared artificially by mixing together linalol, cinnamyl alcohol, amyl salicylate, rhodinol and hydroxycitronellal, with traces of clary sage oil.

HENNA [83-72-7]

CTFA name	Henna; henna extract Known as Jamaica mignonette in the West Indies
History	Once much used by Egyptians and Asiatics for dyeing hair, finger-nails etc. an orange hue. There are well-preserved examples in the British Museum.
Source	
GEOGRAPHIC	Thorny shrub from Northern Africa, China and India.
NATURAL	*Lawsonia inermis*, L., N.O. Lythraceae.
Isolation	The plants are raised in nurseries; the leaves are collected twice a year, and ground into a powder.
Yield	Approx. 16 cwt (810 kg) per acre.
Appearance	Brown powder.
Odour	The flowers have an odour recalling lilac.
Other	The dyeing properties are due to lawsone (2-hydroxynaphtha-1,4-quinone).
Uses	Hair dye.

n-HEPTANAL [111-71-7]

Chemical name	Heptaldehyde Heptanoic aldehyde; heptyl aldehyde; oenanthal; oenanthole; Aldehyde C_7 $CH_3(CH_2)_5CHO$
Isolation	Produced with undecylenic acid when castor oil is distilled under reduced pressure from the decomposition of ricinoleic acid: this is the first step in the synthesis of methyl heptine carbonate.
Appearance	Colourless oily liquid.
Odour	Very powerful oily–fatty with a pungent fermenting fruit character.
Physical	b.p. 153 °C.
Uses	Little used in perfumery, but traces will alter any compound. RIFM Monograph (1975) *FCT*, **13**, 701.

n-HEPTYL ACETATE [112-06-1]

Chemical name	Acetate C_7 n-Hept-1-yl acetate $CH_3COO(CH_2)_6CH_3$
Appearance	Colourless liquid.
Odour	Fruity, fatty–green and slightly floral.
Physical	b.p. 192 °C.

Uses As a topnote in some floral and fruity types.
RIFM Monograph (1974) FCT, **12**, 813.

n-HEPTYL ALCOHOL [111-70-6]

n-Heptan-1-ol; enanthic alcohol; 1-hydroxyheptane
$CH_3(CH_2)_5CH_2OH$; Alcohol C_7

Source
NATURAL Clove oil.
Appearance Colourless liquid.
Odour Light, green–fatty, wine-like note.
Physical b.p. 176 °C.
Uses In perfumery in carnation, jasmine and oriental compositions.
RIFM Monograph (Heptyl Alcohol: Alcohol C_7) (1975) FCT, **13**, 697.

HEPTYL CYCLOPENTANONE

Chemical name 2-Heptylcyclopenten-2-one-1
$C_{12}H_{20}O$
Isolation Synthesized from cyclopentanone and heptyl aldehyde with subsequent hydrogenation.
Appearance Pale yellow liquid.
Odour Fruity-floral, waxy, reminiscent of jasmine.
Uses Principally in jasmine compositions but also in other floral types.
RIFM Monograph (2-n-Heptyl Cyclopentanone) (1975) FCT, **13**, 452.

n-HEPTYL FORMATE [112-23-2]

Chemical name n-Hept-1-yl formate
$HCOO(CH_2)_6CH_3$
Appearance Colourless mobile liquid.
Odour Sharp, fatty–fruity, reminiscent of apple and plum.
Physical b.p. 178 °C.
Uses Occasionally as a topnote ingredient for fruity, woody and elderflower fragrances.
RIFM Monograph (Heptyl Formate) (1978) FCT, **16**, 771.

n-HEPTYL n-HEPTOATE [624-09-9]

Chemical name n-Hept-1-yl n-heptanoate
Heptyl heptylate
$CH_3(CH_2)_5COO(CH_2)_6CH_3$
Appearance Colourless oily liquid.
Odour Green–winy, with a fermenting fruit character.
Physical b.p. 274 °C.

Chemical	Similar properties and uses to aldehyde C_7.
Uses	Combines well with coumarin compounds.

n-HEPTYL PROPIONATE [2216-81-1]

Chemical name	n-Hept-1-yl propanoate $C_2H_5COO(CH_2)_6CH_3$
Appearance	Colourless liquid.
Odour	Ethereal, bitter-wine-like with a waxy nuance.
Physical	b.p. 202 °C.
Uses	In herbaceous and floral-herbaceous compounds.

n-HEPTYL n-VALERATE [5451-80-9]

Chemical name	Heptyl valerianate n-Heptyl-n-pentanoate $CH_3(CH_2)_3COO(CH_2)_6CH_3$
Odour	Fruity–woody, with animalic undertones.
Physical	b.p. 243 °C.
Uses	Blends with lavender, styrax and other herbaceous notes; useful in artificial civet notes.

HEPTYLIDENE ACETONE [10519-33-2]

Chemical name	n-Dec-3-en-2-one Oenanthylidene acetone $CH_3(CH_2)_5CH{=}CHCOCH_3$
Appearance	Colourless mobile liquid.
Odour	Fruity, fatty–green, somewhat jasmine like.
Uses	Occasionally in jasmine and some citrus types.

HERCOLYN [8050-15-5]

Chemical name	Mixture of mainly tetrahydro methylabietate and dihydro methylabietate Hercolyn D (deodorized grade) $C_{21}H_{36}O_2$ and $C_{21}H_{34}O_2$
Isolation	By methyl esterification of hydrogenated acids from American turpentine.
Appearance	Pale straw coloured, very viscous liquid.
Odour	Almost odourless, may have a faint woody or pine character.
Physical	b.p. approx. 366 °C.
Uses	Used extensively as a blender and fixative for many types of low cost fragrances created for household products and detergents. Most suitable for use with pine or herbaceous types but can be used safisfactorily with floral or fruity types. RIFM Monograph (Hercolyn D) (1974) *FCT*, **12**, 931, Special Issue I (Binder, p. 530).

HEXACHLOROPHANE [70-30-4]

Chemical name	Di-(3,5,6-trichloro-2-hydroxyphenyl)methane
CTFA name	Hexachlorophene
	Hexachlorophene, G-11
	$C_{13}H_6Cl_6O_2$
Appearance	White to buff crystalline powder.
Odour	May have a slight phenolic odour.
Physical	Insoluble in water, glycerin, and mineral oils: soluble in 1 in 50 alcohol, 1 in 25 propylene glycol: if heated it may also be soluble in fatty acids and vegetable oils.
Other	Surface bactericide.
Uses	An antiseptic in soaps, shampoos, talcum powder, and baby products: kills up to 95% of the resident bacterial flora on the skin and is therefore used in soaps and lotions for the preparation of the hands of surgeons: used to control staphylococcal infections.
Comments	In 1972 it was declared that hexachlorophane may be a potential health hazard: experiments in America indicated that it could induce brain damage in animals: under normal conditions, there is no evidence to date of actual harm to human beings. It is recommended that any preparation of hexachlorophane should be rinsed off immediately after use. In view of this area of doubt, suitable alternatives should be considered.

HEXADECYL ALCOHOL [36653-82-4]

CTFA name	Cetyl alcohol
	n-Hexadecyl alcohol; 1-hexadecanol
Source	
CHEMICAL	Petroleum.
Isolation	Commercial production from petroleum yields the branched chain primary alcohol.
Appearance	Colourless non-greasy liquid.
Odour	Virtually odourless.
Other	Possesses emollient properties.
Uses	A skin emollient in creams, lotions, suntan oils, bath oils, and after-shave lotions: may also be used in aerosol preparations and has no adverse effects on polystyrene or polypropylene containers.

HEXADECYL ALDEHYDE [629-80-1]

Chemical name	n-Hexadecanal
	Palmitic aldehyde: Aldehyde C_{16} (true)
	$CH_3(CH_2)_{14}CHO$
Odour	Weak, mild, waxy floral.
Physical	b.p. 310 °C, m.p. 34 °C.
Uses	Little used in perfumery.

3a,4,5,6,7,7a-HEXAHYDRO-5(or 6)-ACETOXY-4,7-METHANOINDENE

(6) isomer [5413–60–5]
(5) isomer [2500–83–6]

Appearance	Pale straw or colourless liquid.
Odour	Powerful sweet green floral note.
Physical	s.g. 1.072–1.076; soluble in 3 vols. 70% alcohol.
Uses	In floral-type perfumes for soaps, detergents, and household products, especially when a jasmine theme is required.

3a,4,5,6,7,7a-HEXAHYDRO-5(or 6)-iso-BUTYROXY-4,7-METHANOINDENE

Appearance	Pale straw coloured to colourless liquid.
Odour	Heavy and floral, reminiscent of the jasmine tuberose and gardenia type.
Physical	s.g. 1.025–1.030; soluble in 15 vols. 70% alcohol.
Uses	In all types of floral and bouquet compounds.

HEXAHYDRO HEXAMETHYL CYCLOPENTABENZOPYRAN

[1222–05–5]

Galaxolide
$C_{18}H_{26}O$

Source	
NATURAL	Does not occur naturally.
Isolation	Synthesized from pentamethyl-indane.
Appearance	Colourless viscous liquid.
Odour	Sweet musky.
Uses	One of the most widely used synthetic musks.

RIFM Monograph (1976) FCT, 14, 793.

HEXAHYDRO METHYL IONONE

$C_{14}H_{28}O$

Isolation	Hydrogenation of methyl ionone.
Appearance	Yellow liquid.
Odour	Warm, woody with balsamic undertones.

3a,4,5,6,7,7a-HEXAHYDRO-5-(or 6)-PROPIONOXY-4,7-METHANOINDENE

(6) isomer [17511–60–3]
(5) isomer [102935–86–4]

Appearance	Pale straw to almost colourless liquid.
Odour	Strong woody-earthy floral note.
Physical	s.g. 1.050–1.055; soluble in 10 vols. 70% alcohol.
Uses	In soaps and household products, especially lavender and fougere perfume compounds.

delta-HEXALACTONE [823–22–3]

5-hexanolide; 5-hydroxyhexanoic acid delta lactone; delta-caprolactone; 5-methyl-5-hydroxypentanoic acid lactone; 5-methyl-delta-valerolactone

Chemical name	Tetrahydro-6-methyl-2H-pyran-2-one
Source	
NATURAL	Is a volatile component in coconut oil.
Isolation	Oxidation of 1-substituted cycloalkanes.
Appearance	Colourless liquid.

gamma-HEXALACTONE [695–06–7]

gamma-ethyl-*n*-butyrolactone; tonkalide; gamma-caprolactone; 4-ethyl-4-hydroxybutanoic acid lactone; 1(3H)-furanone, 5-ethyl-dihydro-; 4-hydroxyhexanoic acid, gamma lactone

Chemical name	Hexanolide-1,4
Source	
NATURAL	In various fruits.
Isolation	From ethylene oxide and sodioethylmalonic ester.
Appearance	Colourless liquid.

1,1,2,3,3,6,-HEXAMETHYL-5-ACETYLINDANE [15323–35–0]

Phantolid

Appearance	White crystalline solid.
Odour	Sweet and musk-like.
Physical	m.p. min. 35 °C.
Other	Stable and does not cause discoloration.
Uses	Fixative and blender.
Comments	IFRA GUIDELINE

The Committee recommends for applications on areas of skin exposed to sunshine, excluding bath preparations, soaps and other products which are washed off the skin, to limit 5-acetyl-1,1,2,3,3,6-hexamethyl indan to 10% in fragrance compounds (see remark on phototoxic ingredients in the introduction).

This recommendation is based on test results of RIFM on the phototoxic potential of 5-acetyl-1,1,2,3,3,6-hexamethyl indan, showing no-effects at 10% in human phototoxicity tests (private communication to IFRA).

October 1978, last amendment October 1987
Replacement sheet November 1987

HEX-2-EN-1-AL(*trans*-2-HEXENAL) [6728–26–3]

$C_6H_{10}O$

Isolation	Synthesized from butylaldehyde and vinyl ethyl ether.

Appearance	Colourless liquid.
Odour	Powerful fresh, green-fruity.

RIFM Monograph (1975) *FCT*, **13**, 453.

IFRA GUIDELINE

The Committee recommends that *trans*-2-hexenal should not be used as a fragrance ingredient at a level over 0.01% in fragrance compounds.

This recommendation is made in order to promote Good Manufacturing Practice (GMP), thus avoiding excessive use of this material which has been found to have a sensitizing potential in preliminary tests (private communication to IFRA) and pending further tests.

April 1989

cis-3-HEXENAL [6789-80-6]

cis-beta, gamma-hexylenic aldehyde; 3-hexenal, (Z)-

Source	
NATURAL	Found in various fruits.
Isolation	Oxidation of *cis*-3-hexenol.
Appearance	Colourless liquid.

trans-2-HEXENOL [928-95-0]

2-hexenol, (e)

HEX-3-EN-1-OL (*cis*-3-HEXENOL) [928-96-1]

Leaf alcohol
$C_6H_{12}O$

Source	
NATURAL	Essential oils such as geranium, thyme, and mullberry leaf; also in violet leaves and tea.
Isolation	Synthesized from tetrahydrofuran.
Appearance	Colourless liquid.
Odour	Powerful intensely green, reminiscent of cut grass.
Physical	b.p. 157 °C.
Uses	Adds refreshing green note to many floral and herbaceous compositions.

RIFM Monograph (1974) *FCT*, **12**, 909.

HEXENYL ACETATES [1708-82-3]

$C_8H_{14}O_2$ (*cis*) z[3681-71-8]
(*trans*) E[3681-82-1]

Isolation	The isomers *cis*-3-hexenyl acetate and *trans*-3-hexenyl acetate may be synthesized from the corresponding alcohols and acetic acid.
Appearance	Colourless liquid.

Odour	Powerful, fresh-green, reminiscent of unripe bananas.
Physical	b.p. 169 °C (*cis*-3 isomer): 166 °C (*trans*-2 isomer). RIFM Monograph (*cis*-3-Hexenyl Acetate) (1975) *FCT*, **13**, 454.
Uses	Widely used at low concentrations in delicate floral fragrances such as muguet, lilac, narcisse, etc. where it participates in creating a true-to-nature effect. RIFM Monograph (1975) *FCT*, **13**, 454 (Binder, p. 427).

2-HEXENYL ACETATE [10094–40–3]
[2497–18–9]

Chemical name	*trans*-2-hexenyl acetate; 2-hexen-1-ol acetate 2-hexen-1-yl acetate
Source	
NATURAL	Found in apple, banana, cranberry, grape, peach and pear.
Isolation	Esterification of *trans*-2-hexenol with acetic acid.
Appearance	Colourless liquid.

cis-3-HEXENYL ACETATE

Chemical name	*cis*-beta-gamma-Hexenyl acetate Verdural extra, pipol acetate $C_8H_{14}O_2$
Isolation	By esterification of *cis*-3-hexenol with acetic acid.
Appearance	Colourless mobile liquid.
Odour	Intensively sharp green reminiscent of unripe banana.
Physical	B.P. 169 °C.
Uses	Widely used at low concentrations in delicate floral fragrances such as muguet, lilac, marcisse, etc. where it participates in creating a true-to-nature effect. RIFM Monograph (1975) *FCT*, **13**, 454 (Binder, p. 427).

HEXENYL BUTYRATE [16491–36–4]

Chemical name	*cis*-Hex-3-en-1-yl *n*-butanoate $CH_3CH_2CH_2COO(CH_2)_2CH=CH(CH_2)CH_3$
Appearance	Colourless liquid.
Odour	Fruity green, similar to freshly cut grass or cut privet leaves.
Physical	b.p. 192 °C.
Uses	Gives lift to many floral and herbaceous compounds.

cis-3-HEXENYL SALICYLATE [65405–77–8]

	$C_{13}H_{16}O_3$
Isolation	Synthesized from *cis*-3-hexenol and salicylic acid.
Appearance	Colourless liquid.
Odour	Tenacious, balsamic-sweet.

176 *iso*-Hexenyl tetrahydrobenzaldehyde

Physical   b.p. 271 °C.
                  RIFM Monograph (1979) *FCT*, **17**, 373.

iso-HEXENYL TETRAHYDROBENZALDEHYDE [37677-14-8]

Chemical name   4-(4-methylpent-3-en-1-yl)-cyclohex-3-en-1-carboxaldehyde
                 Myrac Aldehyde
Appearance   Colourless mobile liquid.
Odour   Powerful aldehydic, waxy–citrus.
Uses   Wide range of floral, citrus and pineneedle types.

HEXYL ACETATE [142-92-7]

                 $C_8H_{16}O_2$
Isolation   Synthesized from *n*-hexanol and acetic acid or acetic anhydride.
Appearance   Colourless liquid.
Odour   Sweet, fruity, reminiscent of pears.
Physical   b.p. 160 °C.
                 RIFM Monograph (1974) *FCT*, **12**, 913.

alpha-HEXYL CINNAMIC ALDEHYDE [101-86-0]

                 $C_{15}H_{20}O$
Isolation   Synthesized from benzaldehyde and octanal.
Appearance   Yellow liquid.
Odour   Soft, floral, slightly spicy.
Uses   Extensively in a wide range of floral compounds.
                 RIFM Monograph (Hexyl Cinnamic Aldehyde) (1974) *FCT*, **12**, 915.

n-HEXYL METHYL ACETALDEHYDE [7786-29-0]

                 Methyl hexyl acetaldehyde
Chemical name   2-Methyloctan-1-al
                 $CH_3(CH_2)_5CH(CH_3)CHO$
Appearance   Colourless liquid.
Odour   Delicate, fine flowery, fresh aldehydic character.
Chemical   Isomeric with nonyl aldehyde.
Uses   In cyclamen, lily, lilac, and jasmine creations.

HEXYL METHYL KETONE [111-13-7]

Chemical name   *n*-Octan-2-one (methyl hexyl ketone)
                 2-Octanone
                 $CH_3(CH_2)_5COCH_3$
Appearance   Colourless liquid.
Odour   Pungent, bitter-green reminiscent of sour fruit.
Physical   b.p. 173 °C.
Uses   Principally to reinforce lavender and herbaceous notes.
                 RIFM Monograph (Methyl Hexyl Ketone) (1975) *FCT*, **13**, 861.

HEXYL SALICYLATE [6259-76-3]

	$C_{13}H_{18}O_3$
Isolation	Synthesized from *n*-hexanol and salicylic acid.
Appearance	Colourless, oily liquid.
Odour	Faint, herbaceous–floral.
Physical	b.p. 290 °C.
Uses	As a modifier and blender with more volatile materials.

RIFM Monograph (1975) *FCT*, **13**, 807.

HEXYL VINYL CARBINOL [21964-44-3]

	Anderol: 8-Nonen-7-ol
Source	
NATURAL	Identified in water-fennel oil.
CHEMICAL	Name given to an alcohol isomeric with geraniol and citronellol.
Odour	Powerful oily green.
Physical	b.p. 198 °C.

HEXYLENE GLYCOL [107-41-5]

Chemical name	2-Methyl-pentane-2:4-diol
CTFA name	Hexylene glycol
Other names	2-Methyl-2,4-pentanediol; 2,4-pentanediol,2-methyl $(CH_3)_2COHCH_2CHOHCH_3$
Appearance	Colourless mobile liquid.
Odour	Almost none.
Physical	b.p. 198.3 °C, s.g. 0.924: completely miscible with water.
Uses	A solvent and blending agent in perfumery, and in cosmetics as a pigment-dispersing agent and to improve spreading properties in creams: also a solvent for eosins.

RIFM Monograph (1978) *FCT*, **16**, 777.

2-HEXYLIDENE-CYCLOPENTANONE [17373-89-6]

Appearance	Pale yellow liquid.
Odour	Warm, spicy, jasmine odour.
Physical	s.g. 0.911–0.916: soluble in 4 vols. 70% alcohol; b.p. 240 °C.
Uses	In jasmine compositions in concentrations up to 2%: also particularly useful in floral and fantasy perfumes.

HOMOHELIOTROPIN

Chemical name	3,4-Methylenedioxyphenylacetaldehyde $C_6H_3(CH_2O_2)CH_2CHO$
Appearance	Pale yellow liquid.
Odour	Heliotropin-like.
Physical	b.p. 143–144 °C at 10 mm Hg, m.p. 69 °C, s.g. 1.295: miscible with essential oils and synthetic aromatics.
Chemical	May resinify on exposure.

HOMOMENTHYL SALICYLATE [118-56-9]

Chemical name	3:3:5-Trimethylcyclohexyl salicylate
Appearance	Almost colourless viscous liquid.
Physical	Soluble in ethyl alcohol, mineral and vegetable oils and fatty acid esters.
Other	An oil-soluble sun-screening agent.
Uses	In the preparation of sunscreen oils, creams, lotions and aerosols (concentration 5–8%).

HOMOVANILLIN [5703-24-2]

Chemical name	4-Hydroxy-3-methoxyphenylacetaldehyde $C_6H_3(OH)(OCH_3)(CH_2CHO)$
Appearance	White crystalline solid.
Odour	Floral, resembling vanilla.
Physical	b.p. approx. 300 °C, m.p. 50 °C.

HONEY COMPOUNDS [8028-66-8]

CTFA name	Honey; honey extract May be marketed under the name miel.
Uses	In perfumery with 4-methyl-quinoline or other derivatives of quinoline in combination with phenylacetic acid and esters. This will produce a certain note in perfumes.
Comments	Wax or cire compounds are of a similar composition, but the quinoline derivatives are used in lower percentages.

HONEYSUCKLE [84929-67-9]

Source	Woodbine (*Lonicera periclymenum*)
NATURAL	Twining shrub, which attains a considerable height, N. O. Caprifoliaceae.
Appearance	Possesses flowers with a delicious fragrance.
Uses	Honeysuckle perfumes are not obtained from the flower, but are composed of a mixture of rhodinol, benzyl acetate, linalol, methyl anthranilate, and heliotropin.

HOP OIL [8007-04-3]

CTFA name	Hops oil
Source	
GEOGRAPHIC	A plant cultivated in Germany and in a number of Eastern European countries.
NATURAL	*Humulus lupulus*, L., N. O. Urticaceae.
Isolation	Distillation of the female inflorescences and the lupulin: the

Yield	Up to 1% from the hops and up to 3% from the lupulin.
Appearance	Greenish-yellow to reddish-brown colour.
Odour	Typical harsh, and bitter, odour: the oil obtained from hops preserved by exposure to the fumes of burning sulphur has an unpleasant odour.
Chemical	The following constituents have been identified: dipentene, myrcene, linalol, humulene, free formic acid, heptylic acid, valeric acid, esterified formic, acetic, decylic, nonylic, octylic, oenanthylic, butyric, and probably iso-nonylic acids.
Uses	In the mineral water industry and in brewing: occasionally in tobacco flavours: in perfumery, it may be used in harsh herbal compositions.

strobiles are picked when fully developed and are dried: the colour may be preserved by exposing them to the fumes of burning sulphur.

HORMONES

Source	
NATURAL	Specific secretions of the endocrine glands – adrenal, gonadal, pancreatic, parathyroid, pineal, pituitary, thymic, and thyroid.
Chemical	Have multiple effects upon the human organism.
Uses	Believed that they may be absorbed through the skin when applied in the form of cosmetics, but should only be used on the specific instruction of a qualified medical practitioner.

HUON PINE OIL [94334-02-8]

Source	
GEOGRAPHIC	A tree which grows in Tasmania, New Zealand, Malaysia, and the Archipelagos.
NATURAL	*Dacrydium franklini*, Hook, N. O. Texaceae.
Isolation	Distillation of wood shavings. An oil has been prepared from the leaves of this tree and consists largely of terpenes.
Yield	5%.
Chemical	Consists largely of methyl eugenol (97.5%) with traces of eugenol and cadinine.
Uses	In all preparations containing clove oil.

HYACINTHAL *See* **HYDRATROPIC ALDEHYDE**

HYACINTHIN, a name given to phenylacetaldehyde

HYACINTH OIL [94333-75-2]

Jacinthe oil

Source	
GEOGRAPHIC	A plant cultivated largely in Holland.

Hydratropaldehyde dimethylacetal

NATURAL	*Hyacinthus orientalis*, L., N. O. Lilaceae.
Isolation	Solvent extraction of the blossoms gives the concrete; a second extraction yields the absolute.
Yield	0.01–0.02% as concrete, of which some 50% is obtained as absolute.
Appearance	Dark greenish substance.
Odour	Sharp, penetrating, leaf-like: assumes the fragrance of hyacinths on extreme dilution.
Chemical	Contains phenylethyl alcohol, eugenol, methyl eugenol, benzoic acid, benzyl acetate, benzyl benzoate, benzyl alcohol, cinnamic alcohol, cinnamyl acetate, benzaldehyde, cinnamic aldehyde, methyl o-methoxybenzoate, ethyl o-methoxybenzoate, methyl methylanthranilate, dimethyl hydroquinone and possibly also N-heptanol.
Uses	In green and floral fine fragrances.
Comments	The majority of hyacinth oils on the market are prepared artificially. The following bodies are usually their base: ω-Bromstyrol; generally used as a soap perfume, it has a strong hyacinth odour: Cinnamic alcohol; has distinct fixative qualities, and a milder, sweeter odour: Phenylacetaldehyde; used in fine perfumery, has a smooth, sweet, powerful odour: Styrolyl acetate: used to blend with other oils where a suggestion only of the flower is required: has a sweet, delicate hyacinth–gardenia perfume. RIFM Monograph (1976) *FCT*, **14**, 795 (Hyacinth Absolute).

HYDRATROPALDEHYDE DIMETHYLACETAL [90–87–9]

$C_{11}H_{16}O_2$

Isolation	Synthesized from hydratropaldehyde and methanol.
Appearance	Colourless liquid.
Odour	Powerful, earthy, reminiscent of mushrooms and walnut.

RIFM Monograph (1979) *FCT*, **17**, 819.

HYDRATROPIC ALDEHYDE [34713–70–7]

Chemical name	2-Phenylpropanal alpha-Phenyl propionaldehyde, hyacinthal $C_6H_5CH(CH_3)CHO$
Appearance	Colourless liquid.
Odour	Powerful harsh green–earthy becoming more floral, hyacinth-like on dilution.
Physical	b.p. 204 °C.
Chemical	Stable and does not polymerize.
Uses	In perfumes of the hyacinth, lilac, and rose types.

RIFM Monograph (1975) *FCT*, **13**, 548.

HYDROCINNAMIC ACID [501-52-0]

Chemical name	3-Phenylpropanoic acid
	Benzenepropanoic acid; 3-phenylpropionic acid; beta-phenylpropionic acid; benzylacetic acid
	$C_6H_5CH_2CH_2COOH$
Isolation	Synthesized by the reduction of cinnamic acid with sodium amalgam.
Appearance	White crystalline substance.
Odour	Faint, warm–balsamic.
Physical	b.p. 280 °C, m.p. 49 °C.
Uses	Occasionally as a fixative.

HYDROCINNAMIC ALCOHOL [122-97-4]

Chemical name	3-Phenyl-propan-1-ol
	$C_9H_{12}O$
Source	
NATURAL	Styrax resin.
Isolation	Synthesized by the catalytic hydrogenation of cinnamic aldehyde.
Appearance	Colourless liquid.
Odour	Mild, balsamic–floral.
Physical	b.p. 236 °C.

HYDROCINNAMIC ALDEHYDE [104-53-0]

Chemical name	3-Phenylpropan-1-al
	Phenyl propyl aldehyde
	$C_6H_5CH_2CH_2CHO$
Source	
NATURAL	Ceylon cinnamon oil.
Appearance	Colourless liquid.
Odour	Balsamic, earthy–green.
Physical	b.p. 222 °C.
Chemical	Very stable, and not influenced by alkalis.
Uses	In soap perfumery in red rose, jasmine, lily, lilac, ylang-ylang, sweet-pea, and carnation compounds: also blends with coumarin and heliotropin: may be used in the production of oriental bouquets.

HYDROGEN PEROXIDE [7722-84-1]

	H_2O_2
Source	
CHEMICAL	Presented commercially as a 6% solution, the equivalent of 20 vols. available oxygen.
Appearance	Colourless liquid.
Chemical	A bleach.

HYDROGENATED OILS

Other	Slightly acid taste.
Uses	In nail and skin bleaches, and in peroxide creams.
Comments	May be preserved in 0.1% methyl p-hydroxy-benzoate, which prevents the loss of oxygen for up to 2 years.

HYDROGENATED OILS

Source
NATURAL — Castor oil, palm kernel, cotton seed, sesame, coconut, soya bean, ground nut, and various fish oils.
Isolation — Produced by the catalytic action of hydrogen on the oils.
Appearance — Conversion of the unsaturated glycerides to synthetic stearins results in the liquid oils becoming hard and fatty, or soft and waxy.
Physical — The approximate melting points of hardened coconut are 28 °C at 36 mm Hg and hardened palm kernel 30 °C at 45 mm Hg.
Uses — An important raw material in the manufacture of soaps: also used in edible fats, and in the manufacture of biscuits. Their use in the cosmetic industry is fairly general.

HYDROQUINONE DIMETHYL ETHER [150-78-7]

Chemical name — Dimethyl-hydroquinone
$C_8H_{10}O_2$

Source
NATURAL — Hyacinth oil.
Appearance — White crystalline solid.
Odour — Warm herbaceous–tobacco-like.
Physical — b.p. 213 °C, m.p. 56 °C.
Uses — In fougere, opoponax and amber compositions and may replace coumarin if used carefully: blends well with musk.
RIFM Monograph (Dimethylhydroquinone) (1978) *FCT*, **16**, 715.

p-HYDROXYBENZOIC ACID ESTERS

Sold under a variety of trade names such as nipagin, nipasol, nipabenzyl, and butex.
Appearance — White crystalline solids.
Chemical — Less toxic than benzoic acid.
Other — Preservatives which do not affect the smell, taste, or colour of the material to be preserved have growth-arresting action on *Staphylococcus pyogenes aureus* in an artificial nutrient medium.
Uses — Preservatives in pharmaceutical and cosmetic preparations, firstly, by repressing the development of any microorganisms present in the treated material, subsequently killing them, and then preserving the material from mycoderma, decomposition, putrefaction, fermentation, etc.: finally, they retard rancidity which may be evoked by microbiological influence.

para-HYDROXY BENZYLACETONE [5471-51-2]

Chemical name	Raspberry ketone
	$C_{10}H_{12}O_2$
Isolation	Synthesized from p-hydroxy-benzaldehyde and acetone, with subsequent catalytic reduction.
Appearance	White crystalline substance.
Odour	Typical raspberry odour.

RIFM Monograph (4(p-Hydroxyphenyl)2-Butanone) (1978) *FCT*, **16**, 781.

HYDROXYCITRONELLAL [107-75-5]

Chemical name	7-Hydroxy-3,7-dimethyloctan-1-al
	Muguettine principe, cyclalia, cyclosia, lily aldehyde, laurine, muguet synthetic, etc.
	$C_{10}H_{20}O_2$
Isolation	Synthesized by the hydration of citronellal.
Appearance	Colourless viscous liquid.
Odour	Delicate floral, recalling lily of the valley.
Physical	b.p. 241 °C.
Uses	An excellent fixative: may be used in jasmine, lily, lilac, magnolia, and hyacinth flower oils: a useful modifier: used to be employed in soap perfumery where it blended well with ω-bromstyrene, linaloe, geranium, and palmarosa oils, benzyl acetate and coumarin. Now replaced to a great extent by alternative materials with less skin-sensitizing potential.
Comments	Has been possible to reproduce accurately the odour of muguet, and imitate other floral ottos, previously comparatively crude.

RIFM Monograph (1974) *FCT*, **12**, 921.

IFRA GUIDELINE

The Committee recommends that hydroxycitronellal should not be used as a fragrance ingredient at a level over 5% in fragrance compounds.

This recommendation is made in order to promote Good Manufacturing Practice (GMP) for the use of hydroxycitronellal as a fragrance material due to concern expressed for possible adverse reactions associated with its extensive use. The recommendation is passed on a review of a large number of animal and human test data, indicating a weak sensitizing potential for several qualities of hydroxycitronellal (private communication to IFRA).

HYDROXYCITRONELLAL DIMETHYLACETAL [141-92-4]

Chemical name	Hydroxy acetal
	$(CH_3)_2C(OH)(CH_2)_3CH(CH_3)CH_2CH(OCH_3)_2$

Isolation	Synthesized from hydroxycitronellal and methanol in the presence of hydrogen chloride gas.
Appearance	Colourless, oily liquid.
Odour	Mild, green, fresh floral.
Physical	b.p. 252 °C.
Uses	In perfumery in lilac and orange flower bouquets.

RIFM Monograph (1975) *FCT*, **13**, 548.

HYDROXYCITRONELLAL METHYLANTHRANILATE [89–43–0]

Aurantiol, Aurantine
$C_{18}H_{27}NO_3$

Isolation	Synthesized by the condensation of hydroxycitronellal and methyl anthranilate.
Appearance	Viscous, yellow liquid.
Odour	Floral, very sweet, reminiscent of orange blossoms.
Uses	Most widely used of all Schiff's bases in orange blossom, citrus blends, oriental and chypre types.

RIFM Monograph (1975) *FCT*, **13**, 549.

HYDROXYETHYL CELLULOSE [9004–62–0; 9063–65–4]

CTFA name	Hydroxyethylcellulose
Source	
CHEMICAL	A modified cellulose polymer containing hydroxyethyl sidechains.
Appearance	White, tasteless, free-flowing powder.
Odour	None.
Physical	Dissolves rapidly in both hot and cold water, forming a smooth, viscous solution.
Chemical	Non-ionic.
Other	Tasteless.
Uses	A thickener, binder, emulsion stabilizer, and suspending agent.

(4-HYDROXY-4-METHYL)PENTYL-3-CYCLOHEXENE-1-CARBOXALDEHYDE [31906–04–4]

$C_{13}H_{22}O_2$
Lyral

Isolation	Synthesized by the Diels–Alder condensation of myrcenol and acrolein, forming two isomers.
Appearance	Colourless, viscous liquid.
Odour	Tenacious, sweet floral, with distinct lily of the valley notes.
Uses	Extensively used as a fixative and to give a fine muguet note to perfumes.

HYDROXYQUINOLINE SULPHATE [134-31-6]

Chemical name	8-Hydroxyquinoline sulphate
	8-Quinolinol sulphate; oxine sulphate
	$(C_9H_6NOH)_2 \cdot H_2SO_4$
Appearance	Light yellow, crystalline solid.
Odour	Slight and which is reminiscent of saffron.
Physical	m.p. 175–178 °C: readily soluble in water.
Chemical	Incompatible with alkalis and heavy metal salts.
Other	Fungicide, bactericide, and preservative.
Uses	In skin lotions and deodorant preparations as a bactericide.
Comments	When mixed with an equimolecular mixture of potassium sulphate is known as potassium hydroxyquinoline sulphate: both compounds should be protected from contact with metals and exposure to light.

HYRACEUM

Source	
GEOGRAPHIC	South Africa, Arabia, and Syria.
NATURAL	Hyrax, a small, brown, rabbit-like animal with a short snout, and a small tail: it lives in rock crevices and feeds upon foliage.
CHEMICAL	Probably the faeces and urine.
Isolation	Collected in small masses.
Appearance	Irregular, amorphous masses, which when kneaded with the fingers become plastic: on examination under a microscope, fibres and hairs may be observed.
Odour	Musky and recalling castor.
Uses	Sometimes substituted for castor; not recommended for perfumery.

HYSSOP OIL [84603-66-7]

Source	
GEOGRAPHIC	A culinary herb which grows wild in many Mediterranean countries and is cultivated in Provence and Germany.
NATURAL	*Hyssopus officinalis*, L., N.O. Labiatae.
Isolation	Steam distillation of the flowers and stalks.
Yield	0.5–1%.
Appearance	Oil.
Odour	Sweet and camphoraceous.
Uses	In perfumery to some extent in colognes, but mainly as a flavour.
	RIFM Monograph (1978) *FCT*, **16**, 783.

IANTHONE

Chemical name	iso-Propylidene ionone
	$C_{16}H_{24}O$

Isolation	Condensation of citral with mesityl oxide.
Odour	Dry orris-violet like, with considerable tenacity.

INCENSO MACHO

Source	
GEOGRAPHIC	Probably comes from a plant found in Peru.
NATURAL	Exuded from *Styrax ovatum*, D.C., Styraceae.
Isolation	The resin is collected from the plants.
Appearance	Hard reddish-black resinous mass.
Odour	Reminiscent of Siam benzoin or vanilla.
Chemical	Contains vanillin and benzoic acid.
Uses	In perfumery and as incense.

INDOLE [120–72–9]

Chemical name	2,3-Benzopyrrole
	C_8H_7N
Source	
NATURAL	Neroli oil, jasmine absolute, orange blossom, robinia, wallflower, and some citrus species: also produced in animals by the digestion process.
Isolation	May be isolated from coal tar: synthesized during the manufacture of indigo and also from pyrrol.
Appearance	White crystalline flakes.
Odour	Disagreeable, choking odour: on extreme dilution this acquires a distinct floral character reminiscent of orange blossom and jasmine.
Physical	b.p. 253 °C, m.p. 52 °C.
Chemical	When mixed with other synthetics and exposed to light, becomes reddish-brown.
Uses	In jasmin, neroli, muguet, and lilac fragrances.
	RIFM Monograph (1974) *FCT*, **12**, 925.

alpha-IONONE, *beta*-IONONE α[127–41–3] β[79–77–6]

Chemical name	4-(2,6,6-Trimethylcyclohex-2-en-1-yl)-but-3-en-2-one;
	4-(2,6,6-trimethylcyclohex-1-en-1-yl)-but-3-en-one
	Irisone
	$C_{13}H_{20}O$
History	First synthesized in 1893 by Tiemann and Kruger, as a by-product in the synthesis of irone.
Source	
NATURAL	Boronia absolute.
Isolation	Synthesized from citral and acetone with pseudo-ionone as intermediate: after cyclization, ionone is formed. It has been found that the cyclization of pseudo-ionone may be effected by treatment

	with three times its weight of 60% sulphuric acid, which yields the largest proportion of alpha-ionone. Both alpha- and beta-ionone, and 100% ionone (a mixture of the two ketones without alcohol) are manufactured by various companies under different names.
Appearance	When freshly prepared and 100% pure, both alpha- and beta-ionone are colourless liquids: commercial samples may be pale yellow to greenish.
Odour	Varies depending upon isomeric ratio: alpha-ionone has a sweet odour of violets and beta-ionone has a fruity–woody odour, reminiscent of cedarwood and raspberries: diluted ionone recalls orris, but when mixed with alcohol, it resembles fresh violets.
Physical	alpha-Ionone, b.p. 237 °C; beta-ionone, b.p. 239 °C, used in the synthesis of vitamin A.
Other	May appear to be odourless to an inexperienced worker, as it causes numbing of the olfactory nerves.
Uses	Commercially, the cheapest article is a mixture of the two ketones, and this is widely used in soap perfumery. Alpha-ionone is used for fine perfumery as in violette de parme, and beta-ionone is preferable for jasmine perfumes.
Comments	There are many synthetic substances used in perfumery based upon the chemical structure of ionone: many commercial ionones also contain traces of methyl heptine carbonate. Ionone may cause allergic reactions.

RIFM Monograph (Ionone) (1975) FCT, **13**, 549.

IONYL ACETATE [52210–18–1]

Appearance	Colourless liquid.
Odour	Between linalyl and terpinyl acetate.
Physical	s.g. 0.863–0.867: soluble in 4 vols. 70% alcohol.
Uses	In soap and household toiletry preparations, especially in compounds such as jasmine, rose, lavender, and floral bouquets.

alpha-IRONE [79–69–6]

Chemical name	4-(2,5,6,6-Tetramethylcyclohexen-1-yl)but-3-en-2-one 6-Methylionone $C_{14}H_{22}O$
History	First isolated from orris roots in 1893 by Tiemann and Kruger: the structural elucidation and synthesis were achieved many years later by a Swiss research group.
Source	
NATURAL	Present in concrete orris oil and probably also in the perfume of *Acacia cavenia* and wallflower: the main perfume ingredient of violets.
Isolation	Extracted from orris root oil, or synthesized from dimethyl heptenone via the intermediate pseudo-irone.

Isojasmone

Appearance	Pale yellow to colourless liquid.
Odour	Soft, sweet, characteristic of orris root.
Physical	b.p. 248 °C.
Chemical	Typically contains about 60% gamma-irone, and 40% *cis-cis*-alpha-irone, the latter being the most highly prized by perfumers: the synthetic material comprises about 90% *cis-trans*-alpha-irone and 10% beta-irone; the *cis-trans*-alpha isomer is less well liked than the *cis-cis*-alpha isomer. RIFM Monograph (1975) *FCT*, **13**, 551.

ISOJASMONE [11050-62-7]

2-cyclopenten-1-one, 2-methyl-3-(2-pentenyl)-; 2-hexyl-2-cyclopentanone-1-one and 2-hexylidenecyclopentanone (mixture); 2-hexylidene cyclopentaonone and 2-hexyl-2-cyclopenten-1-one (mixture)

Source	
NATURAL	Does not occur in nature.
Isolation	Prepared from undecylenic acid via undecalactone with polyphosphoric acid.
Physical	GC RIFM no. 74-218; IRC RIFM no. 74-218.
Other	It is a mixture of several isometic ketones.

ISOPULEGOL [89-79-2]

Cyclohexanol, 5-methyl-2-(1-methylethenyl), [1R-(1. alpha., 2. beta., 5. alpha.)]; p-8(9)-menthen-3-ol; 1-methyl-4-isopropenyl-cyclohexan-3-ol

IVA OIL

Source	
GEOGRAPHIC	A plant which grows in high altitudes in the Swiss mountains.
NATURAL	*Achillea moschata*, L., N.O. Compositae.
Isolation	Distillation of the herb.
Yield	0.5% of dry herb.
Appearance	Blue–green liquid, which changes to olive-green in time.
Odour	Intense and aromatic.
Chemical	Contains cineole, *l*-camphor, and probably valeric aldehyde.
Uses	In the manufacture of liqueurs: does not appear to have been employed in the perfume industry.

JABORANDI LEAVES [84696-42-4]

Source	
GEOGRAPHIC	A plant grown in South America.

NATURAL	The leaves from *Pilocarpus jaborandi*, Holmes, *P. microphyllus*, Stapf., and other species of N.O. Rutaceae.
Isolation	A tincture may be prepared 1 in 5 and a liquid extract 1 in 1 with 45% alcohol.
Appearance	An essential oil.
Odour	Resembles rue.
Uses	In hair preparations: the alkaloid pilocarpine, obtained from jaborandi, is used medicinally.

JASMINE [84776-64-7]

Source	
GEOGRAPHIC	A bush native to the East Indies: also cultivated in Southern France, Spain, Algeria, Morocco, India and Egypt.
NATURAL	*Jasminium grandiflorum*, L., N.O. Oleaceae.
Isolation	The concrete is obtained by extraction from the flowers with volatile solvents; further extraction yields the absolute.
Yield	0.2% concrete, of which 50% is obtained as absolute.
Appearance	There are four types of perfume available commercially: 1. That extracted by petroleum ether and used in fine perfumery. 2. That extracted using benzole, yielding a higher percentage: it has a fruity odour, and is consequently cheaper. 3. The discarded chassis flowers are extracted with solvent, and the odour has a more indoloid character, and is used in compound manufacture. 4. The concentrated washings from the enfleurage fats.
Odour	Powerful, honey-like and floral.
	RIFM Monograph (Jasmine Absolute) (1976) *FCT*, **14**, 33.

JASMONE [488-10-8]

Chemical name	3-Methyl-2-(2-pentenyl)-2-cyclopenten-1-one *cis*-Jasmone; 3-methyl-2-(pent-2′-en-1′-yl)-cyclopent-2-en-1-one $C_{11}H_{16}O$
History	First discovered by A. Hesse in the laboratories of Heine & Co. AG. There is some controversy as to the original discovery of its chemical constitution.
Source	
NATURAL	Jasmine absolute, and probably also oils of jonquille, neroli and orange blossom absolute.
Isolation	Synthesized from methyl cyclopentenone and pentenyl chloride.
Appearance	Pale yellow liquid.
Odour	Fruity and celery seed-like: on extreme dilution, this becomes reminiscent of jasmine and cherry blossom.
Physical	b.p. 248 °C.

Chemical	The *cis*-isomer is the naturally occurring material. RIFM Monograph (*cis*-Jasmone) (1979) *FCT*, **17**, 845.

JOCKEY CLUB

History	This name was originally given to a perfume which was said to be an imitation of the fragrance pervading the Downs at Epsom before the Derby.
Comments	Perfumes bearing this name all differ materially in odour value: bouquets may be compounded from jasmine, mimosa, rose, tuberose, orris and bergamot.

JONQUILLE [90064-25-8]

Rush daffodil

Source	
NATURAL	A flower belonging to the narcissus family.
CHEMICAL	The base of jonquille perfumes is phenylethyl phenylacetate.
Isolation	May also contain jasmine, rose, styrolyl acetate, and traces of basil and patchouli oils.
Odour	A good imitation of the odour of jonquille may be obtained with this formulation.

JUNIPERBERRY OIL [8002-68-4]

Oil of Juniper

Source	
GEOGRAPHIC	A bush which grows wild in Italy, Yugoslavia, Hungary and Czechoslovakia.
NATURAL	*Juniperus communis*.
Isolation	Steam distillation of the dried ripe fruits.
Appearance	Colourless to pale green liquid.
Odour	Powerful, characteristic, and turpentine-like.
Physical	Almost insoluble in water; soluble in 4 parts of alcohol and is miscible with benzene, chloroform or amyl alcohol.
Chemical	On exposure to air, and with age, the oil darkens and thickens. The chief known constituents are pinene, cadinene, terpineol, camphene and camphor.
Uses	In perfumery as a toner.
Comments	Should be kept cool, away from air or light. RIFM Monograph (1976) *FCT*, **14**, 333.

KACHI GRASS OIL [84649-81-0]

Inchi grass

Source	
GEOGRAPHIC	A grass which grows in Bangalore and on the high-lying tableland of Mysore.

NATURAL	*Cymbopogon coesius*, Stapf., which is closely allied to *C. martini*, Stapf.
Isolation	Distillation.
Yield	0.5% essential oil.
Appearance	Liquid.
Odour	Similar to ginger-grass.
Chemical	Contains geraniol, perilla alcohol, dipentene, and *l*-limonene.

KAOLIN [1332-58-7]

CTFA name	Kaolin
	Bolus alba; china clay; porcelain clay; white bole; argilla; osmo-kaolin (a superior quality material)
	$Al_2O_3 2SiO_2 H_2O$
Source	
NATURAL	Pegmatite rock.
CHEMICAL	Essentially an hydrated aluminium silicate.
Isolation	Fine white or greyish-white porcelain clay which results from the disintegration of the rock: may be electrolytically purified giving pure hydrated aluminium silicate, known as osmo-kaolin.
Appearance	Purified osmo-kaolin is a smooth, creamy-white colloidal powder.
Uses	In medicines for the treatment of diarrhoea, colitis, and similar bacterial infections: also in the manufacture of toothpastes: in perfumery, in face powders: a superior filtering medium.

KAPUR-KACHRI [93455-95-9]

History	Valued as a perfume in the East, especially by the Arabs and Persians.
Source	
NATURAL	*Hedychium spicatum*, Smith, N.O. Scitamineae.
Isolation	The dried root of the plant.
Uses	As incense in worship.

KARAYA GUM [9000-36-6]

CTFA name	Karaya gum
	Indian tragacanth; gum karaya; kadaya; katilo; kullo; kuteera; sterculia; mucara
Source	
GEOGRAPHIC	A small tree which grows in dense forest near Bombay.
NATURAL	*Sterculia urens*, Roxburgh, N.O. Sterculiaceae.
Isolation	The gum is tapped from the irregular lumps or tears which form on the trees, bagged and sorted into various grades.
Appearance	Liquid gum.
Physical	Swells to form a jelly when dissolved in water; this produces a thick transparent mucilage on dilution.

Chemical	Contains about 80% bassorin.
Uses	Sometimes substituted for tragacanth: can be used as a denture adhesive or as a stabilizer, thickener or texturizer.
Comments	Has cathartic properties.

KARO-KARUNDE [94334-14-2]

Leptactina senegaentica, Hook, f., N.O. Rubiaceae

Source	
GEOGRAPHIC	A shrub native to the Republic of Guinea.
Isolation	The strongly perfumed white flowers are extracted by volatile solvents.
Yield	0.13% concrete which yields about 60–65% absolute.
Appearance	The concrete is orange–red which darkens with age. The decolourized absolute is golden yellow.
Odour	The flowers have a strong odour of ylang-ylang type: the absolute oil has a characteristic jasmine–ylang odour.
Uses	In violet compounds.

KATCHUNG OIL [8002-03-7]

Arachis oil; peanut oil; groundnut oil

Source	
GEOGRAPHIC	A plant which grows in India.
NATURAL	*Arachis hypogoea*, L., N.O. Leguminosae.
Isolation	Pressure on the seeds: there are 60–100 fruits to each plant, and 1–3 brownish-red seeds from each fruit.
Yield	45% fixed oil.
Appearance	Liquid greenish-yellow or almost colourless oil.
Odour	Mild and pleasant.
Physical	Clouds at low room temperature.
Uses	As a cheap substitute for expressed oil of almonds.

KEWDA OIL [91770-47-7]

Source	
GEOGRAPHIC	A plant which grows wild in India, and to a lesser extent in the Andaman Islands and Singapore.
NATURAL	*Pandanus odoratissimus*, L., N.O. Pandanaceae.
Isolation	The blossoms appear between July and December and vary between yellow and white: the best oil is obtained by extraction of the pearly-white flowers: a well-developed flower may weigh about 100 g.
Yield	0.1–0.3%.
Appearance	Light yellow oil.
Odour	Lilac-honey odour.

Chemical	Contains benzyl alcohol, acetate, benzoate and salicylate, geraniol, linalol and linalyl acetate, phenylethyl alcohol, santalol, guaiacol, and ω-bromstyrene.
Uses	In perfumery and may be useful for imparting the green note to rose, lilac, muguet etc.

KIESELGUHR [7631–86–9]

Infusorial earth; siliceous earth; diatomaceous earth; fossil flour SiO_2

Appearance	Available commercially as a fine white or greyish powder.
Odour	Odourless and tasteless.
Uses	A filtering medium; used in dental powders, and sometimes in dusting powders.

KIOU-NOUK [89957–98–2]

Source	
NATURAL	Olibanum, the gum-resin of *Boswellia carterii*, Birdw., and other species of N.O. Burseraceae.
Appearance	Clear, yellowish, semi-liquid resin.
Odour	More aromatic and less sickly.
Chemical	Does not darken with age.
Uses	A fixator in bouquets: blends well with almost any of the essential oils and musks, vanillin, coumarin and heliotropin.

KURO-MOJI OIL

Source	
GEOGRAPHIC	A shrub which grows in the mountainous regions of Japan.
NATURAL	*Lindera fericia*, Bl., Lauraceae.
Isolation	Distillation from the leaves or the whole plant.
Appearance	Light or dark yellow oil.
Odour	Powerful and balsamic.
Chemical	Contains terpineol, carvone, dipentene, limonene, cineol, linalol and geranyl acetate.
Uses	A useful raw material in perfumery.
Comments	The aromatic wood from the plant is employed in the manufacture of toothpicks.

LABDANUM [8016–26–0]

Source	
GEOGRAPHIC	A plant widely distributed in the rocky regions of Southern France, Spain, Portugal and Northern Africa.
NATURAL	Secreted from the glandular hairs on the leaves of several species of the rock rose family, *Cistinae creticus*, L., and *C. ladaniferus*, L.

Isolation	In Crete and Cyprus, the oleo-resinous secretion is collected with a special instrument, called a ladanisterion, resembling a double rake with leathern thongs instead of teeth and used like a whip, the thongs being scraped when fully charged with adhering resin: this is then purified when up to 60% sand may be removed. In Spain, the resin may also be obtained by boiling the branches in water; in France it is extracted directly from the plant by volatile solvents, giving a superior product.
Yield	1–2% volatile oil.
Appearance	A sticky, oleo-resinous substance; when purified may be a yellowish-brown or green liquid: imported as dark brown or green lumps.
Odour	Characteristic, and varies according to the species: a purified specimen has a faintly ammoniacal odour, which may be due to contamination: a good specimen has a powerful, sweet odour, recalling ambergris.
Chemical	Has good fixative properties. The volatile oil has been examined by Masson and the following constituents have been identified: acetophenone and trimethyl-1,5,5-hexanone-6.
Uses	When purified it is an important raw material in perfumery: may be used in many bouquets as a fixator, but is employed principally in the manufacture of artificial ambers: is used in soap perfumery as a fixator for lavender, fern, and chypre compounds. RIFM Monograph (Labdanum oil) (1976) *FCT*, **14**, 335.

LACTIC ACID   [50–21–5] (*d*-form), [10326–41–7] (*dl*-form), [598–82–3] (*l*-form), [79–33–4]

Chemical name	2-Hydroxy-propanoic acid
CTFA name	Lactic acid Ordinary lactic acid is the racemic *dl* form; 2-hydroxypropanoic acid CH$_3$CHOHCOOH
Isolation	Prepared from lactose by fermentation: obtained commercially from sugar by adding decayed cheese containing an abundance of lactic acid bacilli, neutralizing with chalk or lime, filtering and decomposition by sulphuric acid.
Appearance	Colourless, syrupy liquid.
Physical	m.p. 26 °C, s.g. 1.2485.

LANOLIN ANHYDROUS   [8006–54–0] (anhydrous)

CTFA name	Lanolin Adeps lanae; wool fat; lanolin; oesipos; alapurin
Source	
NATURAL	Wool of the sheep, *Ovis aries*.

Isolation	Obtained from the fleece by 'scouring' with soap solution, or by extraction with volatile solvents: the fatty acids are removed and the fat bleached: derivatives may be obtained by the acetylation, ethoxylation, or solvent fractionation of lanolin or lanolin fatty acids or alcohols.
Yield	Fleeces contain from 15–20%.
Appearance	Yellowish, unctuous substance.
Odour	Almost none.
Physical	m.p. 40–44 °C.
Chemical	A complex mixture of esters of high molecular weight, fatty alcohols and acids.
Other	Readily absorbed by the skin: does not tend to go rancid.
Uses	Widely used in toilet preparations, and as a basis for ointments. 'Liquid lanolins', which are more soluble in mineral oil, or 'lanolin oils' may be prepared for use in hair lotions or shampoos.

LARD [61789-99-9]

CTFA name	Lard
	Adeps; axungia porci
Source	
NATURAL	The purified internal fat from the abdomen of the hog, *Sus scrofa*.
Isolation	May be digested at 60 °C for 1 h with 3–5% of gum benzoin to prevent rancidity.
Appearance	Soft, white, unctuous substance.
Chemical	Consists of about 50% olein, palmitin, stearin, laurin, myristin, and a little linolein.
Uses	The basis of most of the fats used in enfleurage.
Comments	Must be kept cool and in tightly capped containers.

LAUREL LEAF OIL [8002-41-3]

Source	
GEOGRAPHIC	A small tree cultivated in the Mediterranean region.
NATURAL	*Laurus nobilis*.
Isolation	Steam distillation of the fresh, green leaves and twigs.
Yield	1.2–1.5% depending upon source of the plant material.
Odour	Fresh, spicy, somewhat camphoraceous.
Uses	In perfumery, especially in masculine creations.
	RIFM Monograph (1976) *FCT*, **14**, 337.

LAURIC DIETHANOLAMIDE [120-40-1]

Appearance	Cream-coloured, waxy solid.
Uses	A foam booster and stabilizer in hair shampoos, and bubble baths: has similar applications to coconut diethanolamide, but gives a closer-type lather.

LAURIC MONOISOPROPANOLAMIDE [142–54–1]

Appearance	Available as cream-coloured waxy flakes.
Other	Dispersible in hot water.
Uses	A foam stabilizer, and a viscosity modifier in shampoos, carpet and upholstery cleaners, and dishwasher detergents.

LAURYL ESTERS

Isolation	Several esters have been prepared from dodecyl alcohol.
Appearance	Liquids.
Odour	Unique characters: the propionic acid ester resembles mushrooms or truffles.
Uses	Fine perfumery.

LAVANDIN OIL [8022–15–9]

One type is known as abrialis.

Source	
GEOGRAPHIC	Flowers which grow on the southern slopes of the lower Alps, and are cultivated in France and Spain.
NATURAL	*Lavandula latifolia fragrans*, Chatenier, a hybrid of N.O. Labiatae.
Isolation	Steam distillation of the dried flowers and stalks, which are collected 3 weeks later in the season than true lavender. The plants have longer stems which makes collection much easier.
Yield	1.0–2.5% depending upon production procedures and plant material: yield per hectare is about five times higher than that of true lavender.
Appearance	Pale straw-coloured oil.
Odour	Lacks the 'bouquet' of true lavender: has a spicy, camphoraceous character.
Chemical	The ester content bears an inverse relationship to the yield of the oil. Commercial samples have an ester content of 20–22%.
Uses	Widely used in lower-priced products, especially in the soap industry: abrialis lavandin is useful in detergents.
Comments	Demand for both oils is increasing and production may reach 600 tons (610 t) per annum.
	RIFM Monograph (1976) *FCT*, **14**, 447.

LAVENDER OIL [8000–28–0]

CTFA name	Lavender oil
	Lavendula officinalis oil; Lavender flowers oil; oil of lavender
Source	
GEOGRAPHIC	Plants native to the lower Alps, but also cultivated extensively in England in Norfolk, and in some parts of Suffolk, Hertfordshire and Kent: cultivated in considerable quantity on the Bridgestowe Estate, Tasmania.

NATURAL	*Lavendula vera*, D.C., N.O. Labiatae: in England the oil is obtained exclusively from two varieties of the cultivated species, dwarf munstead and giant blue.
Isolation	The harvest is collected from 20 July to September: oil content is higher if the cutting is preceded by a few weeks of sunny weather. In France the crop varies with atmospheric conditions and may be particularly influenced by dryness of the air. The inflorescence is cut with the minimum of stem; the oil is obtained by steam distillation.
Yield	1.4–1.6%: may be increased by adopting certain cultural methods, but the weather exerts a constant influence.
Appearance	Pale yellow oil.
Odour	Typically sweet, balsamic, and herbaceous: in France the plants found at higher altitudes bear the finest odour.
Chemical	Composition varies depending upon time of collection: the following constituents have been identified: linalyl acetate, linalyl butyrate, geranyl acetate, linalol, geraniol, with their butyric, capric and valeric acid esters, limonene, caryophyllene, *d*-pinene, coumarin, furfurol, valeric aldehyde, amyl alcohol, ethyl amyl ketone, *d*-borneol and possibly nerol and thymol. Cineol occurs in greater proportion in English lavender oil. There are numerous adulterants, including oils of turpentine, cedar and gurjun, alcohol, resin, terpinyl acetate, and glyceryl triacetate: principal adulterants today are acetylated lavandin and Ho oils.
Uses	In soap perfumery it is blended with other substances such as borneol, bornyl acetate, musk xylene, patchouli, bergamot etc.: quantities of spike-lavender oil may also be added: used extensively in colognes, lavender waters, fougeres, chypres, ambers and many other floral perfumes.
Comments	Sold on its ester content: however, English oil, although it has a lower ester content, is the most esteemed and will fetch a higher price than the finest French oils. RIFM Monograph (1976) *FCT*, **14**, 451.

LAWANG OIL

Source	
GEOGRAPHIC	A tree grown in Indonesia.
NATURAL	A species of Cinnamomum.
Appearance	Oil.
Odour	Clove-type, also reminiscent of sassafras or nutmeg.
Chemical	The odour is probably due to the high eugenol content.

LAWSONE [83-72-7]

Chemical name:	2-Hydroxy-1,4-naphthaquinone
CTFA name	Lawsone 2-Hydroxy-1,4-naphthalenedione

Appearance	Yellow prisms (crystallized from acetic acid).
Chemical	The pigmentary constituent of henna.
Other	Imparts a titian colour to hair.
Uses	In concentrations of 0.05–0.10% in hair shampoos and brilliantines.
Comments	Has a tinting power 100 to 150 times that of henna leaf: can also be used as a sunscreen agent.

LECITHIN [8002-43-5]

CTFA name	Lecithin Phosphatidylcholine
Source	
NATURAL	The naturally occurring mixture of diglycerides of stearic, palmitic and oleic acids, linked to the choline ester of phosphoric acid: occurs in egg yolk and soya beans.
Isolation	Originally obtained from dried egg yolk, but is cheaper to extract from soya beans.
Yield	Present in egg yolk to 16% in the dried state.
Appearance	Blackish-brown when obtained from animal sources, but is a lighter colour when obtained from a vegetable source: turns to golden yellow in solution.
Other	Will keep quite well in vegetable oils.
Uses	In skin creams.

LEMON-GRASS OIL [8007-02-1]

CTFA name	Lemongrass oil
Source	
GEOGRAPHIC	Two species of grasses cultivated in India, China, Brazil, and Guatemala.
NATURAL	*Cymbopogon flexuosus* and *Cymbopogon citratus*.
Isolation	Steam distillation in Singapore of the fresh or partly dried grass, which commences shortly after the rainy season. The stills, housed in bamboo structures, are copper, about 6 ft (2 m) high and 3 ft (1 m) wide: about 700 lb (320 kg) of lemon-grass is placed in 40 gallons (180 litres) of water: distillation takes about 6 h.
Yield	22 oz (624 g) oil is obtained from this method of distillation (on average, 1.8–2.2% depending upon source of plant material.)
Appearance	Pale straw-coloured oil.
Odour	Refreshing and lemon-like: rather bitter.
Chemical	The oil obtained from *Cymbopogon flexuosus*, Stapf., usually has a higher citral content and is more soluble in 70% alcohol. Generally, lemon-grass oil contains 80% citral, together with citronellal, geraniol, methylheptenone, n-decylaldehyde and probably linalol.

Uses	Mainly in low-cost lemon soap perfumes, and is also a cheap odour for hair oils and bath crystals. RIFM Monograph (1976) *FCT*, **14**, 455 and 457.

LEMON OIL [8008-56-8]

CTFA name	Lemon oil
History	The lemon was first introduced to Europe by the Arabs. It was grown in California in 1887.
Source	
GEOGRAPHIC	A small tree cultivated in most Mediterranean countries, principally Sicily, and also in Brazil, USA, Argentina, Israel and West Africa.
NATURAL	*Citrus medica*, L., var. *beta-limonum*, Hooker filius, N.O. Rutaceae, *C. linonum* (L.) Nees.
Isolation	The harvest in Italy is from November to March; in Palermo it is later and may continue until June. The oil is obtained from the peel of the ripe fruit, either by sponge-press or machine. The fruit is first halved, and the inside scraped out. The peels are steeped in cold water for 5 min, drained, and stood aside for 24 h in the winter and 12 h in the summer to increase the turgidity of the cells, thus making the oil easier to express. The oil is produced on an industrial scale by Sandersons at their large factory in Tremestieri, but there are also many small producers in the villages along the coast.
Yield	0.4–0.6% depending upon production procedure.
Appearance	Yellow oil: the machine-pressed oil is darker.
Odour	Typical and refreshing.
Physical	Prone to deterioration if exposed to air or light, and should, therefore, be stored in a well-filled, air-tight container.
Chemical	Contains about 4–6% citral, *d*-limonene and traces of the aldehydes, C_8 and C_9, citronellal, methyl heptenone, phellandrene, terpineol, linalyl and geranyl acetates etc. The terpeneless oil is about 16 times more concentrated, containing about 50% citral and 20% esters in genuine oils.
Uses	Blends well with benzyl acetate, palmarosa, clove, and geranium oils, and is also used in conjunction with citronella and lemon-grass oils as a verbena perfume.
Comments	Principal adulterants are the terpene residues obtained during the manufacture of concentrated oil. RIFM Monograph (Lemon oil Distilled) (1974) *FCT*, **12**, 727 (Lemon oil Expressed) (1974) *FCT*, **12**, 725 (Binder, p. 491). IFRA GUIDELINE The Committee recommends for applications on areas of skin exposed to sunshine, excluding bath preparations, soaps and other

products which are washed off the skin, to limit lemon oil expressed to 10% in the compound (see remark on phototoxic ingredients in the introduction).

This recommendation is based on results of RIFM on the phototoxicity of lemon oil expressed (*Food Cosmet. Toxicol.*, **12**, 725 (1974)), its low bergaptene content (C. K. Shu *et al.* VI. Int. Congress of Essential Oils 1974), and the observed no-effect level of pooled samples in tests using the hairless mouse and the pig (private communication to IFRA).

October 1975, amended October 1978

LIATRIS CONCRETE [68602–86–8]

	Deertongue
Source	
GEOGRAPHIC	A herb indigenous to the Southern States of North America.
NATURAL	*Liatris odoratissima*, Willd., N.O. Compositae.
Isolation	The concrete has been prepared experimentally in Grasse.
Appearance	Dark-green crystalline solid.
Odour	Coumarin type.
Chemical	Constituents are a volatile oil and coumarin.
Uses	In perfumery, and for perfuming smoking, chewing and snuff tobacco.
	RIFM Monograph (1979) *FCT*, **17**, 763 (Deertongue Absolute) (1976) *FCT*, **14**, 743 (Deertongue Incolore).

LILAC PERFUMES

Isolation	Mixtures of synthetics: the absolute has been prepared from the flowers on an experimental scale.
Odour	The synthetic mixtures usually have a predominant odour of terpineol. The odour of the absolute lacks the characteristics of the fresh flower.
Chemical	The other constituents of the synthetic lilac perfumes are phenylethyl alcohol, anisyl alcohol, and hydroxycitronellal, with a small percentage of jasmine.
Uses	Perfumes.

LILY PERFUMES

Isolation	Prepared synthetically.
Odour	A predominating odour of hydroxy-citronellal, also known as lily aldehyde.
Chemical	Usually a combination of linalol, terpineol, and small quantities of jasmine.

LIME OIL DIST./LIME OIL PR. [90063-52-8]

An Italian oil known as limette oil.

Source

GEOGRAPHIC — Trees cultivated mainly in Brazil, Mexico, Italy and the West Indies.

NATURAL — *Citrus aurantifolia, C. Medica* var. *acida/vulgaris*.

Isolation — Montserrat, in the West Indies, is the principal producer: obtained by expression or distillation from the rind of the fruit. The first crop of fruit is collected 5 years after the trees are planted as seedlings, and yield reaches its peak in about the 10th year. In Italy, limette oil is obtained by expression from the peel of *Citrus limetta*, Risso, N.O. Rutaceae. A yellow substance, limettin, is deposited on standing. This is absent from the distilled oil. The genuine article is almost unknown.

Yield — 1.2–2.1% of oil depending upon production procedures and source of raw material: approximate yield per acre is 10 tonnes of fruit.

Appearance — Oil.

Odour — Resembles turpentine, due to the presence of cymene resulting from the decomposition of citral during the distillation process. The pure oil has an intense odour reminiscent of lemon peels.

Uses — At low levels in perfumery in eau-de-colognes: and lime–citrus compositions. Citral is obtained from lime oil and 0.1% may be used to produce freshness in floral compounds: lime oil is also widely used as a flavouring agent.

RIFM Monograph (Lime oil Distilled) (1974) *FCT*, **12**, 729
(Lime oil Expressed) (1974) *FCT*, **12**, 731.

IFRA GUIDELINE

The Committee recommends for applications on areas of skin exposed to sunshine, excluding bath preparations, soaps and other products which are washed off the skin, to limit lime oil expressed to 3.5% in the compound (see remark on phototoxic ingredients in the introduction).

This recommendation is based on results of RIFM on the phototoxicity of lime oil expressed (*Food Cosmet. Toxicol.*, **12**, 731 (1974)), its bergaptene content reported in J.A.O.A.C., **52**, (4) 727 (1969) and the observed no-effect level of pooled samples in tests using the hairless mouse and the pig (private communication to IFRA).

October 1975, amended October 1978

LIMONENE [138-86-3]

Chemical name — 1-Methyl-4-(1-methylethenyl)-cyclohexene
p-Mentha-1,8-diene; cinene; cajeputene; kautschin
$C_{10}H_{16}$

202 Linaloe oil

Source	
NATURAL	Many essential oils such as lemon, orange, and bergamot.
Isolation	Distillation of essential oils.
Appearance	Colourless to pale yellow liquid.
Odour	Fresh, clean, typically citrus.
Chemical	Occurs naturally as two optically active forms, d- and l-limonene: dipentene is the racemic form: d-limonene is easily obtained as some citrus oils contain up to 90%.
Uses	In low cost citrus fragrances.
	RIFM Monograph (d-Limonene) (1975) *FCT*, **13**, 825
	(l-Limonene) (1978) *FCT*, **16**, 809.

LINALOE OIL [8006-86-8]

Source	
GEOGRAPHIC	A tree grown throughout Mexico.
NATURAL	Two species of Burseraceae; linaloe and copal limon, which closely resemble each other.
Isolation	Distillation of the chipped wood: about 5 cwt (250 kg) chips are heated by direct fire in the stills: the process takes about 20 h. The best oil is obtained from trees which are 40–60 years old: the yield may be increased in younger trees by cutting notches in the trunk: the most aromatic part of the wood is the deepest area called the 'heart' wood.
Yield	2.5%.
Appearance	Colourless to pale yelow oil.
Odour	Three different odours may be distinguished; fine, which has a delicate odour, common, which is less delicate, and caraway linaloe, which resembles caraway.
Chemical	Contains 76% linalol with geraniol, terpineol, and methyl heptenol.
Comments	Has almost disappeared commercially due to competition from Brazilian oil and the cheaper Ho oil: nearly all the oil produced comes from the Copal limon.
	RIFM Monograph (Linaloe wood oil) (1979) *FCT*, **17**, 849.

LINALOL [78-70-6]

Chemical name	3,7-Dimethylocta-1,6-octadien-3-ol
	Formerly named licareol; linalool; 2,6-dimethyl-2,7-octadien-6-ol
	$C_{10}H_{18}O$
Source	
NATURAL	Occurs in the free state in cayenne, Mexican and Brazilian linaloe and Ho oils, and in the esters of many essential oils such as lavender, bergamot, petitgrain, clary sage, neroli and basil.
Isolation	Synthesized from pinene or methyl heptenone.
Appearance	Colourless liquid.

Odour	Light refreshing, floral–woody.
Physical	b.p. 198 °C, s.g. 0.86.
Chemical	Isomeric with geraniol and nerol.
Uses	In the preparation of artificial floral oils such as linalol from bois de rose oil, which is used in the composition of lily, lilac, honeysuckle, sweet-pea, rose and neroli fragrances.

RIFM Monograph (Linalool) (1975) *FCT*, **13**, 827.

LINALOOL OXIDE   1365–19–1] [5989–33–3] [14049–11–7] [34995–77–2] [60047–17–8]

2-furanmethanol, 5-ethehyltetrahydro-alpha, alpha, 5-trimethyl-, *cis*; *cis*-trans-2-methyl-2-vinyl-5(2-hydroxy-2-propyl)tetrahydrofuran; *cis*-trans-2-vinyl-2-methyl-5-(1'-hydroxy-1'-methylethyl)tetrahydrofuran

Source	
NATURAL	Found in apricot, grape, papaya, pineapple, strawberry, tomato, thyme, hop oil, cocoa, coffee, tea and many other fruits.
Chemical	Is a mixtue of stereoisomers.
Isolation	From linalool by oxidation.
Physical	GC RIFM no. 77–288; IRC RIFM no. 77–288.

LINALYL ACETATE [115–95–7]

Chemical name	3,7-Dimethyl-1,6-octadien-3-yl acetate Bergamol $C_{12}H_{20}O_2$
Source	
NATURAL	Many essential oils such as lavender, bergamot, petitgrain, clary sage, neroli and also in jasmine and gardenia.
Isolation	Synthesized from dehydrolinalool and acetic anhydride, with subsequent catalytic hydrogenation.
Appearance	Colourless oil.
Odour	Fresh, sweet, fruity, recalling terpeneless bergamot oil.
Physical	b.p. 220 °C, s.g. 0.91: insoluble in water, miscible with alcohol or ether.
Uses	In jasmine, syringa, tilleul, ylang-ylang, and colognes: also invaluable in soap perfumery.

LINALYL ANTHRANILATE

Chemical name	Linalyl *o*-aminobenzoate $NH_2C_6H_4COOC_{10}H_{17}$
Isolation	Difficult to manufacture.
Appearance	Pale straw-coloured viscous liquid.
Odour	Soft, floral, linden blossom, neroli-like.
Physical	b.p. approx. 320 °C.

Chemical	Very unstable.
Uses	In perfumery in synthetic neroli, orange blossom, and jasmine.

RIFM Monograph (1976) *FCT*, **14**, 459.

LINALYL BENZOATE [126-64-7]

$C_6H_5COOC_{10}H_{17}$

Isolation	Difficult to prepare.
Appearance	Almost colourless oily liquid.
Odour	Heavy balsamic floral, recalling the flower of broom and tuberose.
Physical	b.p. 263 °C, s.g. 0.99.
Uses	A blender in such perfumes as tuberose and oriental bouquets.

RIFM Monograph (1976) *FCT*, **14**, 461.

LINALYL iso-BUTYRATE [78-35-3]

$(CH_3)_2CHCOOC_{10}H_{17}$

Appearance	Colourless oily liquid.
Odour	More floral and fresher than the normal butyrate.
Physical	b.p. 230 °C, s.g. 0.89.
Uses	Mainly in lavender bouquets and citrus–cologne types.

RIFM Monograph (1975) *FCT*, **13**, 835.

LINALYL n-BUTYRATE [78-36-4]

$CH_3CH_2CH_2COOC_{10}H_{17}$

Source	
NATURAL	Occurs naturally in French lavender oil.
Appearance	Colourless oily liquid.
Odour	Heavy fruity, sweet bergamot-like.
Physical	b.p. 238 °C, s.g. 0.897.
Uses	In lavender, fougeres, colognes and some floral types.

RIFM Monograph (1976) *FCT*, **14**, 805.

LINALYL CINNAMATE [78-37-5]

$C_{19}H_{24}O_2$

Isolation	Synthesized from cinnamic acid and dehydrolinalool with subsequent catalytic hydrogenation.
Appearance	Colourless to yellow oily liquid.
Odour	Mild, balsamic, lily-jasmine-type.
Physical	s.g. 0.98, b.p. 353 °C.
Uses	In tuberose, rose and in some delicate floral compositions.

RIFM Monograph (1976) *FCT*, **14**, 463.

LINALYL FORMATE [115-99-1]

$HCOOC_{10}H_{17}$

Appearance	Colourless mobile liquid.
Odour	Fresh citrusy, green, bergamot-like.
Physical	b.p. 202 °C; s.g. 0.919.
Uses	Useful to give lift to lemon and bergamot in colognes.

RIFM Monograph (1975) *FCT*, **13**, 833.

LINALYL PHENYLACETATE [7143-69-3]

$C_6H_5CH_2COOC_{10}H_{17}$

Isolation	Difficult to manufacture.
Appearance	Colourless or pale straw-coloured liquid.
Odour	Intensely sweet, mild floral with a honey-like undertone.
Physical	b.p. 317 °C, s.g. 0.97.
Uses	Blends well with coumarin in heavy florals and exotic floral types.

RIFM Monograph (1976) *FCT*, **14**, 465.

LINALYL PROPIONATE [144-39-8]

$C_2H_5COOC_{10}H_{17}$

Appearance	Colourless liquid.
Odour	Fresh, fruity–floral and bergamot-like.
Physical	b.p. 226 °C.
Uses	Used to impart a freshness to colognes, light floral, and herbaceous fragrances.

RIFM Monograph (1975) *FCT*, **13**, 839.

LINALYL n-VALERATE

Chemical name	Linalyl valerianate
	$C_4H_9COOC_{10}H_{17}$
Source	
NATURAL	Oils of sassafras leaves, and French lavender.
Appearance	Colourless oily liquid.
Odour	Rather fruity, citrus–lavender, and suggestive of apricots.
Physical	s.g. 0.90, b.p. 238 °C.
Uses	In lavender, bergamot and some sophisticated pine types.

LIQUIDAMBAR

Sweet gum, Honduras balsam, white Peru balsam

Source	
GEOGRAPHIC	Occurs on the bark of large trees found at high altitudes in the mountains of Honduras.

NATURAL	An oleo-resinous secretion of *Liquidambar styraciflura*, L., N.O. Hamamelideae.
Isolation	Purified by solution in volatile solvents, filtration, and subsequent removal of the solvent.
Appearance	Viscous, transparent, pale amber liquid (after purification).
Odour	Aromatic, and balsamic, sweeter and more aromatic than styrax.
Chemical	Contains cinnamic acid, free and as ester, cinnamyl alcohol, phenylpropyl alcohol, hydrocarbons, and a sesquiterpene: bears some relationship to styrax.
Uses	In soaps and in jasmine compounds.

LIQUID LANOLIN

Lanolin oil, dewaxed lanolin

Source	
CHEMICAL	A form of anhydrous lanolin.
Isolation	Prepared from anhydrous lanolin by extraction and fractionation.
Appearance	Golden-yellow viscous liquid.
Physical	Miscible with mineral and vegetable oils, and fatty acid esters: insoluble in ethyl alcohol, water, and glycols, soluble in isopropyl alcohol and in most aerosol propellants.
Chemical	Consists of low-melting lanolin esters, including those of cholesterol and other sterols.
Other	Spreads easily.
Uses	In skin creams and lotions, when the stickiness of anhydrous lanolin is undesirable: also in cosmetic creams and lotions, and as a water-in-oil emulsifier, or as a stabilizer for oil-in-water emulsions.

LITHIUM STEARATE [4485–12–5]

CTFA name	Lithium stearate
Appearance	Fluffy powder.
Physical	Has excellent adhesion and high oil absorbency: it is water-resistant and non-irritant.
Chemical	In aqueous suspension, it has a mildly alkaline reaction (pH 8.7) and should be used with care.
Uses	In high quality toilet powders; produces a soft, velvety texture.

Litsea cubeba OIL [68855–99–2]

Source	
GEOGRAPHIC	A tree native to China and South East Asia.
NATURAL	*Litsea cubeba*.
Isolation	Steam distillation of the ripe fruit.
Yield	Approx. 2%.
Odour	Intense, citrus-fruity, lemon-like.

LONGOSA OIL

Uses	An important raw material from which citral is obtained: used widely in citrus perfumes. RIFM Monograph (1982) *FCT*, **20**, 731.

LONGOSA OIL

Source	
GEOGRAPHIC	Flowers found on an island a few miles north-west of Madagascar.
NATURAL	*Hedychium flavum*, N.O. Scitaminae.
Isolation	The blossoms are collected during March, early in the morning to avoid the intense heat; extraction with petroleum ether yields the concrete. An oil may be obtained by steam distillation.
Yield	490 g of concrete from 1000 kg of flowers: 50% absolute may be obtained from the concrete: 22% oil is obtained by steam distillation.
Appearance	The concrete is a semi-solid dark-coloured mass: the absolute is dark orange-brown and the oil is straw-yellow.
Odour	The oil has an exotic, peppery fragrance.
Chemical	Contains indole and methyl anthranilate.

LOTOS

History	An ancient Egyptian name for a plant which grows in the Nile: it was described by the Greek historian Herodotus (BC 413).
Source	
GEOGRAPHIC	The River Nile in Egypt.
NATURAL	Appears to be *Nymphea lotus* of Linnaeus, a white waterlily, similar to *Nymphea alba* in England.
Isolation	Lotos perfumes are mixtures of linalol and phenylethyl alcohol with traces of patchouli and undecalactone, fixed with amber and benzoin R.
Uses	Perfumes.

LOVAGE OIL [8016-31-7]

CTFA name	Lovage oil *Levisiticum officinale* oil; oil of lovage
Source	
GEOGRAPHIC	A herb found in Europe.
NATURAL	*Levisticum officinale*, Koch, N.O. Umbelliferae.
Isolation	Steam distillation of the whole plant.
Yield	0.1–0.6% depending upon source of the plant material.
Appearance	Yellow viscous oil.
Odour	Persistent, and spicy, reminiscent of angelica root oil.
Chemical	The chemistry of this oil is not completely known, but terpineol is present.

Uses	A toner, but its use in perfumery is limited: may be employed in some tobacco flavours. RIFM Monograph (1978) FCT, **16**, 813.

LUSTRE PIGMENTS

Source	Nacreous pigments
CHEMICAL	Platelets of mica coated with a thin film of titanium dioxide.
Isolation	Thin transparent platelets of mica coated with a thin film of titanium dioxide: they may also be used with bismuth oxychloride-type pearling agents.
Appearance	Sparkling white powder.
Physical	The reflection of white light gives a brilliance and highlight effect.
Uses	Widely used in cosmetics, in pressed powders, eye make-up preparations and lipsticks.

LYCOPODIUM [84082-56-4]

Source	Club moss spores; lycopodium seed; vegetable sulphur
GEOGRAPHIC	A moss indigenous to Europe, Asia, and North America.
NATURAL	*Lycopodium calvatum*, L., N.O. Lycopodiaceae.
Isolation	The spores of the moss are obtained commercially from Russia. They are collected during July and August by shaking the plants and then removing the impurities by sifting the powder through a fine mesh.
Appearance	Pale yellow, fine, mobile powder.
Odour	None.
Chemical	Contains about 50% fixed oil together with some glyceryl myristate.
Comments	Flammable and has pyrotechnic uses.

MAALI RESIN

Source	
NATURAL	Exuded by an unknown Samoan tree.
Isolation	Distilled to obtain an oil.
Yield	16% oil.
Appearance	Soft, yellowish-white substance which resembles elemi. On distillation a light green oil is obtained, which solidifies at room temperature.
Odour	Balsamic.
Uses	Used by the natives for embalming and for dressing the hair.

MACASSAR OIL

Source	Kusum oil; kon oil; paka oil
GEOGRAPHIC	A plant native to the Sunda Islands in Indonesia.

NATURAL	*Schleichera trijuga*, Willd, N.O. Sapindaceae.
Isolation	Expression of the seeds.
Appearance	Yellow oil, which resembles butter.
Odour	Reminiscent of benzaldehyde.
Physical	Initially melts at 21 °C with complete transparency at 31 °C: miscible with ether, chloroform and vegetable oils.
Uses	In hair oil formulations; in Indonesia, as a dressing for the hair and for removing scurf and eczema.

MACE OIL [8008-45-5]

Source	
GEOGRAPHIC	The nutmeg, cultivated in Indonesia, Sri Lanka and the West Indies.
NATURAL	*Myristica fragrans*.
Isolation	By steam distillation of the arillus of the nutmeg.
Appearance	Essential oil.
Odour	Typical and spicy.
Chemical	Closely resembles oil of nutmeg, but contains a larger proportion of oxygenated constituents.
Uses	True mace oil is used occasionally in soap perfumes of the santal type.
	RIFM Monograph (1979) *FCT*, **17**, 851.

MAGNESIUM CARBONATE [546-93-0]

CTFA name	Magnesium carbonate
	$MgCO_3$
Isolation	Obtained in the light form by precipitating a solution of the sulphate with a solution of sodium carbonate.
Appearance	Flocculent white powder.
Uses	A 'carrier' for perfumes in face powders and talcum powders: also in dental preparations.

MAGNESIUM MYRISTATE [4086-70-8]

COLIPA name	Magnesium myristate
Appearance	Fine, white powder.
Physical	Has a higher specific volume than magnesium or zinc stearates.
Other	Has superior covering power and adhesion.
Uses	In high quality face powders, compressed powder, and talcum powder, without causing an increase in whiteness.

MAGNESIUM PEROXIDE [1335-26-8]

Magnesium dioxide
Mg_2O_2

Isolation	Obtained by adding a solution of hydrogen peroxide to calcined magnesia. After standing, the residue is filtered out, washed and dried.
Appearance	White, tasteless, odourless powder.
Physical	Insoluble in water.
Chemical	Contains between 15 and 20% of true magnesium peroxide.
Uses	In dentifrices.

MAGNESIUM STEARATE [557-04-0]

CTFA name	Magnesium stearate Octadecanoic acid magnesium salt
Appearance	Fine, white, fluffy powder.
Physical	Insoluble in water.
Other	Has good covering, adhesive and lubricating properties.
Uses	In proportions from 5–10% as a constituent of face powders, talcum powders, and dusting powders: also as a mould lubricant in the manufacture of tablets: may be used to give opacity to cream and lotion shampoos.

MAGNOLIA

Source	
GEOGRAPHIC	Trees and shrubs native to China, Japan, and North America.
NATURAL	The flowers have given rise to some artificial oils.
Isolation	The artificial oils are usually jasmine, neroli, and rose with traces of ionone and benzaldehyde.
Uses	The artificial oils are employed in perfumery.

MAIZE STARCH

Source	
GEOGRAPHIC	A plant cultivated in subtropical countries.
NATURAL	*Zea mays*, L., N.O. Graminaceae.
Isolation	The cells which contain the starch grains are ruptured by grinding the fruits to a pulp. A weak solution of alkali is used to wash them out and dissolve the gluten. Further washing and straining purifies the starch, which is dried. The granules separate into angular fragments which are powdered.
Appearance	Powder.
Uses	In face powders, rouges, and other cosmetics: however, the starch obtained from rice is preferred.

MALE FERN RESIN [84650-05-5]

Source	
GEOGRAPHIC	A fern plant indigenous to Europe.
NATURAL	*Dryopteris felix-mas*, Schott, N.O. Felicineae.

Isolation	The rhizomes are collected in the autumn, dried and the resin obtained by volatile solvents from the dried rhizomes.
Appearance	Greenish-brown semi-liquid.
Odour	Heavy and persistent.
Chemical	The principal constituent is filmarone to which the male fern owes its properties as a vermifuge.
Uses	An excellent fixative base.

MALTOL [118-71-8]

3-Hydroxy-2-methyl-4H-pyran-4-one; larixinic acid
$C_6H_6O_3$, veltol, corps praline

Source	
NATURAL	Beechwood tar: occurs naturally in malt and caramel.
Appearance	White crystals.
Odour	Intense, sweet, fruity, caramel-like: when diluted smells of strawberries and pineapples.
Physical	Freely soluble in hot water, chloroform, soluble in alcohol; m.p. 164 °C.
Uses	A flavouring agent, imparting a 'freshly baked' odour and flavour to bread and cakes. In perfumery at low levels to impart a rich sweet character to some pine needle types and in rose compositions. RIFM Monograph (1975) *FCT*, **13**, 841.

MANDARIN OIL [8008-31-9]

Source	
GEOGRAPHIC	A tree found in Calabria.
NATURAL	*C. reticulata blanco, Citrus madurensis*, Laur., Rutaceae.
Isolation	The peels of the fruit are cold-pressed by machine.
Yield	0.7–0.8%: one tree produces about 1000 fruits per annum, and these yield about 400 g oil.
Appearance	Bright golden-yellow oil.
Odour	Refreshing and fragrant, reminiscent of sweet orange oil.
Chemical	Contains methyl methylanthranilate with large proportions of *d*-limonene.
Uses	In perfumery in colognes and floral compounds.

MARJORAM OIL [8015-01-8]

History	The ancient Egyptians grew sweet marjoram as a pot herb; has been used in perfumes and unguents from earliest times.
Source	
GEOGRAPHIC	A plant cultivated in Germany, France, Spain, Tunisia, Hungary and Egypt.
NATURAL	The culinary herb, *Marjorana hortensis*.
Isolation	Steam distillation of the fresh, dried leaves and flowering tops.

Yield	0.5–3% depending upon production procedure and source of the raw material.
Appearance	Colourless to slightly yellow oil.
Odour	Tenacious, and reminiscent of nutmeg and mint: may be confused with thyme and organum oils.
Chemical	Consists largely of terpenes.
Uses	In perfumery as a toner, especially in herbal-spicy creations: occasionally for perfuming hair pomades.
Comments	There is some confusion surrounding this oil, probably due to the vernacular names of the parent plants. RIFM Monograph (Marjoram oil, Spanish) (1976) *FCT*, **14**, 467 (Marjoram oil, Sweet) (1976) *FCT*, **14**, 469.

MAROCAINE OIL

Source	
NATURAL	An unidentified species of pelargonium.
Isolation	Distillation.
Appearance	Semi-solid at normal temperatures.
Odour	Recalls a mixture of Algerian geranium and rose otto.
Chemical	Contains a large proportion of stearoptene.

MARSH ROSEMARY OIL [90063-39-1]

Source	
GEOGRAPHIC	A herb grown in Silesia.
NATURAL	*Ledum palustre*, L., N.O. Ericaceae.
Yield	0.33%.
Appearance	Lemon-yellow oil.
Odour	Reminiscent of coriander and worm seed oils.

MASTIC [68991-39-9]

Balsam tree; pistachia galls; mastiche; mastix; lentisk

Source	
GEOGRAPHIC	A tree indigenous to the countries bordering the Mediterranean.
NATURAL	The broad-leaved variety of *Pistacia lentiscus*, L., N.O. Anacardiaceae.
Isolation	Obtained from an incision in the bark of the stem and branches of the trees.
Yield	2% volatile oil.
Appearance	A hard resinous exudate which occurs as rounded, irregular, or pear-shaped tears of a white or pale yellow substance: may be either opaque or glassy.
Odour	Balsamic, and recalls turpentine.
Physical	Almost insoluble in water, nearly completely soluble in alcohol.

Uses	A useful fixative in honeysuckle, lavender, fern, mimosa and sweet-pea perfumes.
Comments	Also used in tooth cements, chewing gums and incense.

MATE ABSOLUTE [68916-96-1]

CTFA name	Mate extract
	Paraguay tea extract
Source	
GEOGRAPHIC	A tree cultivated in most of South America.
NATURAL	*Ilex paraguayensis.*
Isolation	The concrete is obtained by extraction of the leaves. Further extraction yields the absolute.
Yield	Depends upon the raw material and isolation procedure.
Odour	Rich and herbaceous.
Uses	In perfumery in fantasy creations.

MATICO OIL [84696-39-9]

Source	
NATURAL	*Piper augustifolium*, R. et P., N.O. Piperaceae.
Isolation	Distillation of the leaves.
Appearance	Yellowish-brown liquid.
Odour	Intense and reminiscent of pepper, cubebs, and mint.
Uses	In perfumery to the extent of about 0.1% in carnation compounds which are based upon pimento and iso-eugenol.

MAWAH OIL [90082-51-2]

Source	
GEOGRAPHIC	Kenya and South Africa, where there are some 500 acres under cultivation.
NATURAL	*Pelargonium graveolens*, Ait. At Njoro.
Isolation	Distillation.
Odour	Fine, geranium odour.
Uses	In perfumery.
Comments	Kenyan geranium oil is an article of commerce.

MAY CHANG OIL [97553-37-2]

GEOGRAPHIC	A tree native to Tonquin: attains a height of about 30 ft (9 m).
NATURAL	*Litsaea citrata*, Bl., Lauraceae.
Isolation	Distillation from the flowers.
Appearance	Oil.
Odour	Pleasant, and reminiscent of bois de rose and coriander.

214 Meadow sweet

Chemical	Contains citral, *d*-linalol, terpineol, geraniol, and limonene, with traces of saturated aliphatic aldehydes.
Uses	The flowers may be used for perfuming tea.
Comments	If this oil was produced commercially, it could be usefully employed in the perfume industry.

MEADOW SWEET

Spirea ulmaria, L., N.O. Rosaceae

Source	
GEOGRAPHIC	A common plant in meadows, hedgerows, and on the banks of streams in England.
Isolation	An oil may be obtained by distillation from the flowers.
Appearance	Oil.
Odour	The flowers possess an almond-like fragrance.
Chemical	The oil contains salicylic aldehyde, methyl salicylate, heliotropin, and vanillin.

MECCA BALSAM

Balm of Gilead

Source	
GEOGRAPHIC	A tree found in Arabia and southern Nubia: however, the balsam is collected only from the trees in the valleys near Mecca.
NATURAL	*Balsamodendrom opobalsamum*, Engl., N.O. Burseraceae.
Isolation	Obtained by making incisions in the trees, through which minute quantities are exuded; the twigs are then collected and boiled in water.
Appearance	Yellowish viscid liquid.
Odour	Fragrant.
Chemical	Contains about 30% volatile oil.
Uses	As a medicine, and in perfumery for obtaining Arabian blends.

MEERSCHAUM

The name means 'sea-foam' in German.

Source	
GEOGRAPHIC	Asia Minor and the USA.
NATURAL	Occurs in alluvial deposits.
Chemical	A hydrous form of magnesium silicate, $Mg_2Si_3O_8 2H_2O$.
Appearance	Irregular greyish-white nodular masses.
Other	Has good absorbent properties.

MELALEUCA OILS [85085-48-9]

Tea tree oil

Source	
GEOGRAPHIC	A tree which grows in Australia.

NATURAL	*Melaleuca alternifolia*, N.O. Myrtaceae.
Isolation	Distillation of the leaves.
Yield	Nearly 2%.
Appearance	Bright yellow oil.
Odour	Aromatic.
Chemical	Contains pinene, terpinene, cymene, terpineol, and 8% cineol: the oil obtained from *M. bracteata*, F. v. Mull, consists of methyl eugenol, together with eugenol, cinnamic acid, cinnamic aldehyde, cinnamyl cinnamate, and *l*-phellandrene.
Uses	In medicated soaps and dentifrices, and is excellent as a perfume in deodorants and disinfectants due to its germicidal properties and the absence of toxicity or irritation.
	RIFM Monograph: *Melaleuca alternifolia* oil: Tea tree oil (1988) *FCT*, **26**, 407
	Melaleuca leucadendron oil: Cajeput oil (1976) *FCT*, **14**, 701.

MELILOT [8023–73–2]

Source	
GEOGRAPHIC	Throughout Europe and Western Asia.
NATURAL	*Melilotus officinalis*, Desse, a clover-like annual or biennial of N.O. Leguminosae.
Odour	A fragrance more noticeable when the plant is dry; resembles sweet woodruff and appears to be due to derivatives of coumarin.
Uses	A flavouring in Gruyère cheese.

MELISSA OIL [84082–61–1]

	Balm oil
Source	
GEOGRAPHIC	A herb native to the Mediterranean countries and also cultivated in the USA.
NATURAL	*Melissa officinalis*, L., N.O. Labiatae.
Isolation	Distillation from the leaves and tops of the plants.
Yield	0.11%.
Appearance	Brown oil.
Comments	It is believed that the commercial product is obtained by distilling lemon oil over the herb. In some cases, citronellal is added.

Mentha citrata OIL *See* BERGAMOT MINT OIL

MENTHOL [89–78–1]

Chemical name	5-Methyl-2-(1-methylethyl)-cyclohexanol
CTFA name	Menthol
	l-Menthol
	$C_{10}H_{20}O$

Menthone

Source	
NATURAL	Peppermint oil.
Isolation	Obtained by freezing peppermint oils, or synthesized from thymol, piperitone or citronellal.
Appearance	Colourless, acicular crystals.
Odour	Refreshing, typically peppermint, but slightly pungent.
Physical	l-form, m.p. 43 °C, b.p. 216 °C, slightly soluble in water, very soluble in alcohol, chloroform or ether.
Uses	The most commonly used forms are the naturally occurring l-menthol, its d-isomer and the d,l-racemate, which may be used in perfumery to enhance the cooling properties in solid eau-de-colognes: in medicine, l-menthol is an important flavouring material: useful in dental creams to obtain freshness.

RIFM Monograph (Menthol, Racemic) (1976) FCT, **14**, 473.

MENTHONE [14073–97–3]

5-Methyl-2-(1-methylethyl)cyclohexanone
l-Menthone
$C_{10}H_{18}O$

Source	
NATURAL	The characteristic ketone found with menthol and menthyl acetate in oil of peppermint.
Isolation	Synthesized by the oxidation of l-menthol: may also be obtained from the distillation of peppermint oil.
Appearance	Colourless liquid.
Odour	Refreshing-minty reminiscent of menthol with slightly woody undertones.
Physical	b.p. 207 °C, m.p. −6 °C: slightly soluble in water, soluble in organic solvents.
Chemical	'Menthone' always means l-menthone. However, it should be noted that there exist also d-menthone, d- and l-isomenthone and the corresponding racemates.
Uses	At low levels in geranium, rose and lavender compositions.

RIFM Monograph (Menthone, Racemic) (1976) FCT, **14**, 475.

MENTHYL ACETATE [89–48–5]

Chemical name	5-Methyl-2-(1-methylethyl)cyclohexanol acetate
CTFA name	Menthyl acetate
	$CH_3COOC_{10}H_{19}$
Source	
NATURAL	Peppermint oil.
Appearance	Colourless liquid.
Odour	Mild sweet, herbaceous–minty character.
Physical	b.p. 227 °C, s.g. 0.927: slightly soluble in water, miscible with alcohol or ether.

Uses In perfumery in colognes and lavender fragrances: also useful in flavours to give a 'cool' effect.
RIFM Monograph (1976) *FCT*, **14**, 477 (laevo)
(1976) *FCT*, **14**, 473 (racemic).

MENTHYL SALICYLATE [89-46-3]

Chemical name 2-Hydroxybenzoic acid 5-methyl-2-(1-methylethyl)-cyclohexyl ester
CTFA name Menthyl salicylate
$HOC_6H_4COOC_{10}H_{19}$
Appearance Colourless oily liquid.
Odour Almost odourless, or slightly fruity.
Physical Insoluble in water, miscible with most organic solvents; b.p. 322 °C.
Other Has filtering properties for ultraviolet light.
Uses 10% is used in sun-tan oils to be effective.
Comments Keep well closed and away from light.

MENTHYL *iso*-VALERATE [89-47-4]

Chemical name 3-Methylbutyric acid 5-methyl-2-(1-methylethyl)-cyclohexyl ester
Isovaleric acid *p*-menth-3-yl ester
$(CH_3)_2CHCH_2COOC_{10}H_{19}$
Source
NATURAL American peppermint oil.
Appearance Colourless oily liquid.
Odour Sweet–herbaceous, somewhat minty–rooty character with a cooling, faintly bitter taste.
Physical s.g. 0.907, b.p. 241 °C.
Uses Little used in fragrances.
RIFM Monograph (1982) *FCT*, **20**, 735.

p-METHOXYACETOPHENONE [100-06-1]

Chemical name *p*-Acetylanisole
Acetanisole, anisyl methyl ketone, helional
$CH_3COC_6H_4OCH_3$
Source
NATURAL Its closely allied hydroxy derivative, paenol, has been identified in nature.
Appearance Crystalline solid.
Odour Intense-flowery with a slightly harsh hay-like note.
Physical b.p. 258 °C, m.p. 38–39 °C, s.g. 1.08.
Uses Mainly in woody–floral, fougere, trefle, and mimosa types for soaps, detergents and other household products.
RIFM Monograph (1974) *FCT*, **12**, 927.

o-METHOXYCINNAMIC ALDEHYDE [1504-74-1]

Chemical name	3-(o-Methoxyphenyl)acrolein $CH_3OC_6H_4CH=CHCHO$
Source	
NATURAL	Cassia oil.
Appearance	Pale yellow crystalline solid.
Odour	Warm, sweet–spicy resembling cassia.
Physical	b.p. 295 °C (decomp.), m.p. 46 °C.
Chemical	Its isomer, p-coumaric aldehyde methyl ether also occurs in the oils of tarragon and chlorocodon. Colours human skin intensely yellow.
Uses	Rarely used in perfumes due to discolouring properties. RIFM Monograph (1975) FCT, **13**, 845.

p-METHOXYPHENYLACETALDEHYDE [5703-26-4]

	p-Anisyl acetaldehyde $CH_3OC_6H_4CH_2CHO$
Odour	Sweet floral; green resembling that of hawthorn.
Physical	b.p. 255 °C.
Uses	In soap perfumes and in lilac and bouquet compounds.

p-METHYL ACETOPHENONE [122-00-9]

Chemical name	Methyl p-tolyl ketone 4-Acetyltoluene, Melilot ketone $CH_3COC_6H_4CH_3$
Source	
NATURAL	Rosewood oil.
Isolation	Synthesized by the Friedel–Crafts reaction from toluene and acetic anhydride in the presence of aluminium chloride.
Appearance	Colourless needles, or opaque, crystalline mass.
Odour	Harsh, pungent but with a warm, sweet and floral character.
Physical	b.p. 228 °C, m.p. 28 °C.
Uses	In perfumery and blends well with terpineol, benzyl formate, iso-butyl salicylate, oakmoss, patchouli and rose: effectively replaces coumarin in foin coupe perfumes. RIFM Monograph (1974) FCT, **12**, 933.

METHYL-n-AMYLKETONE [110-43-0]

Chemical name	2-Heptanone Ketone C-7 $C_7H_{14}O$
Source	
NATURAL	Clove oil and cinnamon bark oil.
Isolation	Hydrogenation of heptyne or from ethyl butyl acetoacetate.

Appearance	Colourless liquid.
Odour	Penetrating fruity odour which is rather 'chemical'; on dilution it is reminiscent of pear and lavender.
Physical	b.p. 152 °C: very slightly soluble in water, soluble in alcohol or ether.

RIFM Monograph (1975) FCT, **13**, 847
RIFM Monograph (1975) FCT, **13**, 845.

METHYL ANISATE [121–98–2]

Chemical name	Methyl p-methoxybenzoate $CH_3OC_6H_4COOCH_3$
Appearance	White crystalline solid.
Odour	Sweet, aromatic, delicately floral, resembling linden blossom and magnolia.
Physical	b.p. 256 °C, m.p. 48–49 °C.
Uses	In perfumery in lilac, sweet pea, hawthorn, mimosa, and linden blossom compositions: employed to give 'body' to fancy perfumes.

RIFM Monograph (1976) FCT, **14**, 481.

METHYL ANTHRANILATE [134–20–3]

Chemical name	Methyl o-aminobenzoate 2-Aminobenzoic acid methyl ester $C_8H_9NO_2$
Source	
NATURAL	Flower oils of neroli, ylang-ylang, jasmine, wallflower, and tuberose.
Isolation	Synthesized from anthranilic acid and methanol.
Appearance	Colourless liquid or a white crystalline mass: has a characteristic violet–blue fluorescence in solution.
Odour	Orange flower, with a musty, dry–floral note.
Physical	b.p. 237 °C, m.p. 24 °C. Slightly soluble in water, freely soluble in alcohol or ether.
Chemical	When condensed with hydroxy-citronellal, an intensely yellow viscous liquid is produced called aurantiol.
Uses	Widely used in sweet floral and orange blossom types; better used in coloured soaps as it turns white soaps a reddish-brown with age.

RIFM Monograph (1974) FCT, **12**, 935.

p-METHYL BENZALACETOPHENONE [4224–96–8]

Chemical name	1-Phenyl-3-p-tolylprop-1-en-3-al (benzylidene-p-methyl acetophenone) $CH_3C_6H_4COCH{=}CHC_6H_5$
Appearance	Colourless viscous liquid.
Odour	Warm, sweet, heavy floral type, reminiscent of tuberose.
Uses	Occasionally used in tuberose, gardenia and mimosa fragrances.

METHYL BENZOATE [93–58–3]

Chemical name	Benzoic acid methyl ester Niobe oil $C_8H_8O_2$
Source	
NATURAL	Oils of clove, ylang-ylang and tuberose.
Isolation	Synthesized by passing dry HCl gas through a solution of benzoic acid in methyl alcohol.
Appearance	Colourless oily liquid.
Odour	Pungent, sweet floral with strong fruity undertones resembling blackcurrant; similar to ylang-ylang and tuberose on dilution.
Physical	b.p. 200 °C, m.p. -12.3 °C: insoluble in water, miscible with alcohol or ether.
Uses	Widely in perfumery with vetivert, oakmoss resin, and labdanum as a base for peau d'espagne perfumes: also important in the preparation of artificial ylang-ylang oil. RIFM Monograph (1974) FCT, **12**, 937.

METHYL n-BUTYRATE [623–42–7]

Chemical name	Methyl n-butanoate Butanoic acid methyl ester $CH_3(CH_2)_2COOCH_3$
Appearance	Colourless liquid.
Odour	Penetrating, diffusive, sweet-ethereal, fruity character.
Physical	b.p. 103 °C, m.p. -84.8 °C: soluble in about 60 parts of water, miscible with alcohol or ether.
Uses	In the manufacture of artificial rum and fruit essences and for its topnote effect in fragrance compositions. RIFM Monograph (Methyl Butyrate) (1982) FCT, **20**, 741.

METHYL CAPRINATE [110–42–9]

Chemical name	Methyl n-decanoate Methyl caprate $CH_3(CH_2)_8COOCH_3$
Appearance	Colourless oily liquid.
Odour	Oily, winy and slightly fruity.
Physical	b.p. 228 °C, m.p. -18 °C.
Uses	Traces in perfumery in sophisticated bouquets.

METHYL CAPRYLATE [111–11–5]

Chemical name	Methyl n-octanoate
CTFA name	Methyl caprylate $CH_3(CH_2)_6COOCH_3$

Appearance	Colourless liquid.
Odour	Powerful winy-fruity, orange-like character.
Physical	b.p. 193 °C, m.p. −40 °C, s.g. 0.887.
Uses	As methyl caprinate.

METHYL CELLULOSE [9004-65-7]

CTFA name	Methylcellulose
	Cellulose, methyl ether; there are many trade names.
Appearance	A non-fermentable, cream, fibrous or powdered solid.
Odour	Odourless and tasteless.
Physical	Forms a mucilage when stirred into boiling water and then cooled. These are stable to all conditions except temperature change.
Chemical	Unaffected by weak alkalis or electrolytes.
Other	Methyl and ethyl cellulose are graded according to their viscosity.
Uses	In toiletries as a binding agent for toothpaste, or for thickening and stabilizing cosmetic creams.

METHYL CHAVICOL [140-67-0]

Chemical name	p-Methoxyallylbenzene
	Iso-anethole, estragole, esdragol
	$C_{10}H_{12}O$
Source	
NATURAL	Oils of tarragon, fennel, anise bark and bay.
Appearance	Colourless oily liquid.
Odour	Aromatic, sweet herbaceous, anise–fennel like.
Physical	b.p. 216 °C, s.g. 0.97.
Uses	In fougere, lavender, chypre and aromatic types for soaps and men's toiletries.
	RIFM Monograph (1976) *FCT*, **14**, 603.

METHYL CINNAMATE *trans* [103-26-4]

Chemical name	Methyl 3-phenylacrylate
	Cinnamic acid methyl ester
	$C_{10}H_{10}O_2$
Source	
NATURAL	Oils of galanga, watara, alpinia and basil.
Isolation	Synthesized from cinnamic acid and methanol.
Appearance	White crystalline solid.
Odour	Balsamic, fruity, heavy amber odour: recalls strawberry in dilution.
Physical	b.p. 263 °C, m.p. 38 °C.
Uses	A useful fixative and blender in soap and detergent perfumery, especially of the carnation, oriental, citrus cologne and amber types.
	RIFM Monograph (1975) *FCT*, **13**, 849.

6-METHYL COUMARIN [92-48-8]

	Toncarine
Appearance	Colourless crystalline solid.
Odour	Persistent, herbaceous, tonka-like, more fig- or date-like than coumarin.
Physical	m.p. 75 °C, b.p. 305 °C.
Uses	No longer used as a fragrance ingredient due to its photoallergenic properties.

RIFM Monograph (1976) *FCT*, **14**, 605
(1979) *FCT*, **17**, 257.

IFRA GUIDELINE

The Committee recommends that 6-methyl coumarin should not be used as a fragrance ingredient.

This recommendation is based on the potential for inducing photoallergic reactions of this material (K. H. Kaidbay and A. M. Kligman (1978), *Contact Dermatitis*, **4**, (5) 277; D. L. J. Opdyke (1979), *Food Cosmet. Toxicol.*, **17**, 275). It should be noted that coumarin, when tested under the same conditions as the above material, showed no potential for inducing photoallergic reactions (private communication to IFRA).

March 1978, amended October 1978

METHYL DIHYDROJASMONATE [24851-98-7]

Chemical name	Methyl 1-oxo-2-pentylcyclopentane-3-acetate
	Hedione
	$C_{13}H_{22}O_3$
Source	
NATURAL	Tea.
Isolation	Synthesized from 2-pentyl-cyclopenten-2-one and ethyl malonate.
Appearance	Colourless to pale yellow liquid.
Odour	Pleasant, floral, jasmine-like.
Physical	b.p. approx. 300 °C.
Uses	Extensively used in a wide range of floral compositions.

METHYL DODECYL ALDEHYDE [10522-20-0]

Chemical name	3-Methyldodecan-1-al (methyl duodecyl aldehyde)
	$CH_3(CH_2)_8CH(CH_3)CH_2CHO$
Isolation	Produced by condensing methyl nonyl ketone (from rue oil) with bromoacetic ester in benzene solution in the presence of zinc.
Appearance	Colourless oily liquid.
Odour	Waxy–fatty, in dilution becoming more refreshing and citrusy.
Physical	b.p. 255 °C.
Uses	In minute traces in the bouquets of violet, mimosa and lilac type perfumes.

METHYL FORMATE [107–31–3]

	$HCOOCH_3$
Appearance	Colourless mobile liquid.
Odour	Very diffusive ethereal–winy type.
Physical	b.p. 32 °C, m.p. −99 °C, s.g. 0.98. Its vapours form explosive mixtures with air.
Uses	Mainly as a fumigant for dried fruits and tobacco.

METHYL FUROATE [611–13–2]

Chemical name	Methyl furan-2-carboxylate
	Methyl pyromucate
	$C_4H_3OCOOCH_3$
Isolation	Furoic acid, from which the ester is made, comes from furfural obtained from oat hulls by the action of steam and acid in a suitable enclosed vessel.
Appearance	Yellow, mobile, volatile liquid.
Odour	Fruity–winy and berry-like recalling a mixture of ethyl acetate, lactate and benzoate.
Physical	b.p. 181 °C.
	RIFM Monograph (1979) FCT, **17**, 869.

METHYL HEPTENOL [4630–06–2]

Chemical name	6-Methylhept-5-en-2-ol
	$(CH_3)_2C{=}CH_2CH_2CH_2CH(OH)CH_3$
Source	
NATURAL	Cayenne and Mexican linaloe oils.
Appearance	Colourless oily liquid.
Odour	Green, oily-herbaceous reminiscent of unripe berries.
Physical	b.p. 180 °C, s.g. 0.85.
Uses	In perfumery in citrus colognes, rose bouquets and oriental-type perfumes.
Comments	May also contain some 6-methylhept-6-en-2-ol, depending upon source and method of preparation.
	RIFM Monograph (1978) FCT, **16**, 817.

METHYL HEPTENONE [110–93–0]

Chemical name	6-Methylhept-5-en-2-one
	$C_8H_{14}O$
Source	
NATURAL	Essential oils of bois de rose, palmarosa, citronella, lemongrass and *Litsea cubeba*.
Isolation	As a by-product from processing of lemongrass and litsea oils. Synthesized from isoprene and acetone.

Appearance	Colourless mobile liquid.
Odour	Pungent, herbaceous, grassy–green with green–fruity note.
Physical	b.p. 174 °C, m.p. −67.1 °C, s.g. 0.86.
Uses	To boost citrus notes in cheap soap and household products fragrances.

RIFM Monograph (1975) FCT, 13, 859.

METHYL HEPTOATE [106–73–0]

Chemical name	Methyl n-heptanoate
	$CH_3(CH_2)_5COOCH_3$
Appearance	Colourless mobile liquid.
Odour	Fruity-green, waxy berry-like note.
Physical	b.p. 174 °C, m.p. −55.8 °C.
Uses	In traces for modifying the odour of perfume bouquets.

RIFM Monograph (Methyl Heptanoate) (1988) FCT, 26, 381.

METHYL HEPTYLENE CARBINOL

Chemical name	2-Methyloct-1-en-7-ol
	$CH_2\!=\!C(CH_3)CH_2CH_2CH_2CH_2CH(OH)CH_3$
Appearance	Colourless oil.
Odour	Powerful, fatty–fruity, oily–herbaceous.
Physical	b.p. 186 °C, s.g. 0.851.
Uses	Occasionally in perfumery in compositions of the rose type and as a reinforcer of lavender notes for soap.

METHYL n-HEPTYL KETONE [821–55–6]

Chemical name	Nonan-2-one
	$CH_3CO(CH_2)_6CH_3$
Source	
NATURAL	Rue oil, Algerian oil, clove and coconut oils.
Appearance	Colourless liquid.
Odour	Fatty–herbaceous with a fruity floral aspect.
Physical	b.p. 194 °C, m.p. −7.8 °C.
Uses	Occasionally in perfumery in lavender and citrus cologne types.

RIFM Monograph (Methyl Heptyl Ketone: 2-Nonaone) (1988) FCT, 26, 393.

p-METHYL HYDROCINNAMIC ALDEHYDE [5406–12–2]

Chemical name	3-p-Tolylpropan-1-al
	$CH_3C_6H_4CH_2CH_2CHO$
Source	
NATURAL	Cinnamon bark oil.

Appearance	Colourless oily liquid.
Odour	Powerful sweet, warm–floral, heliotrope-type.
Physical	b.p. 122 °C at 15 mm Hg.
Uses	Occasionally used in floral, balsamic, oriental and spicy compositions.

IFRA GUIDELINE

The committee recommends that p-methylhydrocinnamic aldehyde should not be used as a fragrance ingredient at a level over 1% in fragrance compounds.

This recommendation is based on test results of RIFM, showing sensitization reaction at 20% and no sensitization reaction when tested at 2% (private communication to IFRA).

November 1987

METHYL HYDROGENATED ROSINATE   [8050–13–3]; [8050–15–5]

CTFA name	Methyl hydrogenated rosinate Hercolyn D
Source	
CHEMICAL	The ester of methyl alcohol and the hydrogenated mixed long chain acids derived from rosin.
Appearance	Very viscous amber-coloured liquid.
Odour	Faint woody-piny note.
Physical	b.p. approx. 366 °C.
Uses	As a blender and fixative in low cost fragrances.

RIFM Monograph (Methyl Ester of Rosin) (1974) *FCT*, **12**, 939.

METHYL p-HYDROXYBENZOATE   [99–76–3]

Chemical name	Methyl paraben, methyl parahydroxybenzoate
Appearance	Fine, white crystalline solid.
Odour	None.
Physical	Soluble in water 1 to 500 after being dissolved by boiling: soluble in alcohol, acetone, chloroform, ether, vegetable oils and glycols.
Chemical	The methyl, ethyl, propyl, butyl and benzyl esters of p-hydroxybenzoic acid are used as preservatives.
Uses	The methyl ester is used as a protective in aqueous preparations in concentrations of 0.1 to 0.2%: when used in emulsified creams and lotions it is employed together with the propyl ester which is less soluble in water (1 in 2000), but soluble in alcohol, organic solvents, propylene glycol, and vegetable and mineral oils: the propyl ester is used in concentrations of from 0.02 to 0.2% as a protective of the oil or oil/wax phase of the product: a concentrated solution of the mixed esters, in propylene glycol, may be added to emulsions.

METHYL IONONES [1335–46–2]

$C_{14}H_{22}O$

Isolation
Synthesized by the condensation of citral and methyl ethyl ketone with subsequent cyclization of the resulting pseudo-methyl ionone: there are six positional isomers and commercial products are often a mixture of isomers.

Appearance
Pale yellow liquid.

Odour
Recalls iron: the alpha-normal form has a mild sweet odour and the beta-normal form has a weak, uninteresting odour; alpha-isomethyl ionone has a creamy, tobacco-like odour and is the most familiar to perfumers; beta-isomethyl ionone has a musky, animalic odour.

Physical
There are six positional isomers:

alpha-*n*-Methyl ionone, 5-(2,6,6-trimethylcyclohex-2-en-1-yl)-pent-4-en-3-one, b.p. 238 °C, s.g. 0.92; [7779–30–8]
alpha-iso-Methyl ionone, 4-(2,6,6-trimethylcyclohex-2-en-1-yl)-3-methylbut-3-en-2-one, b.p. 230 °C, s.g. 0.93; [127–51–5]
beta-*n*-Methyl ionone, 5-(2,6,6-trimethylcyclohex-1-en-1-yl)pent-4-en-3-one, b.p. 242 °C, s.g. 0.93; [127–43–5]
beta-iso-Methyl ionone, 4-(2,6,6-trimethylcyclohex-1-en-1-yl)-3-methylbut-3-en-2-one, b.p. 232 °C, s.g. 0.92; [79–89–0]
gamma-*n*-Methyl ionone, 5-(2-methylene-6,6-dimethylcyclohex-1-yl)pent-4-en-3-one, b.p. 252 °C;
gamma-iso-Methyl ionone, 4-(2-methylene-6,6-dimethylcyclohex-1-yl)-3-methylbut-3-en-2-one, b.p. 255 °C.

Uses
Widely used and indispensable in modern sophisticated perfumery. RIFM Monograph (1975) *FCT*, **13**, 863.

METHYL JASMONATE [1211–29–6]

$C_{13}H_{20}O_3$

Source
NATURAL   Jasmine absolute.

Isolation
Synthesized from muconic acid by the intermediate oxycyclopentenyl acetic acid.

Appearance
Colourless liquid.

Odour
Powerful herbaceous floral with notes of jasmine absolute.

Physical
b.p. approx. 300 °C.

METHYL LAURATE [111–82–0]

Chemical name   Methyl *n*-dodecanoate
CTFA name   Methyl laurate
Methyl laurinate; dodecanoic acid, methyl ester
$CH_3(CH_2)_{10}COOCH_3$

Appearance	Colourless slightly oily liquid.
Odour	Intense oily–fatty note, which becomes violet–floral in nature on extreme dilution.
Physical	b.p. 262 °C, m.p. 5 °C, s.g. 0.873.
Uses	In traces in perfumes of the violet, jasmine and mimosa type.

METHYL MYRISTATE [124-10-7]

Chemical name	Methyl n-tetradecanoate
CTFA name	Methyl myristate
	Tetradecanoic acid, methyl ester
	$CH_3(CH_2)_{12}COOCH_3$
Appearance	Waxy–crystalline solid at ordinary temperatures.
Odour	Weak oily orris-like.
Physical	b.p. 295 °C (m.p. 22 °C may vary depending upon the presence of related methyl esters).
Uses	Useful in violet bouquets especially in perfuming powders.

METHYL-beta-NAPHTHYLACETALDEHYDE

Chemical name	2-(2-Naphthyl)propanal (2-beta-naphthylpropionaldehyde)
	$C_{10}H_7CH(CH_3)CHO$
Odour	Harsh, heavy–sweet, orange-blossom-type.
Physical	m.p. 53 °C.

METHYL beta-NAPHTHYL ETHER [93-04-9]

Chemical name	2-Methoxynaphthalene
	beta-Naphthol methyl ether, yara-yara, nerolin I
Appearance	White shiny crystals.
Odour	Intensively sweet, orange-blossom-like.
Physical	b.p. 274 °C, m.p. 73 °C: soluble in alcohol.
Uses	Widely employed in low cost fragrances for household products, soaps and detergents.

METHYL alpha-NAPHTHYL KETONE [941-98-0]

Chemical name	Acetonaphthone
	Oranger liquid
	$C_{10}H_7COCH_3$
Isolation	Synthesized by condensing naphthalene with acetyl chloride in the presence of aluminium chloride and monochlorobenzene.
Appearance	Colourless or pale yellow oily liquid.
Odour	Sweet floral orange blossom, but is sweeter and smoother than the naphthol ethers.
Physical	b.p. 303 °C, m.p. 55–56 °C, s.g. 1.14.

Uses	In perfumery in orange blossom compounds or eau-de-colognes. RIFM Monograph (1982) *FCT*, **20**, 755.

METHYL-*beta* NAPHTHYL KETONE [93–08–3]

	Oranger crystals, cetone-D $C_{12}H_{10}O$
Isolation	Synthesized from naphthalene and acetyl chloride in the presence of aluminium chloride (Friedel–Crafts reaction).
Appearance	White crystalline solid.
Odour	Fruity–floral orange-blossom-type, more powerful and pleasant than the alpha-isomer.
Physical	b.p. 300 °C, m.p. 55 °C.
Uses	Widely used in perfume compositions in orange blossom, sweet pea, honeysuckle, jasmine and many other floral types. RIFM Monograph (1975) *FCT*, **13**, 867.

METHYL NONYL ACETALDEHYDE [110–41–8]

	2-Methylhendecan-1-al; methyl nonyl acetic aldehyde; 2-methyl-hendecan-1-al; methyl nonyl acetic aldehyde; aldehyde MNA $CH_3(CH_2)_8CH(CH_3)CHO$
Source	
NATURAL	Algerian oil of rue (*R. montana*).
Isolation	Synthesized (by Darzen's glycidic reaction) from methyl-*n*-nonyl ketone and monochloroacetate.
Appearance	Colourless liquid.
Odour	Powerful and diffusive, slightly orangey with dry ambre overtones.
Physical	b.p. 232 °C.
Uses	Widely used in traces to give fresh aldehydic topnote to a wide range of floral, amber, tabac and citrus compositions. RIFM Monograph (1973) *FCT*, **11**, 485 and (1973) *FCT*, **11**, 1081.

METHYL NONYLATE [1731–84–6]

Chemical name	Methyl *n*-nonanoate Methyl pelargonate $CH_3(CH_2)_7COOCH_3$
Appearance	Colourless oily liquid.
Odour	Fruity–winy, faintly nut-like.
Physical	b.p. 214 °C, s.g. 0.88.
Uses	In traces as a topnote in fruity and floral compositions.

METHYL NONYLENATE [111–79–5]

Chemical name	Methyl-2-nonenoate Neo-folione, Beauvertate $C_{10}H_{18}O_2$

Isolation	By Claisen condensation of heptaldehyde and methylacetate using sodium methylate.
Appearance	Colourless oily liquid.
Odour	Powerful earthy vegetable green, suggestive of violet leaves.
Uses	At low levels in floral compositions, particularly for soap and detergents. Often used as a replacement ingredient for methyl heptine carbonate. RIFM Monograph (1976) FCT, **14**, 811, Special Issue III (Binder, p. 564).

METHYL NONYL KETONE [112–12–9]

Chemical name	Hendecan-2-one Undecanone-2; Rue ketone $CH_3CO(CH_2)_8CH_3$
Source	
NATURAL	French oil of rue, Spanish oil and to a lesser extent in Algerian rue.
Isolation	Separated from oil of rue.
Appearance	Colourless oily liquid.
Odour	Fatty–oily, slightly herbaceous and fainty orris-like.
Physical	b.p. 225 °C, m.p. 15 °C.
Uses	In proportions of no more than 0.5% in compounds such as sweet-pea, lavender, verbena and other herbaceous types. RIFM Monograph (1975) FCT, **13**, 869.

METHYL NONYNOATE [111–80–8]

Chemical name	Methyl-2-nonynoate Methyl octin(e) carbonate: methyl octyne carboxylate
Appearance	Colourless oily liquid.
Odour	Powerful, green violet-leaf like; on dilution it becomes more vegetable green and cucumber-like.
Physical	b.p. 220 °C.
Chemical	More stable and more pungent in the pure state than methyl-heptine carbonate.
Uses	In green floral compositions at extremely low levels. RIFM Monograph (Methyl 2-Nonynoate: Methyl Octine Carbonate) (1975) FCT, **13**, 871, Special Issue II (Binder, p. 565). IFRA GUIDELINE The Committee recommends that methyl heptine carbonate and methyl octine carbonate should not be used as fragrance ingredients at a total level greater than 0.05%, individually or in combination, in fragrance compounds. This recommendation is based on extensive data from sensitization studies conducted by RIFM and others on the substances, individually and in combination, and on evidence of crossreactivity (private communication to IFRA).

METHYL OCTYL ACETALDEHYDE [19009-56-4]

Aldehyde MOA; 2-Methyldecanal
$C_{11}H_{22}O$

Isolation By the reduction of alpha-methylene decanal.
Appearance Colourless liquid.
Odour Powerful aldehydic reminiscent of orange peel.
RIFM Monograph (1976) *FCT*, **14**, 609.

METHYL OCTYNOATE [111-12-6]

Chemical name Methyl oct-2-ynoate, methyl heptin(e) carbonate
Vert de violette; folione
History Discovered by Moureu and Delange.
Isolation Heptaldehyde must first be obtained by the destructive distillation of castor oil.
Appearance Colourless oil.
Odour Powerful, penetrating vegetable, green-foliage note, develops a violet fragrance on extreme dilution.
Physical b.p. 217 °C, s.g. 0.93.
Uses At extremely low levels due to its strong skin sensitizing properties.
RIFM Monograph (Methyl 2-Octynoate: Methyl Heptine Carbonate) (1979) *FCT*, **17**, 375.
IFRA GUIDELINE
The Committee recommends that methyl octine carbonate and methyl heptine carbonate should not be used as fragrance ingredients at a total level greater than 0.05%, individually or in combination, in fragrance compounds.

This recommendation is based on extensive data from sensitization studies conducted by RIFM and others on the substances, individually and in combination, and on evidence of crossreactivity (private communication to IFRA).

October 1990

3-METHYL-2-PENTYL-CYCLOPENT-2-EN-1-ONE [1128-08-1]

Dihydro jasmine
Isolation Synthesized from heptanal and butenone.
Appearance Colourless to faintly yellow liquid.
Odour Intensely floral, jasmine-type.
Physical Soluble in 2 vols. of 70% alcohol.
Uses Widely used in perfumery in concentrations of 3–5% in floral and fantasy compounds and as a modifier in fougere, lavender and citrus type perfumes.

p-METHYL PHENYLACETALDEHYDE [104-09-6]

Chemical name *p*-Tolylacetaldehyde, syringa aldehyde
$CH_3C_6H_4CH_2CHO$

Appearance	Colourless oily liquid.
Odour	Powerful, refreshing, fruity, sweet-floral note.
Physical	b.p. 210 °C.
Uses	To give sweet–green notes to floral compositions such as lilac, hyacinth, rose, sweet pea etc.

RIFM Monograph (*p*-Tolylacetaldehyde) (1978) *FCT*, **16**, 877.

METHYL PHENYLACETATE [101-41-7]

$C_9H_{10}O_2$

Isolation	Synthesized from phenylacetic acid and methanol.
Appearance	Colourless liquid.
Odour	Powerful, floral, honey-like, somewhat musky character.
Physical	b.p. 220 °C.
Uses	Widely used in perfumery in floral and oriental fragrances and in low cost rose, jasmine and non-floral types for soaps and household products.

RIFM Monograph (1974) *FCT*, **12**, 941.

METHYL PHENYL CARBINOL [1321-27-3]

Chemical name	Phenyl methyl carbinol, 1-phenyl-ethanol alpha-Phenylethyl alcohol; styrallyl alcohol $C_6H_5CHOHCH_3$
Appearance	Colourless liquid.
Odour	Earthy–green reminiscent of gardenia and hyacinth.
Physical	b.p. 204 °C.
Uses	Occasionally in lilac and mimosa types.

RIFM Monograph (Styralyl Alcohol) (1974) *FCT*, **12**, 995.

METHYL PHENYL CARBINYL ACETATE [93-92-5]

Chemical name	1-Phenylethyl acetate Phenyl ethylene acetate; phenyl methyl carbinyl acetate; gardiniol II; styrallyl acetate
Source	
NATURAL	Leaves and oil of the gardenia flower.
Isolation	Produced from styrolyl alcohol and acetic acid.
Appearance	Mobile, colourless liquid.
Odour	Powerful, green floral gardenia-type.
Physical	b.p. 214 °C.
Uses	In perfumery in lilac, lily of the valley, and hyacinth type perfumes: as a modifier in floral bouquets.
Comments	Should be well blended with the ionones and hydroxycitronellal to obtain a really fine perfume: the powerful gardenia odour makes its use difficult.

RIFM Monograph (1976) *FCT*, **14**, 611.

METHYL PHENYLGLYCIDATE [37161-74-3]

Chemical name	Methyl 2,3-epoxy-3-phenyl-propanoate
	Hexadecyl aldehyde, fraesol
Appearance	Colourless liquid.
Odour	Mild fruity pineapple, strawberry-like.
Physical	b.p. 257 °C.
Uses	In traces to obtain a new note in fancy bouquets.

METHYL PHENYLPROPIONATE [103-25-3]

Chemical name	Methyl 3-phenylpropanoate
	Methyl hydrocinnamate
	$C_6H_5CH_2CH_2COOCH_3$
Appearance	Colourless mobile liquid.
Odour	Powerful, winy–floral with sweet balsamic undertones.
Physical	b.p. 239 °C, s.g. 1.05.
Uses	As a modifier in jasmine, rose, lilac, carnation and many other fragrance types.

METHYL PROPIONATE [554-12-1]

Chemical name	Methyl propanoate
	Propanoic acid methyl ester
	$C_2H_5COOCH_3$
Appearance	Colourless mobile liquid.
Odour	Ethereal rum-like character.
Physical	b.p. 80 °C.
Uses	In traces as parts of diffusive fruity topnotes.
	RIFM Monograph (1982) *FCT*, **20**, 765.

METHYL ROSINATE [68186-14-1]

CTFA name	Methyl rosinate
	Abalyn; rosin acid, methyl ester; methyl abietate
Source	
CHEMICAL	Methyl ester of acids recovered from rosin.
Appearance	Pale straw-coloured very viscous liquid.
Odour	Almost odourless, faintly piney–woody.
Physical	b.p. 365 °C (decomposition).
Uses	Extensively as a cheap fixative, blender or diluent.

METHYL SALICYLATE [119-36-8]

Chemical name	Methyl o-hydroxybenzoate
	2-Hydroxybenzoic acid methyl ester, oil of wintergreen
	$C_8H_8O_3$

Source	
NATURAL	Principal constituent of wintergreen oil and birch bark oil: also occurs in many other volatile oils, including cassie, spicewood, rue and tuberose.
Isolation	Synthesized by condensing salicylic acid and methyl alcohol in the presence of sulphuric acid.
Appearance	Colourless oily liquid.
Odour	Pungent, sweet rather medicinal odour, reminiscent of wintergreen oil.
Physical	b.p. 223 °C, m.p. −9 °C: slightly soluble in water, miscible with alcohol or glacial acetic acid.
Uses	Widely in medicine and as a flavouring agent in dental creams: in perfumery in synthetic cassie, tuberose and chypre. RIFM Monograph (1978) *FCT*, **16**, 821.

2-METHYL UNDECANOL [10522-26-6]

Chemical name	2-Methyl hendecanol Methyl nonyl ethyl alcohol; 'secondary duodecyl alcohol' $CH_3(CH_2)_8CH(CH_3)CH_2OH$
Appearance	Colourless oily liquid.
Odour	Mild, sweet, waxy character.
Uses	Occasionally in lily, rose, freesia, peony types etc. RIFM Monograph (Aldehyde C_{12} MNA) (1973) *FCT*, **11**, 485 (1973) *FCT*, **11**, 1081.

METHYL VINYL ETHER/MALEIC ANHYDRIDE COPOLYMERS
[9011-16-9]

CTFA name	PVM/MA Copolymer
Source	
CHEMICAL	A copolymer of methyl vinyl ether and maleic anhydride.
Appearance	Fine, white, free-flowing powder.
Physical	In water or aqueous/alcoholic solutions it forms high viscosity gels when neutralized with a suitable alkali, such as sodium hydroxide or an amine base, to pH 6.0–7.0.
Uses	In the preparation of perfumes, deodorants and hair dressing gels: also as a stabilizing or thickening agent in creams and lotions.

METOL [55-55-0]

Chemical name	*p*-Hydroxy-*N*-methylaniline sulphate
CTFA name	*p*-Methylaminophenol sulfate *p*-Methylaminophenol sulphate $(CH_3NHC_6H_4OH)_2H_2SO_4$
Appearance	Crystals which discolour in air.

234   Microcrystalline wax

Physical	m.p. about 260 °C (decomp.): soluble in 20 parts cold, 6 parts boiling water, slightly soluble in alcohol but insoluble in ether.
Uses	A photographic developer which may be used as a hair dye.
Comments	Keep well closed and away from light.

MICROCRYSTALLINE WAX [63231-60-7]

CTFA name	Microcrystalline wax
Source	
CHEMICAL	A long-chain hydrocarbon derivative of petroleum.
Isolation	Derived from petroleum.
Appearance	Microscopically small and irregular crystals.
Physical	Differs from paraffin wax in its crystalline structure (microscopic examination of paraffin wax reveals large well-formed crystals.)
Chemical	There are several microcrystalline waxes available with differing hardness and melting points.
Uses	Used to control the melting point in creams, lipsticks, solid brilliantines and thixotropic preparations.

MIGNONETTE [91721-98-1]

Reseda odorata

Source	
GEOGRAPHIC	Egypt.
NATURAL	Cultivated around the Mediterranean.
Isolation	The perfume is extracted from the flowers by volatile solvents to yield an essential oil.
Odour	The oil has a powerful odour of violet leaves.
Uses	In perfumery.
Comments	A synthetic resembling reseda may be obtained with ethyl decine carbonate (used in conjunction with orris, methyl ionone, basil, phenylacetaldehyde, dimethyl acetal and santal oils) to produce mignonette compounds.

MILFOIL OIL [84082-83-7]

CTFA name	Yarrow extract
Source	
NATURAL	*Achilles millefolium*, L., N.O. Compositae (commonly known as yarrow).
Isolation	Distillation of the fresh flowers.
Appearance	Dark blue liquid.
Odour	Reminiscent of bornyl acetate.
Chemical	Principal constituent is cineole, together with valeric acid, eugenol, pinene, thujone, borneol, camphor, caryophyllene and azulene.

MIMOSA CONCRETE/MIMOSA ABSOLUTE [8031-03-6]

Source
GEOGRAPHIC A tree, native to Australia, but grown in southern France and Italy.
NATURAL *Acacia dealbata* or *A. floribunda*, N.O. Leguminosae.
Isolation The concrete is obtained from the flowers and twig ends by extraction with volatile solvents: a second extraction yields the absolute.
Yield 1% concrete of which about 25% is obtained as absolute.
Odour Powerful and green–floral.
Chemical Artificial mimosa perfumes are produced from methyl acetophenone, terpineol, hydroxycitronellal, etc.
Uses In perfumery in fine fragrances: mimosa perfumes are all obtained from the natural flower absolutes in the south of France.
RIFM Monograph (1975) *FCT*, **13**, 873, Special Issue II (Binder, p. 571).

MONARDA OIL [91772-89-3]

Source
GEOGRAPHIC Plants native to the USA.
NATURAL Different species of Monarda, N.O. Labiatae, most importantly *M. didyma*, L.
Isolation Distillation.
Appearance Pale yellow liquid.
Odour Sweet and balsamic odour, reminiscent of lavender and ambergris.
Comments *M. punctata*, L., is a North American perennial (also known as horsemint) with a phenolic–minty odour.

MORPHOLINE [110-91-8]

Chemical name Tetrahydro-4H-1,4-oxazine
CTFA name Morpholine
Diethylamine oximide; diethylene imidozide
C_4H_9NO
Appearance Colourless, mobile, hygroscopic liquid.
Odour Slightly ammoniacal.
Physical b.p. 128.9 °C, m.p. −4.9 °C: miscible with water with evolution of some heat; also miscible with acetone, benzene, ether, alcohol and ethylene glycol.
Chemical Reacts as a secondary amine in the same way as diethanolamine, of which it is the anhydride.
Uses Useful emulsifying agent in combination with fatty acids.
Comments Can be corrosive to human skin.

MOSKENE [116-66-5]

1,1,3,3,5-Pentamethyl-4,6-dinitroindane; 1H-indene, 2,3-dihydro-1,1,3,3,5-pentamethyl-4,6-dinitro-

Source	
NATURAL	Does not occur in nature.
Isolation	From p-cymene plus isobutene followed by nitration.
Physical	GC RIFM no. 71–51; IRC RIFM no. 71–51.

MUSCONE [541-91-3]

Chemical name	3-Methylcyclopentadecanone-1
Other names	Muskone; methylexaltone
	$C_{16}H_{30}O$
Source	
NATURAL	Animal musk.
Isolation	Obtained from diacetyldodecane by intramolecular condensation and subsequent catalytic reduction.
Appearance	Colourless to white crystalline solid.
Odour	Soft, sweet and extremely tenacious musky type.
Physical	m.p. 33 °C, b.p. approx. 328 °C.
Uses	With natural musk tinctures in fine fragrances.

3-Methylcyclopentadecanone: RIFM Monograph (1982) *FCT*, **20**, 749, Special Issue VI.

MUSHROOMS

Source	
NATURAL	Many different types.
Odour	May be rather unpleasant, but the following have odours of some interest:

Hygrophora hyacenthinus – jasmine;
Tricholoma aurantium – cucumber;
Clitocybe geotrope – lavender;
Lactary glyciosmos – bergamot;
Lactory camphoratus – melilot;
Hygrophora agathosmus – cherry laurel;
Psatella arvensis – aniseed.

MUSK [8001-04-5]

Tonquin musk from Tibet is highly esteemed: has several local names suggestive of its character; 'lard' musk, 'vegetable-oil' musk, 'snake's-head' musk;
Blue skin musk, or 'blue pile';
Grain musk which consists of the dried secretion and hairs etc. which have been removed from the musk pod;
Carbardine musk, both Chinese and Russian, from the 'Kabarga' musk deer found in the Atlas mountains;
Yunnan musk which occurs in 'pig-faced pods'; Assam and Nepal musk which occurs in very small pods.

Source	
GEOGRAPHIC	A species of deer from the mountains of Nepal, China and Tibet.
NATURAL	Comes from the internal pouch between the hind legs of the male musk deer, *Moschus moschiferus*, L., N.O. Ungulata and other species. The musk pod is a round sac approx. 1.5 in (4 cm) in diameter, smooth on one side and covered with dark hairs on the other concentrically arranged around a nearly central orifice. There are also plant sources of musk materials, which are macrocyclic ketones or lactones.
CHEMICAL	Synthesis is possible.
Isolation	The dried secretion from the preputial follicles: the animal is killed and the gland, with an outer covering of hair and hide, is completely removed and dried either in the sun, on a heated stove, or immersed in hot oil: after removing most of the attached hair and hide, the musk pods are soaked in water, to soften them, and the agreeable perfume now becomes apparent.
Yield	The best yield is obtained from 10-year-old deer, when 0.5–2 oz (14–56 g) may be collected.
Odour	Pleasant and agreeable.
Physical	About 50% of unadulterated musk will dissolve in water: about 10–20% is soluble in strong alcohol.
Chemical	The odiferous principle is a ketone (muscone.)
Uses	The most indispensable natural raw material in perfumery: should be added in small quantities to fortify the basic odour: has both subtlety and diffusive power: in China it is highly esteemed as a medicine.
Comments	Should be handled with great care: camphor or benzaldehyde, in alcoholic solution, are useful for removing the odour from the hands: is liable to adulteration with earth or dried blood.

MUSK AMBRETTE [83–66–9]

Chemical name	1,3-Dinitro-4-methoxy-2-methyl-5-tert-butyl-benzene $C_{12}H_{16}N_2O_5$
Isolation	Synthesized by alkylating *m*-cresol methyl ether with subsequent nitration.
Appearance	Yellow crystalline solid.
Odour	Persistent, sweet, heavy, floral, musky.
Physical	m.p. 83 °C.
Uses	Once the most popular synthetic nitromusk: safety restrictions have severely limited its use in fragrance compositions. RIFM Monograph (1975) *FCT*, **13**, 875, Special Issue II (Binder, p. 572). IFRA GUIDELINE The Committee recommends that Musk Ambrette should not be used in fragrance compounds for cosmetics, toiletries and other products which under normal conditions of use will come into

contact with the skin. This includes rinse-off products. For other applications, Musk Ambrette should not be used as a fragrance ingredient at a level over 4% in fragrance compounds. This restriction should not be exceeded, irrespective of the end-use concentration.

This recommendation is based on results of animal experiments on abraded skin under UV exposure showing photosensitivity (I. E. Kochevar, G. L. Zalar, J. Einbinder and L. C. Harber (1979), *J. Invest. Dermatol.*, 73, 144) and on reports of light-dependent skin reactions related to the use on damaged skin of after-shave products with a high content of Musk Ambrette (G. J. Raugi, F. J. Storrs and W. G. Larsen (1979), *Contact Dermatitis*, 5, 251; V. J. Giovinazzo, L. C. Harber, R. B. Armstrong and I. E. Kochevar (1980), *J. Am. Acad. Dermatol.*, 3 (4), 384). It has been confirmed that certain dermatological patients are photosensitive to Musk Ambrette (E. Cronin (1984), *Contact Dermatitis*, 11, 88). Other studies have indicated that Musk Ambrette is showing neurotoxic effects (P. S. Spencer, M. C. Bischoff-Fenton, O. M. Moreno, D. L. Opdyke and R. A. Ford (1984), *Toxicol. Appl. Pharmacol.*, 75, 571), can penetrate human skin and is only slowly excreted (private communication to IFRA).

June 1981, last amendment October 1985

MUSK KETONE [81–14–1]

Chemical name	3,5-Dinitro-2,6-dimethyl-4-tert-butylacetophenone
	Musk C
	$C_{14}H_{18}N_2O_5$
Isolation	Manufactured by the Friedel–Crafts acetylation of tert-butylxylene and subsequent nitration.
Appearance	Pale yellow crystalline solid.
Odour	Very tenacious, a warm, sweet, powdery, musky smell, with a mild animal character.
Physical	m.p. 137 °C.
Uses	Highly appreciated in fine perfumery and has the widest application.
	RIFM Monograph (1975) *FCT*, 13, 877, Special Issue II (Binder, p. 574).

MUSK PLANTS

Source	
NATURAL	White musk mallow, Spikenard, Sumbul, Ambrette, Musk thistle, and *Mimulus moschatus*, Douglas, N.O. Scrophulariaceae.
Isolation	The odour may come from either the flowers, roots or the whole plant.
Odour	All plants exhale a musky odour.

MUSK TINCTURE

Source
GEOGRAPHIC   Musk deer from the mountains of Nepal, China and Tibet at an altitude of 1500 m.
NATURAL   The glandular secretion contained in an internal pouch between the hind legs of the male musk deer, *Moschus moschiferus*.
Isolation   The musk is treated with alcohol which extracts the odiferous matter from the secretion: the tincture is obtained by filtration.
Appearance   A paste-like substance.
Odour   Typically erogenous and animalic.
Uses   Only in very expensive perfumes.
Musk Tonquin Tincture: RIFM Monograph (1983) *FCT*, **21**, 865.

MUSK RAT OIL [8023-83-4]

Source
GEOGRAPHIC   A rodent related to the beaver, largely found in North America.
NATURAL   *Castor fibre*, L.
Isolation   The musk rats are trapped and their pelts, known as musquash fur, are removed.
Appearance   Light yellow oil.
Odour   Musky.

MUSK XYLENE [81-15-2]

Chemical name   2,4,6-Trinitro-1,3-dimethyl-5-tert-butylbenzene
Musk xylol
$C_{12}H_{15}N_3O_6$
Isolation   Produced by the nitration of alkylated *m*-xylene.
Appearance   Pale yellow crystalline solid.
Odour   Tenacious, sweet, musky and slightly harsher in character than musk ketone.
Physical   m.p. 114 °C.
Uses   Widely used in soap, detergent and household products: in perfumery in a wide range of fragrance types.
RIFM Monograph (1975) *FCT*, **13**, 881, Special Issue II (Binder, p. 577).

MYRRH [8016-37-3]

CTFA name   Myrrh extract
Gum-resin myrrh
History   One of the first gum-resins mentioned in history: of considerable value as a perfume and used with spices: it is thought that the earliest references were not to true myrrh, but to labdanum.

Myrtenal

Source	
GEOGRAPHIC	Wild trees or shrubs of Somalia, Ethiopia and Saudi Arabia.
NATURAL	*Commiphora myrrha*, Holmes, and other species of N.O. Burseraceae.
Isolation	The gum-resin is exuded between the cortical layers of the tree and falls to the ground: it is collected in goat-skin bags: the resinoid is obtained by extraction of the crude gum: steam distillation will produce the oil: may be prepared in the Grasse houses by the volatile solvent process from a purified resinoid.
Yield	Steam distillation yields approx. 5% essential oil.
Appearance	Oleo-resinous solid.
Odour	Attractive, slightly spicy and aromatic.
Uses	As a fixative in oriental compounds: recommended in honeysuckle and broom compounds: used in liquid dentifrices as a tincture (1 in 5 in 90% alcohol).
Comments	A carminative and topically astringent to mucous membrane. Myrrh oil: RIFM Monograph (1976) *FCT*, **14**, 621 (Binder, p. 581).

MYRTENAL [564–94–3]

Chemical name	2-Pinen-10-al
Appearance	Colourless liquid.
Odour	Powerful, sweet, herbaceous smell reminiscent of dry hay and bayleaf.
Physical	s.g. 0.99, b.p. 199 °C.
Uses	In the preparation of simulated essential oils, where it imparts the natural dry hay character.
Comments	Has an intense odour and should be handled as a 10% solution: recommended to be stored in containers sealed under nitrogen. 2-Formyl-6,6-Dimethylbicyclo (3.1.1) hept-2-ene: RIFM Monograph (1988) *FCT*, **26**, 329, Special Issue VII.

MYRTENOL [515–00–4]

Chemical name	2-Pinen-10-ol
Appearance	Colourless liquid.
Odour	Woody, warm, herbaceous with a mild camphoraceous character.
Physical	When exposed to air it develops hay-like overtones, due to the formation of traces of myrtenal: soluble in 5 vols. or more of 60% alcohol; s.g. 0.98, b.p. 224 °C.
Uses	In perfumery in spruce, lavender, citrus and bay type fragrances for men's colognes.

MYRTENYL ACETATE [35670–93–0]

Chemical name	2-Pinen-10-yl acetate
Appearance	Colourless mobile liquid.
Odour	Refreshing, sweet herbaceous and slightly warm, spicy.

Physical	Soluble in 5 vols. or more of 70% alcohol; s.g. 0.99.
Uses	In perfumery to impart a natural fruitiness to essential oils, particularly to bergamot: also employed in herbal or citrus type perfume compounds.

RIFM Monograph (1988) *FCT*, **26**, 389, Special Issue VII.

MYRTENYL iso-BUTYRATE [29021–37–2]

Chemical name	2-Pinen-10-yl 2-methylpropionate
Appearance	Colourless liquid.
Odour	Heavy, woody, pine-type.
Physical	s.g. 0.963–0.968: soluble in more than 5 vols. of 80% alcohol.
Uses	In perfumery as a modifier: blends well with ionones, patchouli, galbanum and vetivert.

MYRTENYL FORMATE [72928–52–0]

Chemical name	2-Pinen-10-yl formate
Appearance	Colourless liquid.
Odour	Strong pine-type, linseed odour.
Physical	Soluble in 5 vols., or more, of 70% alcohol.
Uses	In perfumery to enhance the galbanum-type note: also in simulation of essential oils of rosemary and galbanum.

MYRTLE OIL [8008–46–6]

CTFA name	Myrtle extract
Source	
GEOGRAPHIC	A small tree or bush from the shores of the Mediterranean.
NATURAL	*Myrtus communis*, L., N.O. Myrtaceae.
Isolation	Steam distillation of the blossoms, leaves and twigs.
Yield	0.5–1.0% depending upon source of the plant material.
Appearance	Yellow to green oily liquid.
Odour	Distinct, spicy and camphoraceous.
Chemical	Contains pinene, borneol, iso-borneol, cineole, myrtenol, geraniol and nerol.
Uses	In perfumery to impart a spicy–herbaceous component to many natural-type creations: also in Eastern bouquets, and in certain soap compounds.

RIFM Monograph (1983) *FCT*, **21**, 869.

beta-NAPHTHOL [135–19–3]

Chemical name	2-Hydroxynaphthalene
CTFA name	2-Naphthol
	2-Naphthalenol; beta-hydroxynaphthalene
	$C_{10}H_7OH$
Appearance	Colourless to buff crystalline solid.

Physical	b.p. 286 °C, m.p. 122 °C, s.g. 1.22: darkens with age: only slightly soluble in water, soluble in glycerol or olive oil.
Chemical	Has antiseptic and germicidal properties.
Uses	5% dissolved in eau-de-cologne may be used as a deodorant.

NARCISSUS CONCRETE/NARCISSUS ABSOLUTE [8023-75-4]

Source	
GEOGRAPHIC	A plant cultivated in the Mediterranean litorral, Morocco and Egypt.
NATURAL	*Narcissus poeticus* (a white flower).
Isolation	The perfume may be extracted by enfleurage and the volatile solvent method.
Yield	0.2–0.3% concrete from which approx. 30% is obtained as absolute.
Odour	Hay-like and spicy; on dilution becomes typical of the flower.
Chemical	Artificial narcissus is a mixture of phenylethyl alcohol, terpineol, methyl ionone, anisic aldehyde, benzyl acetate and linalol, with up to 10% of phenylacetic aldehyde or *p*-cresol.
Uses	In perfumery in fantasy-type creations.
Comments	Very dilute solutions of *p*-cresylacetate, iso-butyrate or phenylacetate possess an odour which resembles narcissus: artificial narcissus oil is improved on its addition. RIFM Monograph (1978) *FCT*, **16**, 827, Special Issue IV.

NAULI 'GUM'

Source	
GEOGRAPHIC	A tree which grows in the Solomon Islands.
NATURAL	*Canarium commune*, L., N.O. Burseraceae (the 'Java almond tree').
Isolation	The oleo-resinous secretion is collected and the oil obtained by steam distillation.
Yield	Approx. 10% volatile oil.
Appearance	Soft, yellow, oleo-resinous substance.
Chemical	Contains 34% anethole.

Nelumbium speciosum

	Known as the sacred bean in India.
History	This aquatic plant of the water lily family was held sacred in ancient times by the Egyptian worshippers of Isis.
Source	
GEOGRAPHIC	Found in India, Tibet, China and Japan.
NATURAL	Possesses beautiful, fragrant, rose-pink or white flowers which resemble gigantic tea roses.
Isolation	The leaf stalks, which are spiral fibres, are extracted and used as

Odour	Delightful fragrance. wicks to burn before idols. The leaves are used as plates for offerings.
Uses	Widely employed in religious invocations and ceremonies.

NEROL [106-25-2]

Chemical name	3,7-Dimethyl-2,6-octadien-1-ol *cis*-3,7-Dimethylocta-2,6-dien-1-ol; 2,6-dimethyl-2,6-octadien-8-ol $C_{10}H_{18}O$
History	Heine first synthesized nerol in 1902.
Source	
NATURAL	Essential oils of neroli, champaca, wallflower, ylang-ylang, and *Helichrysum angustifolium*, N.O. Compositae.
Isolation	Manufactured with its isomers, geraniol and linalol, from myrcene which is hydrochlorinated to a mixture of chlorides: these are converted to esters by treatment with sodium acetate and hydrolysed to a mixture of alcohols: fractionation produces nerol: by successive removal and addition of hydrogen iodide, nerol may be obtained from geraniol.
Appearance	Colourless oily liquid.
Odour	Fresh, sweet rose–neroli odour.
Physical	b.p. 227 °C, s.g. 0.88.
Uses	Widely used in perfumery in orange blossom, rose, and many other floral and citrus cologne types. RIFM Monograph (1976) *FCT*, **14**, 623 (Binder, p. 584).

NEROLIDOL [7212-44-4]

Chemical name	3,7,11-Trimethyl-1,6,10-dodecatrien-3-ol *cis*-3,7,11-Trimethyldodeca-1,6,10-trien-3-ol $C_{15}H_{26}O$
Source	
NATURAL	Orange flower oil and Peru balsam oil: the main component of cabreuva oil.
Isolation	Synthesized from linalool via geranyl acetone, which is reacted with acetylene and partially reduced to yield nerolidol: may be obtained from cabreuva oil by distillation.
Appearance	Colourless liquid.
Odour	Mild, peculiar and lily-like.
Physical	b.p. 276 °C, s.g. 0.88.
Chemical	The synthetic product will yield farnesol when treated with acetic anhydride.
Uses	In perfumery, particularly in muguet, lilac, jasmine and orange blossom compounds. RIFM Monograph (1975) *FCT*, **13**, 887, Special Issue II (Binder, p. 586).

NEROLI OIL [8016-38-4]

Source	Neroli Portugal, neroli bigarade
GEOGRAPHIC	A tree grown in Morocco, Algeria, Egypt and Southern France.
NATURAL	The sweet and bitter orange tree, or *Citrus aurantium*, subspecies *amara*, and also *Citrus bigardia*.
Isolation	Steam distillation of the fresh, hand-picked blossoms: is largely produced in Vallauris in the south of France; other oils may also be distilled in Algeria, Sicily, Spain and Syria, but these do not possess such a delicate bouquet: a synthetic product may be manufactured.
Yield	0.08–0.1%.
Appearance	Pale yellow liquid.
Odour	Sweet, spicy, orange blossom odour.
Physical	Darkens on exposure to light.
Chemical	Contains limonene, linalol, esters of geraniol, linalol, and nerol, phenylethyl alcohol, methyl anthranilate and indole.
Uses	Indispensable in perfumery in the preparation of eau-de-cologne.
Comments	The terpeneless neroli oil is about three times as concentrated and very expensive. Neroli oil, Tunisian: RIFM Monograph (1976) *FCT*, **14**, 813, Special Issue III (Binder, p. 587).

NEROLIN BROMELIA [93-18-5]

Chemical name	beta-Naphthol ethyl ether Nerolin II; Bromelia $C_{12}H_{12}O$
Isolation	Synthesized from beta-naphthol and ethanol in the presence of sulphuric acid.
Appearance	White crystalline solid.
Odour	Soft floral.
Physical	m.p. 37 °C, b.p. 282 °C, s.g. 1.06.
Uses	Widely used as a sweetener and floralizer in many fragrance types. Beta-Naphthyl Ethyl Ether: RIFM Monograph (1975) *FCT*, **13**, 883, Special Issue II (Binder, p. 582).

NEROLIN YARA-YARA [93-04-9]

Chemical name	beta-Naphthol methyl ether Nerolin I, Yara-yara $C_{11}H_{10}O$
Isolation	Synthesized from beta-naphthol and dimethyl sulphate.
Appearance	White crystalline solid.
Odour	Tenacious warm, intensely sweet, floral, orange blossom acacia-like.

Physical	m.p. 73 °C, b.p. 274 °C.
Uses	Extensively in low cost fragrances for soaps, detergents and household products.

Beta-Naphthyl Methyl Ether: RIFM Monograph (1975) *FCT*, **13**, 885, Special Issue II (Binder, p. 583).

NERYL ACETATE [141-12-8]

Chemical name	2-*cis*-3,7 Dimethyl-2,6-octadien-1-yl acetate
	$C_{12}H_{20}O_2$
Source	
NATURAL	Helichrysum, neroli and petitgrain oils.
Isolation	Synthesized from nerol and acetic anhydride.
Appearance	Colourless oily liquid.
Odour	Sweet, floral–rose odour.
Physical	b.p. 231 °C, s.g. 0.91.
Uses	As a sweetener in fine perfumery, and in neroli, jasmine and other floral types.

RIFM Monograph (1976) *FCT*, **14**, 625 (Binder, p. 589).

NERYL *iso*-BUTYRATE [2345-24-6]

	$(CH_3)_2CHCOOC_{10}H_{17}$
Appearance	Oily.
Odour	Sweet and of the rose type.
Physical	b.p. 130 °C at 3 mm Hg, s.g. 0.89.
Uses	In perfumery in rose, neroli and fancy bouquets.

RIFM Monograph (1988) *FCT*, **26**, 391, Special Issue VII.

NERYL *n*-BUTYRATE [999-40-6]

	$CH_3CH_2CH_2COOC_{10}H_{17}$
Appearance	Colourless liquid.
Odour	Leafy-floral, fruity orange-like.
Physical	b.p. 240 °C, s.g. 0.89.
Uses	In perfumery as a modifier in lilac, lily, neroli, rose and other floral types.

NERYL FORMATE [2142-94-1]

	$HCOOC_{10}H_{17}$
Appearance	Colourless liquid.
Odour	Penetrating herbaceous-green, orange–rose-type.
Physical	b.p. 225 °C, s.g. 0.92.
Uses	Occasionally in delicate leafy-floral-type fragrances.

RIFM Monograph (1976) *FCT*, **14**, 627 (Binder, p. 590).

NERYL PHENYLACETATE

	$C_6H_5CH_2COOC_{10}H_{17}$
Appearance	Liquid.
Odour	Rich honey-rose-type.
Physical	b.p. 327 °C, s.g. 0.92.
Uses	Occasionally in perfumery as a modifier and fixative to introduce heavy, sweet character to perfumes.

NERYL PROPIONATE [105-91-9]

	$C_2H_5COOC_{10}H_{17}$
Odour	Sweet, fruity and 'jam-like'.
Appearance	Colourless oily liquid.
Physical	b.p. 233 °C.
Uses	Useful in perfumery, particularly in violet compositions to produce new and original shades: also valuable in reproducing the odour of honeysuckle, lily, neroli and jasmine.
	RIFM Monograph (1976) *FCT*, **14**, 629 (Binder, p. 592).

NERYL n-VALERATE [10522-33-5]

Chemical name	Neryl valerianate
	$CH_3CH_2CH_2CH_2COOC_{10}H_{17}$
Appearance	Colourless oily liquid.
Odour	Herbaceous, reminiscent of hops and clary sage.
Physical	b.p. 258 °C.
Uses	As a modifier in heavy floral and oriental type fragrances.

NGAI CAMPHOR [92201-48-4]

History	Used by the natives in religious ceremonies and in the preparation of Indian ink.
Source	
GEOGRAPHIC	A shrubby plant native to India, Tonquin, and the Malaysian Archipelago.
NATURAL	*Blumea balsamifera*, D.C., N.O. Compositae.
Isolation	Distillation.
Chemical	Principal constituents are *l*-borneol and *l*-camphor.

NIAOULI OIL

	May be called 'Gomenol'
Source	
GEOGRAPHIC	A tree found throughout New Caledonia; also occurs in Annam.
NATURAL	*Melaleuca viridiflora*, Brongn et Grio, N.O. Myrtaceae.

Isolation	Distillation of the leaves.
Appearance	Yellow oil.
Chemical	Principal constituent is cineole.
Other	Generally resembles cajuput oil.

NIGELLA OIL [90064-31-6]

Source	
NATURAL	*Nigella damascena*, L., often described as 'black caraway'.
Isolation	Distillation from the crushed seeds.
Yield	0.37%.
Appearance	Yellow oil with blue fluorescence.
Odour	The crushed seeds have a strawberry smell; the oil recalls ambrette seeds.
Chemical	The fluorescence is due to damascine.

NIKKEL OIL [97659-68-2]

Source	
GEOGRAPHIC	A tree indigenous to Japan.
NATURAL	*Cinnamomum laureirii*, Nees, N.O. Lauraceae.
Isolation	Distillation from the leaves and young twigs: an essential oil may be obtained from the stems and roots of the tree.
Appearance	Bright yellow liquid.
Odour	Lemon–cinnamon.
Chemical	Principal constituents are citral, linalol and cineol: the essential oil contains a high proportion of cinnamic aldehyde.

NITROBENZENE [98-95-3]

	Mirbane, nitrobenzol
	$C_6H_5NO_2$
Appearance	Pale yellow mobile liquid.
Odour	Strong, pungent and harsh resembling benzaldehyde and bitter almonds.
Physical	b.p. 211 °C, m.p. 6 °C: marginally soluble in water, freely soluble in alcohol, benzene, ether and oils.
Chemical	Inflammable and poisonous.
Uses	No longer used in fragrances due to its toxic nature.
Comments	Rapidly absorbed through the skin.
	IFRA GUIDELINE
	The Committee recommends that nitrobenzene should not be used as a fragrance ingredient.
	This recommendation is based on the high acute toxicity and the recognized skin toxicity effects of this material (H. E.

Christensen, Toxic Substances List, *Natl Inst. for Occupational Safety and Health*, (1972), p. 369).

June 1974

NITRO-CELLULOSE [9004-70-0]

Chemical name	Cellulose nitrate
CTFA name	Nitrocellulose
	Pyroxylin; collodion cotton; soluble gun cotton; collodion wool; colloxylin; xyloidin; celloidin.
Appearance	A fine, white fibrous substance which resembles cotton in appearance.
Physical	Available in varying degrees of viscosity as it may be damped down by industrial spirit, or iso-propyl alcohol: soluble in methanol or acetone.
Uses	In the preparation of fine manicure enamels.
Comments	Should be stored in air-tight containers: highly flammable.

2,6-NONADIENAL [557-48-2]

Violet leaf aldehyde; cucumber aldehyde
$C_9H_{14}O$

Source	
NATURAL	Violet leaves.
Isolation	Obtained when nonadienol is oxidized.
Appearance	Colourless to pale yellow liquid.
Odour	Intense, powerful vegetable, green character reminiscent of violet leaves and cucumbers on dilution.
Physical	b.p. 187 °C.
Uses	As a trace ingredient in violet and cucumber bases.

RIFM Monograph (1982) *FCT*, **20**, 769, Special Issue VI.

2,6-NONADIENOL [7786-44-9]

Violet leaf alcohol, cucumber alcohol
$C_9H_{16}O$

Source	
NATURAL	Violet leaf oil.
Isolation	Synthesized from hexenyl chloride and acrolein by the Grignard reaction with subsequent allyl rearrangement.
Appearance	Colourless liquid.
Odour	Intense herbaceous, green with a distinct violet–cucumber note.
Physical	b.p. 196 °C.
Uses	In traces in violet, cucumber and narcisse bases.

RIFM Monograph (1982) *FCT*, **20**, 771, Special Issue VI.

gamma-NONALACTONE [104-61-0]

Chemical name	4-Hydroxynonanoic acid lactone
	Nonyl lactone, aldehyde C_{18} (so called), coconut aldehyde; Prunolide
	$C_9H_{16}O_2$
Isolation	Synthesized by cyclization of nonenoic acid or by reacting *n*-hexanol with acrylic acid.
Appearance	Colourless liquid.
Odour	Creamy, sweet, soft, coconut odour.
Uses	Most frequently used lactone in perfumery, finds a wide range of application in most fragrance types.
Comments	This is a lactone and the name aldehyde C_{18} is misleading.
	RIFM Monograph (1975) *FCT*, **13**, 889, Special Issue II (Binder, p. 593).

iso-NONYL ACETATE [58430-94-7]

Chemical name	3,5,5-Trimethyl hexylacetate
	Vanoris, Inonyl acetate
	$C_{11}H_{22}O_2$
Isolation	By acetylation of trimethylhexanol.
Appearance	Colourless liquid.
Odour	Refreshing, light fruity–woody and slightly green character.
Uses	Widely used in lavender, pine, fougere and some floral types for soap, detergent, airfresheners and household products.
	RIFM Monograph (1974) *FCT*, **12**, 1009, Special Issue I (Binder, p. 716).

n-NONYL ACETATE [143-13-5]

Chemical name	*n*-Nonan-1-yl acetate
CTFA name	Nonyl acetate
	Acetic acid, nonyl ester acetate C-9
	$CH_3COO(CH_2)_8CH_3$
Appearance	Colourless liquid.
Odour	Slightly harsh, earthy, fruity–green note.
Physical	b.p. 212 °C, s.g. 0.88.
Uses	Occasionally in perfumery in gardenia, rose, orange-blossom, and lavender compounds.
	Acetate C-9: RIFM Monograph (1973) *FCT*, **11**, 97
	(1973) *FCT*, **11**, 1079 (Binder, p. 24).

n-NONYL ALCOHOL [143-08-8]

Chemical name	*n*-Nonan-1-ol; Alcohol C_9
	Pelargonic alcohol
	$CH_3(CH_2)_7CH_2OH$

250   n-Nonyl aldehyde

Source	
NATURAL	Occurs as an ester in sweet orange oil.
Appearance	Colourless liquid.
Odour	Peculiar oily-floral odour, recalling rose and neroli on dilution.
Physical	b.p. 213 °C, m.p. −5 °C.
Uses	In perfumery in synthetic rose ottos, e.g. rose d'orient: also employed in the preparation of artificial jasmine and neroli oils.

Alcohol C_9: RIFM Monograph (1973) *FCT*, **11**, 103
(1973) *FCT*, **11**, 1079 (Binder, p. 41).

n-NONYL ALDEHYDE [124-19-6]

Chemical name	n-Nonanal, pelargonic aldehyde; Aldehyde C_9 $CH_3(CH_2)_7CHO$
Source	
NATURAL	Oils of rose, tangerine, cinnamon, ginger, lemon and orris.
Isolation	Synthesized from castor oil by fractional distillation which produces undecylenic acid: this is converted into potassium nonylate, nonylic acid, nonyl alcohol, and finally nonyl aldehyde.
Odour	Powerful, diffusive, fatty–floral type odour.
Physical	b.p. 191 °C, m.p. 6 °C. Has a tendency to polymerize on prolonged storage.
Uses	Extensively at low concentrations in perfumery in the preparation of rose, lily, citrus, orange-blossom and many other fragrance types to impart a natural petal-like effect.

Aldehyde C_9: RIFM Monograph (1973) *FCT*, **11**, 115
(1973) *FCT*, **11**, 1080 (Binder, p. 52).

NONYL iso-BUTYRATE [10522-34-6]

$(CH_3)_2CHCOOCH_2(CH_2)_7CH_3$

Appearance	Colourless liquid.
Odour	Delicate, fruity, orange–rosy type.
Physical	s.g. 0.86, b.p. 220 °C.
Uses	As a modifier in oriental compositions.

gamma-n-NONYL-gamma-BUTYROLACTONE [7370-36-7]

Chemical name	gamma-Nonyl lactone Tridecanolide 1:4
Appearance	Colourless oily liquid.
Odour	Weak, fatty and slightly musky.
Uses	Rarely used for tuberose notes.

NONYL PHENOL ETHOXYLATES [9016-45-9]

Isolation	Condensation products of nonyl phenol and ethylene oxide.
Physical	Available as a series of compounds having different ethylene oxide

Uses	Those prepared with up to 9 molecules of ethylene oxide are used as emulsifying agents and as solubilizers for oils and perfumes in bath preparations.

NOOTKATONE [4674–50–4]

$C_{15}H_{22}O$

Source	
NATURAL	Grapefruit juice and oil.
Isolation	From orange oil by the tert-butyl chromate oxidation of valencene.
Appearance	Pale yellow to colourless liquid.
Odour	Intense, citrus, grapefruit-type.
Uses	Occasionally at low levels and in conjunction with d-limonene in citrus colognes grapefruit and other citrus themes.

IFRA GUIDELINE

The Committee recommends that nootkatone used as a fragrance ingredient should be at least 98% pure with a melting point of at least 32 °C. Lower-purity nootkatone, with a minimum purity of 86%, as determined by GLC, should not exceed a level of 1% in a compound and should be used only in conjunction with at least four times the amount of d-limonene.

This recommendation is based on test results of RIFM showing absence of sensitizing potential in a sample of nootkatone of 98% purity and in a mixture of 1 part nootkatone of 86% purity with four parts d-limonene tested at 4% (private communication to IFRA).

The GLC determination of the purity of nootkatone uses the following parameters:
    glass capillary column, SP-2100
    column temperature 75–220 °C, programmed at 4 °C/min
(private communication to IFRA).

October 1980

NOPYL ACETATE [128–51–8]

Chemical name	2-(2'-Acetoxyethyl)-6,6-dimethyl-2-norpinene $CH_3COOC_{11}H_{17}$
Appearance	Colourless liquid.
Odour	Sweet woody–fruity character similar to linalyl acetate.
Physical	b.p. 234 °C.
Uses	As a versatile low-cost alternative to linalyl acetate in floral bouquets, citrus colognes, lavender, fougeres and many other types of perfumes for soaps and household products. RIFM Monograph (1974) *FCT*, **12**, 943, Special Issue I (Binder, p. 595).

NUTMEG OIL [8008-45-5]

CTFA name	Nutmeg oil
	Myristica fragrans oil; Myristica oil; oil of mace
Source	
GEOGRAPHIC	A tree found in the Molucca Islands; is also grown in Java, Sumatra, Penang, and Sri Lanka.
NATURAL	*Myristica fragrans*, van Houtt, N.O. Myristicaceae.
Isolation	The nutmegs are exported and the oil is distilled from the small, damaged seeds: the mace (arillus) yields a concrete oil on treatment with heat and pressure.
Yield	7–15% essential oil and about 20% concrete oil.
Appearance	The essential oil is a yellow liquid.
Odour	Highly aromatic.
Chemical	Mace yields a concrete oil which melts at 45 °C, and contains 12% volatile oil, together with resin and glyceryl myristinate. The following constituents have been identified in both nutmeg and mace oils: eugenol, iso-eugenol, terpineol, borneol, linalol, geraniol, safrole, an aldehyde, terpenes and free acids.
Uses	The expressed oil of nutmeg may be used in perfumes to obtain a spicy odour: it imparts a special characteristic to lavender water. The essential oil is employed as a flavour in dental creams in combination with peppermint, methyl salicylate, and cloves: may also be used as a tobacco flavour, when blended with cassia, cloves, vanilla, coumarin, etc.

RIFM Monograph (1976) *FCT*, **14**, 631 (Binder, p. 596).

NUANUA OIL

Source	
GEOGRAPHIC	A tree or shrub which grows in Samoa.
NATURAL	A species of *Nelitris*, Gartn, of the Rubiaceae family.
Isolation	Steam distillation from the leaves.
Appearance	Greenish-yellow liquid.
Odour	Reminiscent of ambergris.

OAKMOSS RESINOID/OAKMOSS ABSOLUTE [9000-50-4]

Mousse de Chene

History	It is believed that in ancient Egypt oakmoss was an ingredient in breadmaking as it has been discovered in the tombs of the Pharaonic period. Although it does not occur in Egypt, it was probably imported from the Islands of Archipelago.
Source	
GEOGRAPHIC	A lichen found in the mountainous regions of France, Czechoslovakia, Herzegovina, and Italy (Piedmont).
NATURAL	The lichen, *Evernia prunastri*, Ach., and *E. furfuracea*, Ach.,

Isolation	belongs to N.O. Parmeliaceae and is found primarily on oak trees: it also grows on spruce and sometimes on fruit trees.
The moss is pressed into large bales and transported to the perfumery works, where extraction by volatile solvents produces the concrete. The volatile oil may then be extracted with acetone, which produces a resin, wax, and chlorophyll: may also be produced by direct steam distillation of the lichen.
Yield	2–4% resinoid, about half of which may be obtained as absolute.
Appearance	The colour of the lichen varies according to its host, being greenish when obtained from oak, but greyish black from spruce. The absolute is a colourless liquid.
Odour	Generally earthy, woody, and slightly phenolic: however, there are many different types due to the diverse botanical origins, place of origin and collection: it is also further complicated by the fact that most of the constituents of oakmoss react upon the solvent.
Chemical	A phenol ('lichenol') has been isolated from oakmoss: further examination has shown this to be the ethyl ester of everninic acid.
Uses	At low levels in perfumery, it is an excellent fixative for all bouquets: in small quantities used for perfuming face powders in combination with jasmine, tuberose, and orange blossom absolutes: will strengthen and improve soap perfumes but will discolour white soap.
Comments	The following rules should be followed when blending oakmoss:
To obtain freshness: bergamot, lemon, linalol, linalyl acetate, and citronellyl formate may be used;
For lightness: alpha ionone, and hydroxy-citronellal;
To obtain a floral blend: jasmine, rose, cassie, and orange blossom;
To tone: ylang, coriander, sweet orange, tarragon, sassafras, and vanillin;
For stability and depth: patchouli, vetivert, coumarin, and musk.
Oakmoss concrete (*Evernia* spp.): RIFM Monograph (1975) *FCT*, **13**, Special Issue II (Binder, p. 599).
IFRA GUIDELINE
The Committee recommends that Oak moss absolute and resinoid (concrete) obtained from *Evernia prunastri* should not be used as fragrance ingredients at levels over 3% in fragrance compounds.
This recommendation is made in order to promote Good Manufacturing Practice (GMP) for the use of Oak moss absolute and resinoid (concrete) as fragrance ingredients. It is based on the known sensitizing potential of Oak moss and its derivatives reported in the literature and supported by RIFM test data (private communication to IFRA).
September 1988, amended July 1989

OCIMENYL ACETATE

Chemical name	*cis*-Ocimenylacetate
$C_{12}H_{18}O_2$

254   Octalactone

Isolation	Synthesized by the acetylation of ocimenol.
Appearance	Colourless liquid.
Odour	Fresh, herbaceous, slightly citrus-like.
Uses	In artificial lavender and bergamot compositions.
	RIFM Monograph (1976) FCT, **14**, 819, Special Issue III (Binder, p. 602).

OCTALACTONE [104–50–7]

Chemical name	*gamma-iso*-Octalactone
	$C_8H_{14}O_2$
Isolation	Synthesized from pentanol and acrylic acid.
Appearance	Colourless oily liquid.
Odour	Intense odour of coconut–coumarin with a slightly spicy undertone.
Physical	b.p. 223 °C.
Uses	Occasionally used as a modifier for coumarin and as a sweetener for heavy floral and oriental types.
	RIFM Monograph (1976) FCT, **14**, 821.

delta-OCTALACTONE [698–76–0]

	5-hydroxyoctanoic acid lactone; tetrahydro-6-propyl-2H-pyran-2-one; 5-propyl-5-hydroxypentanoic acid lactone; delta-propyl-delta-valerolactone; 2H-pyran-2-one, tetrahydro-6-propyl-
Source	
NATURAL	Found in various fruits and animal products.
Isolation	Lactonization of 5-hydroxyoctanoic acid.

n-OCTYL ACETATE [112–14–1]

Chemical name	*n*-Oct-1-yl acetate
	Acetate C-8
	$C_{10}H_{20}O_2$
Source	
NATURAL	Seeds from wild parsnip.
Isolation	Synthesized from *n*-octanol and acetic acid.
Appearance	Colourless liquid.
Odour	Fruity, slightly fatty, waxy with a green-apple-like characteristic smell.
Physical	b.p. 208 °C.
Uses	Traces are useful in perfumery in synthetic rose, neroli, cassie, and jasmine etc.
	Acetate C-8: RIFM Monograph (1974) FCT, **12**, 815, Special Issue I (Binder, p. 22).

n-OCTYL ALCOHOL [111–87–5]

Chemical name	n-Octan-1-ol
CTFA name	Caprylic alcohol
	1-Octanol; n-octanol; Alcohol C_8
	$CH_3(CH_2)_6CH_2OH$
Source	
NATURAL	Probably in cow-parsnip oil.
Isolation	May be synthesized.
Appearance	Colourless liquid.
Odour	Powerful, fresh, waxy, orange–rose character.
Physical	b.p. 195 °C.
Uses	In perfumery, for modifying floral bouquets: and many single floral types.

Alcohol C_8: RIFM Monograph (1973) *FCT*, **11**, 101
(1973) *FCT*, **11**, 1079 (Binder, p. 39).

n-OCTYL ALDEHYDE [124–13–0]

Chemical name	n-Octan-1-al
	Aldehyde C_8
	$CH_3(CH_2)_6CHO$
Source	
NATURAL	Lemon, rose, and orange flower oils.
Odour	Powerful harsh fatty becoming sweeter and more orange-like on dilution.
Physical	b.p. 170 °C, s.g. 0.80.
Uses	Traces are used in perfumery to impart a special note to jasmine, orange, rose, and citrus types.

Aldehyde C_8: RIFM Monograph (1973) *FCT*, **11**, 113
(1973) *FCT*, **11**, 1080 (Binder, p. 51).

n-OCTYL BENZOATE [94–50–8]

Chemical name	n-Oct-1-yl benzoate
	$C_6H_5COO(CH_2)_7CH_3$
Appearance	Colourless liquid.
Odour	Weak, oily-balsamic character.
Physical	b.p. 305 °C.
Uses	Occasionally in perfumery as a fixative.

n-OCTYL iso-BUTYRATE [109–15–9]

Chemical name	n-Oct-1-yl 2-methylpropanoate
	$(CH_3)_2CHCOO(CH_2)_7CH_3$
Odour	Green-herbaceous slightly reminiscent of galbanum.

n-Octyl n-butyrate

Physical   s.g. 0.87.
Uses   Occasionally in rose, geranium, fougere and chypre compositions. Octyl Isobutyrate: RIFM Monograph (1975) *FCT*, **13**, 893, Special Issue II (Binder, p. 605).

n-OCTYL n-BUTYRATE [110-39-4]

Chemical name n-Oct-1-yl n-butanoate
               $CH_3CH_2CH_2COO(CH_2)_7CH_3$
Source
NATURAL   Rhizome of the male fern.
Appearance   Oil.
Odour   Heavy fruity-green, vegetable character.
Physical   b.p. 244 °C, s.g. 0.86.
Uses   Little used in perfumery, but may be employed as a modifier in chypre, fougere, lavender and new-mown hay types.

n-OCTYL FORMATE [112-32-3]

Chemical name n-Oct-1-yl formate
               $HCOO(CH_2)_7CH_3$
Appearance   Colourless mobile liquid.
Odour   Powerful, fruity-green note.
Physical   b.p. 200 °C, s.g. 0.880.
Uses   In traces in perfumery as a modifier in orris, rose and citrus notes. Octyl Formate: RIFM Monograph (1979) *FCT*, **17**, 883, Special Issue V.

OCTYL METHYL KETONE [693-54-9]

Chemical name n-Decan-2-one
               Methyl octyl ketone; heptyl acetone
               $CH_3(CH_2)_7COCH_3$
Appearance   Colourless oily liquid.
Odour   Powerful orange-citrus and slightly floral.
Physical   b.p. 209 °C, m.p. 14 °C.
Uses   In combination with oakmoss, heliotropin, and vanillin, to enhance these notes.

OCTYL PROPIONATE [142-60-9]

Chemical name n-Oct-1-yl propanoate
               $C_2H_5COO(CH_2)_7CH_3$
Source
NATURAL   Distilled oil from the dried fruits of *Pastinaca satica*, L.
Appearance   Colourless liquid.
Odour   Oily–waxy and fruity herbaceous odour.

Physical	b.p. 226 °C, s.g. 0.87.
Uses	In perfumery to obtain a novel effect in carnation, jonquille, and floral–herbaceous types.

OCTYL iso-VALERATE [7786-58-5]

Chemical name	n-Oct-1-yl 3-methylbutanoate
	Octyl valerianate
	$(CH_3)_2CHCH_2COO(CH_2)_7CH_3$
Appearance	Colourless liquid.
Odour	Intense fatty–fruity herbaceous odour.
Physical	b.p. 245 °C.
Uses	Mainly in flavouring essences.

OLEA FRAGRANS, See OSMANTHUS CONCRETE/ OSMANTHUS ABSOLUTE

OLEIC ACID [112-80-1]

Chemical name	9-Octadecenoic acid
CTFA name	Oleic acid
	$C_{17}H_{33}COOH$
Isolation	Saponification of oleins, such as tallow or lard: the soap is then decomposed with a mineral acid: the solid stearic and palmitic acids are removed by cooling and expression.
Appearance	Colourless to yellow, syrupy liquid.
Odour	Slight odour of tallow.
Physical	Solidifies at approx. 4 °C to a crystalline mass: practically insoluble in water, soluble in alcohol, chloroform, benzene or ether.
Chemical	Oxidizes in air.
Other	Has skin-penetrating properties.
Uses	In cosmetics in small proportions to blend with the oil phase of cosmetic emulsions.

OLEYL ALCOHOL [143-28-2]

Chemical name	(Z)-9-Octadecen-1-ol
CTFA name	Oleyl alcohol
	cis-9-Octadecen-1-ol
	$C_{18}H_{35}OH$
Appearance	Colourless, viscous or oily liquid.
Odour	Faint, fatty odour.
Physical	m.p. 13–19 °C, has a bland taste: soluble in ether and alcohol, insoluble in water.
Chemical	Several grades are available, differing in degree of chemical purity, including the proportion of other fatty alcohols present, such as stearyl and palmityl alcohol.

Uses — In cosmetics, especially lipsticks, as an emollient, emulsion stabilizer, superfatting agent, or pigment dispersing material: may also be employed to improve the texture of creams and lotions.

OLIBANUM RESINOID/OLIBANUM OIL [8050-07-5]

Frankincense, gum thus (describes the oleo-resin obtained from the trunks of *Pinus palustris*, N.O. Coniferae, of Central America)

Source
GEOGRAPHIC — A tree which grows wild in Somalia, Ethiopia and Saudi Arabia.
NATURAL — *Boswellia carterii*, Birdw., and other species of N.O. Burseraceae.
Isolation — The oleo-resin is obtained by incisions in the inner bark of the tree: it is extracted and the resinoid obtained by steam distillation: an alcoholic extract (1 in 2) may be prepared by reducing the gum resin to a powder before maceration. The Grasse perfume manufacturers use the volatile solvent process.
Yield — 50% resinoid and 5–10% volatile oil.
Appearance — Colourless to pale yellow ovate tears: may be purchased as a colourless liquid from any of the Grasse manufacturers.
Odour — Highly balsamic and spicy.
Chemical — About 60% of the gum resin is soluble in 90% alcohol; the remaining portion consists mainly of calcium and magnesium arabinate, with some bassorin.
Uses — Important in perfumery as a fixative: used in face powders, perfumes, and heavy bouquets: also in the preparation of incense and fumigating mixtures.
Olibanum Absolute: RIFM Monograph (1978) *FCT*, **16**, 835, Special Issue IV
Olibanum Eum: RIFM Monograph (1978) *FCT*, **16**, 837, Special Issue IV.

OLIVIL [2955-23-9]

Chemical name — Tetrahydro-3-hydroxy-2-(4-hydroxy-3-methoxyphenyl)-4-((4-hydroxy-3-methoxyphenyl)methyl)-3-furan-methanol
Source
NATURAL — Obtained from the resin from wild olive.
Isolation — The crystalline constituent of the resin is obtained.
Appearance — The monohydrate is in the form of crystals which form a thick vapour on heating: soluble in hot water, alcohol, acetic acid, or fatty oils.
Odour — Resembles vanillin.
Uses — Used to fumigate sickrooms in Italy.

OPOPONAX RESINOID/OPOPONAX OIL [8021-36-1]

Bisabol, perfumed bdellium, sweet myrrh; in Somaliland, as habbak hadi (the gum of the hadi tree)

Orange oil, bitter 259

Source	
GEOGRAPHIC	A plant which grows wild in Somalia and Ethiopia.
NATURAL	*Commiphora erythroea*, var. *glabrescens*, Engler, N.O. Burseraceae.
Isolation	The oleo-resin is formed when the bark of the tree is damaged: the resinoid is prepared using the volatile solvent process: the essential oil is obtained by distillation.
Yield	Approximately 20% resinoid and 5–10% oil.
Appearance	Appears as yellowish-brown lumps on the bark of the tree: a pale yellow essential oil.
Odour	Balsamic and reminiscent of olibanum.
Chemical	Consists of gum, resin, and essential oil.
Uses	The resinoid is occasionally used in perfumery as a fixative. The oil is employed in many different perfume types.

IFRA GUIDELINE

The Committee recommends only the use of opoponax preparations which are obtained from opoponax gums (the exudation from the bark of *Commiphora erythraea*, var. *glabrescens*, Engler) by extraction with suitable solvents or steam distillation.

These recommendations are based on test results of RIFM with samples of resinoids, concretes, absolutes and oils of opoponax of defined botanical origin obtained by steam distillation or solvent extraction using methanol, ethanol or benzene (private communication to IFRA).

March 1978

ORANGE FLOWER CONCRETE/ORANGE FLOWER ABSOLUTE
[8030-28-2]

Source	
GEOGRAPHIC	A tree cultivated in France, Spain, Algeria, and Morocco.
NATURAL	The bitter orange tree, *Citrus aurantium*, and *Citrus bigaradia*.
Isolation	The concrete is obtained by extraction of the flowers. The absolute is then obtained by dissolving the concrete in alcohol, 'freezing out' the waxy materials, and filtration.
Yield	0.3% concrete, from which approximately 30% absolute is obtained.
Odour	Warm, slightly spicy, bitter orange-type.
Uses	In perfumery in expensive colognes and fine floral fragrances.

RIFM Monograph (1982) *FCT*, **20**, 785, Special Issue VI.

ORANGE OIL, BITTER
[68916-04-1]

Source	
GEOGRAPHIC	A tree cultivated in Morocco, Egypt, Sicily, Spain and France.
NATURAL	The ripe fruit of *Citrus aurantium*, var. *amara*, L., also known as the bitter orange tree.
Isolation	Cold-pressed from the peel either by hand or by machines in a similar way to that for lemon oil: contains 2–5% citric acid, which can be extracted.

Yield	0.1–0.4%.
Appearance	Orange oil.
Odour	Bright, bitter odour of oranges.
Chemical	Principal constituent is limonene, with small quantities of linalol, terpineol, citral, citronellal, methyl anthranilate, nonyl alcohol, and decyl aldehyde.
Uses	Occasionally in perfumery to add a freshness.

RIFM Monograph (1974) *FCT*, **12**, 735.

IFRA GUIDELINE

The Committee recommends for applications on areas of skin exposed to sunshine, excluding bath preparations, soaps and other products which are washed off the skin, to limit bitter orange oil to 7.0% in the compound (see remark on phototoxic ingredients in the introduction).

This recommendation is based on results of RIFM on the phototoxicity of bitter orange oil (*Food Cosmet. Toxicol.*, **12**, 735 (1974)), its low bergapten content (C. K. Shu *et al.* VI. Int. Congress of Essential Oils 1974) and the observed no-effect level of pooled samples in tests using the hairless mouse and the pig (private communication to IFRA).

ORANGE OIL, SWEET [8008-57-9]

Source	
GEOGRAPHIC	A tree cultivated on large plantations in Brazil, South Africa, and America: also grown in most of the Mediterranean countries.
NATURAL	The sweet orange tree, *C. aurantium*, var. *dulcis*, L., *C. sinensis* (L.) *Osbeck* very similar to the bitter orange tree.
Isolation	Obtained by cold-pressing the fresh peels, either by hand or by machine.
Yield	0.3–0.5%.
Appearance	Orange oil.
Odour	Sweet, fruity odour of oranges.
Chemical	Contains limonene, small quantities of linalol, terpineol, citral, citronellal, methyl anthranilate, nonyl alcohol, and decyl aldehyde.
Uses	In perfumery, especially in creations of the citrus-type: may be used in the manufacture of eau-de-cologne, replacing neroli oil in part: the terpeneless oil, in combination with terpeneless petitgrain oil, makes an excellent base for eau-de-colognes.
Comments	Sweet orange oil is more important than bitter orange oil. Orange Oil Expressed: RIFM Monograph (1974) *FCT*, **12**, 733 (Binder, p. 607).

ORIGANUM OIL [8007-11-2]

GEOGRAPHIC	Many different species of a herb which grows wild in the Middle East and Spain, but is cultivated in France and the Balkans.

NATURAL	The following species have been identified:
	Algerian — *Origanum floribundum*, Mumby, and *Origanum cinereum*, de Noe;
	Cretan — *Origanum onites*, L.O. maru, L.;
	Cyprian — *Origanum majoranoides*, Willd., *Origanum bevani*, Holmes;
	Smyrnian — *Origanum smyrnoem*, L.;
	Triestian — *Origanum birtum*, Lk.;
	Europe generally — *Origanum vulgare*, L. (or wild marjoram).
Isolation	Steam distillation of the dried herb.
Yield	1–3% (the fresh herb gives a lower yield).
Appearance	Varies, being either a white to yellowish, or red liquid.
Odour	Similar to that of thyme, being intensely phenolic and spicy.
Chemical	Composition varies, but the following constituents have been identified: carvacrol, thymol, linalol, cedrol, cymene, pinene, and origanene.
Uses	In perfumery, especially in spicy, natural-type compounds.
	RIFM Monograph (1974) *FCT*, **12**, 945, Special Issue I (Binder, p. 611).

ORRIS OIL [8002-73-1]

CTFA name	Orris root extract
	Orris root oil; extract of florentine iris.
	Iris oil
Source	
GEOGRAPHIC	A plant cultivated mainly in and around Florence, but also in other parts of Europe, and Morocco: the most renowned district for the cultivation of iris is the Florentine hills: the largest proportion of the commercially produced oil comes from Grasse, either from rhizomes which are grown there, or from those imported from Tuscany.
NATURAL	The rhizomes of three species of iris; *I. germanica*, *I. pallida*, and *I. florentina*, N.O. Iridaceae.
Isolation	The rhizomes are cleaned, decorticated, and artificially warmed and dried, and are then exported to Milan, Grasse, London, Germany etc. The oil occurs as both concrete and liquid. The former is obtained by steam distillation of the powdered rhizomes. The liquid is obtained by distilling orris roots with cedarwood oil. Orris resin may be obtained from the rhizomes by extraction with a volatile solvent such as acetone, petroleum ether or benzole.
Yield	0.1–0.2% oil.
Appearance	Solid fat at normal temperatures.
Odour	Oily, woody, and violet-like.
Physical	m.p. 45 °C.
Chemical	The concrete oil contains 85% odourless myristic acid, 15% irone (the violet-like ketone), and traces of the esters of myristic and

	oleic acids, furfural, benzoic, nonylic, and decyclic aldehydes. The presence of the following saturated aliphatic acids has been shown: octanoic, nonanoic, decanoic, undecanoic, dodecanoic, and tridecanoic.
Uses	Orris oil is an important, though expensive, perfume material which may be used for blending with ionone as a base for violet compositions: also employed in other floral ottos, or in opoponax, lavender sophisticated bouquet perfumes: orris resin is an excellent fixative for perfumes, and high-class soaps.
	Orris absolute: RIFM Monograph (1975) *FCT*, **13**, 895, Special Issue II (Binder, p. 612).

OSMANTHUS CONCRETE/OSMANTHUS ABSOLUTE [68917-05-5]

Source	Olea fragrans, Loureiro, Thumb, *Osmanthus fragrans*
GEOGRAPHIC	An evergreen tree native to Japan, China, and India.
NATURAL	*Osmanthus fragrans*, N.O. Oleaceae, grows to a height of 20 ft and is cultivated for its flowers called 'Kwei Hwa' or 'Mo Hsi'.
Isolation	The concrete may be obtained by extraction from the flowers; further extraction yields the absolute.
Yield	0.1–0.2% concrete, of which some 70% is obtained as absolute.
Odour	Both products possess a distinct, exotic, floral odour reminiscent of plums and raisins.
Uses	In perfumery in modern compositions, or to flavour tea and confectionary.

PALMAROSA OIL [8014-19-5]

	Turkish geranium oil; Indian grass oil; rusa oil
History	Palmarosa oil entered Europe via Turkey for many years, hence its other name.
Source	
GEOGRAPHIC	A grass ('Rusa grass') grown in India, Brazil, and Central America.
NATURAL	*Cymbopogon martini*, Stapf., *motia*.
Isolation	The grass is harvested just before it flowers and the oil obtained by steam distillation of the fresh or dried grass in copper or iron cauldrons: the grass is trampled down in the cauldrons which contain about 1 ft (30 cm) of water; the natives build these into stone fireplaces on the banks of streams, principally in the Bombay Presidency and the central provinces.
Yield	Approx. 1.5%.
Appearance	Colourless to pale yellow liquid. An orange oil may be obtained from *Cymbopogon intermedius*, Stapf.
Odour	Sweet and floral recalling rose–geranium: the oil from *Cymbopogon intermedius*, Stapf., has an odour recalling vetivert.
Chemical	Contains 75–90% of the finest commercial geraniol, also geranyl acetate and caproate, dipentene, methyl heptenone, and farnesol.

Uses	An excellent raw material in soap perfumery, being used extensively in many perfumes: is comparatively cheap, and blends well with clove, cassia, and patchouli oils, terpineol, benzyl acetate, coumarin, and musk: it gives an excellent result when blended with Algerian geranium oil. At one time it was used as an adulterant of Bulgarian rose otto, being prepared by shaking with gum-arabic solution, followed by subsequent exposure to the sun, which lightened the colour of the oil.
Comments	The closely related *Cymbopogon martini*, Stapf., *sofia* is used for producing ginger-grass oil, which is less highly esteemed. RIFM Monograph (1974) *FCT*, **12**, 947, Special Issue I (Binder, p. 614).

PALMITIC ACID [57–10–3]

Chemical name	Hexadecanoic acid
CTFA name	Palmitic acid
	Hexadecylic acid; cetylic acid
	$CH_3(CH_2)_{14}COOH$
Source	
NATURAL	Tallow: occurs in association with other fatty acids, particularly palm oil.
Isolation	Obtained from tallow and sold commercially as stearic acid. It is, however, a eutectic mixture which contains an almost equal weight of palmitic acid: a purer product may be obtained by vacuum distillation.
Appearance	White crystalline substance.
Physical	m.p. 63 °C. The degree of crystallinity may be standardized by a setting point curve: insoluble in water, sparingly soluble in cold alcohol or petroleum ether, freely soluble in hot organic solvents.
Uses	In cosmetics with stearic acid, particularly in the preparation of foundation creams and shaving creams.

PANDANUS OIL

Source	
GEOGRAPHIC	A plant native to India, Arabia and Persia.
NATURAL	The flowers of *Pandanus odoratissimus*, L., N. O. Pandanaceae, which retain their fragrance when dry.
Isolation	Distillation with sandalwood.
Appearance	Oil.
Odour	Pleasant and honey-like.

PARAFFIN LIQUID [8012–95–1]

CTFA name	Mineral oil
	Liquid paraffin, white mineral oil
Isolation	Obtained from petroleum by distillation.

Appearance	Colourless, clear, oily liquid.
Odour	None.
Physical	b.p. 370 °C.
Chemical	Sold according to its specific gravity and viscosity.
Uses	In medicine as a lubricant and mild aperient: in the preparation of cold cream, cleansing creams and other emulsified preparations: imparts the gloss in lipsticks, and may also be used in suntan oils, and brilliantines: it may be mixed with vegetable oils and fatty acid esters to use as the oil phase of cosmetic emulsified products.
Comments	When used in cosmetic preparations it should comply with the sulphuric acid limit test for carbonizable substances, and the limit test for sulphur compounds, given in the British Pharmacopoeia.

PARAFFIN MOLLE OR SOFT PARAFFIN [8012–95–1]

Source	
CHEMICAL	Obtained from the petroleum fraction which distils between 360 °C and 390 °C, m.p. 38–56 °C: a mixture of various paraffin hydrocarbons.
Isolation	Obtained from the residue after crude petroleum has been distilled and fractionated.
Appearance	Semi-solid yellow or white substance.
Uses	Widely employed in pharmacy as an ointment base: used in cosmetics in the preparation of toilet and massage creams.

PARAFFIN WAX [8002–74–2]

CTFA name	Paraffin
	Hard Paraffin; pyroparaffin; protoparaffin
	C_nH_{2n+2}
Source	
NATURAL	Crude tarry oil.
Isolation	Pyroparaffin is prepared from the crude tarry oil obtained by the destructive distillation of shale. Protoparaffin is obtained from the oil from Pennyslvanian oilfields.
Appearance	Colourless, semi-transparent body.
Odour	None.
Physical	Pyroparaffin solidifies at 43–50 °C: protoparaffin solidifies at 48–57 °C: hard paraffin is insoluble in water or alcohol but soluble in benzene, chloroform or ether.
Uses	In the preparation of toilet articles.

PARSLEY SEED OIL, PARSLEY HERB OIL [8000–68–8]

CTFA name	Parsley seed oil
	Parsley oil herb

Source	
GEOGRAPHIC	A plant native to Asia, but cultivated in France and the Balkan countries.
NATURAL	The parsley plant, *Petroselinum sativum*, N. O. Umbelliferae.
Isolation	Steam distillation of the ripe seeds or leaves.
Yield	Approx. 7% from the seeds and 0.02–3% from the leaves.
Appearance	Colourless or yellow rather viscid liquid.
Odour	Herbaceous and spicy, typical of the fresh plant.
Physical	Very slightly soluble in water.
Chemical	Chief constituents are apiol, terpene and possibly l-pinene.
Uses	In perfumery as a modifier.
	RIFM Monograph (1975) *FCT*, **13**, 897, Special Issue II (Binder, p. 615) (Parsley herb oil) (1983) *FCT*, **21**, 871 (Parsley seed oil).

PARSNIP OIL [90082–39–6]

Source	
GEOGRAPHIC	A wild plant which may also be cultivated in Eastern Europe, Central America, and South East Asia.
NATURAL	The parsnip, *Pastinaca sativa*, N. O. Umbelliferae.
Isolation	Steam distillation of the fruit, flowers and roots.
Yield	Approximately 2–3% depending upon source of the plant material.
Odour	The oil from the fruits contains octyl butyrate, which gives it a characteristic odour. The oil from the umbels has an odour reminiscent of ambrette seeds: that from the roots, which contains vanillin, is reminiscent of vetivert.
Uses	Occasionally in spicy herbal-type perfumes.

PATCHOULI OIL [8014–09–3]

Source	
GEOGRAPHIC	A small shrub cultivated in India, the Philippines, Java, Sumatra, and Singapore.
NATURAL	*Pogostemon patchouli*, Pellet, and *P. cablin*, Benth.
Isolation	Oil content is at its highest in the first three pairs of leaves: it is, therefore, suggested that the first cutting should take place as soon as five pairs of leaves have developed. The cut leaves are dried in the sun, allowed to ferment slightly, and the oil is obtained by steam distillation: possible to obtain four different grades of oil by varying the distillation process, known as 'ordinary', 'medium', 'special', and 'extra-special', the last two being a close approximation to the oils distilled in Europe. In England and France, very fine oils may be produced from imported leaves. Patchouli may also be extracted with volatile solvents, resulting in a good, cheap resinoid.
Yield	Up to 3% depending upon quality of the leaves.
Appearance	Yellowish to greenish viscid liquid.

Odour	Powerful, woody, and balsamic.
Physical	m.p. 56 °C. Patchouli camphor separates as odourless crystals from very old oil: practically insoluble in water, soluble in ether.
Chemical	The following constituents have been identified: 'patchouli alcohol', cinnamic aldehyde, benzaldehyde, eugenol, an alcohol and a ketone.
Uses	An important material in perfumery, being employed as a fixative in amber, chypre, and fougere-type perfumes: used in toilet powder and soap perfumes, blending well with geranium, palmarosa, clove, and cassia oils.
Comments	In Singapore, patchouli leaves are known locally as 'tilam wangi' and 'dhalum wangi': they may be subject to adulteration with wild patchouli: patchouli oil may be adulterated with cubeb, and cedar oils.

Patchouli oil: RIFM Monograph (1982) *FCT*, **20**, 791, Special Issue VI.

PEARL ESSENCE OR NATURAL PEARL ESSENCE [73-40-5]

CTFA name	Guanine Crystalline fish guanine $C_5H_5N_5O$
Source	
NATURAL	A silvery substance obtained from the scales or body of several types of fish.
Isolation	Obtained from a suspension of the organic nacreous material derived from the fish: used in dispersion in a suitable vehicle, being prepared for cosmetics with different characteristics. Formulators may obtain special effects by blending different grades together.
Appearance	Leafy, brilliant pearlescent, crystalline particles.
Physical	The smaller the particles, the more brilliant the pearlescent effect, as a result of the simultaneous reflection of light from several layers. An increase in the particle size gives greater opacity and covering power: insoluble in water and most organic solvents.
Chemical	A natural purine. The particles are destroyed by alkalis.
Other	The crystalline structure may be destroyed by milling so gentle stirring to incorporate the natural pearl essences into cosmetic preparations is required.
Uses	As a dispersion in cellulose nitrate lacquer for nail lacquers, in castor oil or mineral oil for use in lipsticks, eye colours, and other make-up stick preparations: may be dispersed in water, alcohol, or propylene glycol for use in products which are non-oily and non-fatty.

PEAU D'ESPAGNE

Isolation	A piece of leather material, such as tanned kid or chamois, is immersed in a specially formulated perfume mixture for about 10

	days, removed, drained, and spread on a sheet of glass: after drying, a second mixture is brushed on, and the leather is pressed under a weight and allowed to dry.
Appearance	Liquid.
Odour	Variable, depending upon the materials used in the mixture.
Uses	Used to treat leather which is made up into fancy goods, producing so-called Spanish leather, or perfumed skin: sold encased in fancy silk, or satin envelopes, and placed in contact with gloves, writing paper, etc.
Comments	A specially formulated perfume mixture containing rectified oil of birch tar, sandalwood oil, bergamot oil, petitgrain oil, lavender oil, and clary sage oil: may be varied by the use of patchouli, labdanum, and oakmoss.

PECTINS [9000-69-5]

CTFA name	Pectin Citrus pectin
Source	
NATURAL	Ripe fruits and vegetables.
Isolation	Derived from pectose.
Physical	Have a molecular weight ranging from 20 000 to 400 000: readily soluble in water and are good emulsifiers in essential oils: may be diluted with under 70% strength alcohol, forming a gel.
Chemical	More stable than many mucilages: possess all the characteristics of colloids. However, they should only be used with acid or neutral substances as they are decomposed by alkaline carbonates.
Uses	Widely employed in the manufacture of jam and other foodstuffs, and are beginning to be used in cosmetics.

PENNYROYAL OIL [8007-44-1]

Source	
GEOGRAPHIC	A flowering herb cultivated in Japan, Spain, Morocco, and Tunisia.
NATURAL	*Mentha pulegium* and other related varieties.
Isolation	Steam distillation of the fresh, slightly dried herb.
Yield	1.5–2%.
Appearance	Oil.
Odour	Warm, fresh, and minty.
Uses	In perfumery in the reproduction of certain essential oils. Pennyroyal oil, Eurafrican: RIFM Monograph (1974) *FCT*, **12**, 949, Special Issue I (Binder, p. 617).

PEONOL [4136-21-4]

Chemical name	2-Hydroxy-4-methoxyacetophenone Paeonol; peony ketone $CH_3COC_6H_3(OCH_3)OH$

Source	
NATURAL	*Paeonia montana*.
Appearance	Colourless or white needles.
Odour	Rather pungent, warm, aromatic and remotely hay-like.
Physical	m.p. 50 °C.
Uses	In new-mown hay and certain heavy floral types.

PEPPER OIL [8006–82–4]

Source	
GEOGRAPHIC	The pepper vine, native to Southern India, but now cultivated in Indonesia, South East Asia and Brazil: the principal commercial centre is Singapore.
NATURAL	*Piper nigrum*, L., N. O. Piperaceae.
Isolation	The vines are grown up trees: the unripened berries are removed and the oil obtained by steam distillation.
Yield	Approx. 2%.
Appearance	Greenish-yellow volatile oil.
Odour	Intense, spicy, and peppery, reminiscent of cubebs.
Chemical	Contains phellandrene, dipentene, and probably caryophyllene.
Uses	In perfumery in masculine compositions.

Black pepper oil: RIFM Monograph (1978) *FCT*, **16**, 651, Special Issue IV.

PEPPERMINT OIL ARVENSIS [68917–18–0]

Source	Mint oil or cornmint oil
GEOGRAPHIC	A plant native to Japan and China, but also cultivated in Brazil, Paraguay, Korea, and Taiwan: in Japan, the main centres for the industry are Hokkaido, Okayama, and Hiroshima.
NATURAL	Two varieties of the herb, *Mentha arvensis*, N. O. Labiatae, known as Akamaru and Aomaru.
Isolation	The herb is collected in September, tied in bundles, and hung up to dry for about 3 weeks: the essential oil is obtained by steam distillation of the dried herb; the crude distillate may be used to produce menthol, which is crystallized out in metal containers placed in a freezing mixture.
Yield	Each acre yields approximately 1.25 tons (1.27 t) of dried herb, from which the yield of oil is 1–2% depending upon quality of the herb. Plants which are 2–3 years old yield the highest percentage: the yield of menthol is about 45%.
Appearance	Oil.
Odour	Powerful, slightly sweet, and minty.
Chemical	Principal constituents are menthol, menthone, menthyl acetate, and menthyl iso-valerate.
Uses	Little used in perfumery; is possibly employed in soap perfumery,

and to impart cooling properties to toilet waters: used as a flavouring agent in medicine and in dental preparations.
Cornmint oil: RIFM Monograph (1975) *FCT*, **13**, 771, Special Issue II (Binder, p. 258).

PEPPERMINT OIL PIPERITA [8006-90-4]

CTFA name	Peppermint oil
	Mentha piperita oil
Source	
GEOGRAPHIC	An English herb, now cultivated throughout the world, mainly in America, Spain, France, Italy, and Germany: however, the premier oil is obtained from the English product.
NATURAL	The herb, known as English Mitcham, or *Mentha piperita*, N. O. Labiatae, is cultivated in two varieties; black mint, with a purplish stem, and white mint, with a greenish stem. Large quantities of American peppermint oil are rectified in England.
Isolation	The crop is harvested at the end of August (just before the buds open) and partially dried; the oil is obtained by direct steam distillation, which takes 4–6 h.
Yield	0.1–1% depending upon source of the herb and production procedure.
Appearance	Oil.
Odour	Powerful, fresh, and minty.
Chemical	Principal constituents of all peppermint oils are menthol, menthone, methyl acetate, and methyl iso-valerate: however, different hydrocarbons are found in oils from different sources.
Uses	Widely used as a flavouring agent in a variety of consumable articles.
Comments	The intrinsic value lies in the odour and flavour which make the English Mitcham oil the most expensive.

PERILLA OIL [90082-61-4]

Source	
GEOGRAPHIC	A plant cultivated in Japan, China, and India.
NATURAL	*Perilla frutescens*.
Isolation	Steam distillation of the flowers and leaves.
Yield	0.1–0.15%.
Odour	Powerful and oily, reminiscent of cumin.
Uses	In traces in perfume compositions.

RIFM Monograph (1988) *FCT*, **26**, 397, Special Issue VII.

PERSIC OIL [8023-98-1]

CTFA name	Peach kernel oil
	Oleum amygdalae persic; expressed peach oil

Source	
NATURAL	Peach and apricot kernels.
Isolation	Expression of the kernels.
Appearance	Light yellow liquid.
Physical	Soluble in chloroform or ether, but only slightly soluble in alcohol.
Uses	In cosmetics as a basis for skin creams, and glycerine and cucumber creams.

PERU BALSAM OIL [8007–00–9]

CTFA name	Balsam Peru
	Peruvian balsam; Indian balsam; China oil; black balsam; Honduras balsam; Surinam balsam
Source	
GEOGRAPHIC	A tree which grows in vast forests in Central America.
NATURAL	The oleo-resinous secretions from the trunk of *Myroxylon pereirae*, Klotzsch, N. O. Leguminosae, which attains a height of about 50 ft (15 m).
Isolation	Indians work the trees for the balsam in November or December, after the last rains. Three exudations are collected, the first two on rags, the third from the wounded bark, which is first removed and boiled in water: the commercial product is a mixture of all three in definite proportions. The oil is obtained by steam distillation of the balsam.
Yield	Approx. 50%.
Appearance	The balsam is an oleo-resinous exudate. The oil may be obtained in dilution with an equal volume of alcohol, as a water-white viscous oil.
Odour	Balsamic, sweet, and vanilla-like.
Physical	Insoluble in water, partly soluble in ether, soluble in alcohol or chloroform.
Chemical	After clarification, the balsam resembles black–brown treacle; it contains 30% Peruresinotannol, and the benzoic acid and cinnamic esters of benzyl alcohol, esters of benzoic and cinnamic acids, free cinnamic acid, coumarin, vanillin, and nerolidol.
Other	Has soothing qualities.
Uses	No longer used as a fragrance ingredient due to its skin sensitizing properties and has now been replaced by synthetic alternatives.
Comments	The balsam sometimes causes irritation to the skin.
	RIFM Monograph (1974) *FCT*, **12**, 953, Special Issue I (Binder, p. 620)
	IFRA GUIDELINE
	The Committee recommends that Peru balsam (the exudation from *Myroxylon pereirae* (Royle) Klotzsch) should not be used as a fragrance ingredient. Only preparations produced by methods which give products not showing a potential for sensitization (e.g. distillation) should be used.

This recommendation is based on test results of RIFM on the sensitizing potential of Peru balsam (D. L. Opdyke (1974), *Food Cosmet. Toxicol.*, **12**, 951) and absence of sensitizing reactions in tests with several samples of Peru balsam oil (D. L. Opdyke (1974), *Food Cosmet. Toxicol.*, **12**, 953), as well with compounds containing Peru balsam oil (private communication to IFRA).

PETITGRAIN CITRONNIER OIL [8014–17–3]

Lemon petitgrain

Source
NATURAL   The lemon tree, *Citrus medica*, L., var. *beta-limonum*, Hooker filius, N. O. Rutaceae.
Isolation   Distillation from the leaves and twigs.
Odour   Distinct and pleasant.
Chemical   Contains 24% citral together with camphene, limonene, linalol, geraniol, and esters.
Uses   In cologne-type perfumes.
Comments   Similar products may be obtained from the lime and mandarin. Lemon Petitgrain Oil: RIFM Monograph (1982) *FCT*, **16**, 807, Special Issue VI, Lemon Petitgrain oil: RIFM Monograph (1978) *FCT*, **16**, 807.

PETITGRAIN MANDARIN OIL [8014–17–3]

Mandarin petitgrain

Source
GEOGRAPHIC   A tree cultivated in Algeria, Spain, France, and Sicily.
NATURAL   The mandarin orange tree, *Citrus mandurensis*, Laur.
Isolation   Distillation of the leaves and twigs, the nitrogenous constituents of the oil increasing as the distillation proceeds: this causes the oil to become heavier and sink to the bottom of the receiver.
Odour   Distinctive, and predominated by methyl methyl anthranilate and a suggestion of fine thyme.
Chemical   Stable and persistent: the following constituents have been identified: traces of free acids, phenols, pyrrol, and other non-saponifiable bases; 1% methyl anthranilate, 56% methyl methyl anthranilate, 42% pinene, camphene, dipentene, limonene, p-cymene, 2% terpenic alcohols (one-third esters), and traces of sesquiterpenes.
Uses   In both perfumes and soap perfumery.

PETITGRAIN OIL [8014–17–3]

Source
GEOGRAPHIC   A tree cultivated in the south of France, Morocco, Sicily, Algeria, southern Spain, and the forests of Paraguay.

NATURAL	The bitter orange tree, *Citrus aurantium*, var. *amara*, L., N. O. Rutaceae.
Isolation	Steam distillation of the leaves, twigs, and unripe fruit: the finest oil comes from Europe: in the south of France, the leaves only are used producing a fine oil.
Yield	0.5–1%: in France, 500 kg leaves are required to produce about 1 kg essence.
Odour	The terpeneless oil produced in France has an exceptionally fine odour and is sometimes sold as synthetic neroli. The oil from Paraguay has a sweet, pleasant bouquet.
Chemical	The following constituents have been identified: nerol, geraniol, geranyl acetate, linalol, linalyl acetate, terpineol, dipentene, limonene, pinene, camphene, furfural, pyrrol and possibly methyl anthranilate.
Uses	An important component in perfumery, being the basis of cologne compounds for soaps: is also used in perfumes for skin creams.
Comments	The most common adulterants are Paraguay oil, linalol, and linalyl acetate.

Petitgrain Paraguay Oil: RIFM Monograph (1982) *FCT*, **20**, 801, Special Issue VI.

alpha-PHELLANDRENE [99–83–2]

Chemical name	*p*-Mentha-1,5-diene
	2-Methyl-5-(1-methylethyl)-1,3-cyclohexadiene; 5-isopropyl-2-methyl-1,3-cyclohexadiene
	$C_{10}H_{16}$
Source	
NATURAL	Oil of *Eucalyptus dives*.
Isolation	Distillation and fractionation of the residues left in the separation of piperitone from the oil of *Eucalyptus dives*.
Appearance	Colourless mobile oil.
Odour	Fresh, slightly resinous, citrus character.
Physical	b.p. 175 °C, s.g. 0.843: there are *d*- and *l*-forms.
Uses	In perfumery for artificial citrus, pepper and geranium bases.
Comments	Can be irritating, and is absorbed through the skin.

RIFM Monograph (1978) *FCT*, **16**, 843.

PHENOL [108–95–2]

Chemical name	Carbolic acid
CTFA name	Phenol
	Liquid phenol; phenic acid; phenylic acid; phenyl hydroxide; hydroxybenzene; oxybenzene
	C_6H_5OH
Source	
NATURAL	Coal tar.

Isolation	Fractional distillation of coal tar, with subsequent purification.
Appearance	Detached, colourless crystals or a white crystalline mass.
Odour	Characteristic.
Physical	m.p. 42 °C: partially soluble in water, very soluble in alcohol, chloroform, ether, glycerol and volatile and fixed oils.
Comments	Poisonous and caustic: it should be kept well closed and away from light.

PHENYLACETALDEHYDE [122-78-1]

Chemical name	Benzeneacetaldehyde Hyacinth aldehyde; phenylacetic aldehyde; alpha-toluic aldehyde; alpha-tolualdehyde; hyacinthin C_8H_8O
Source	
NATURAL	Narcissus absolute, neroli, and rose oils.
Isolation	Synthesized by isomerizing styrene oxide.
Appearance	Viscous, colourless liquid.
Odour	Intense pungent, green, floral, hyacinth type.
Physical	b.p. 206 °C: polymerization may be prevented with the addition of 5% ethyl alcohol when freshly made: slightly soluble in water, soluble in alcohol or ether.
Uses	In perfumery in many types of floral fragrances and as the base of most hyacinths: and in the preparation of special types of rose compounds: only small quantities are employed due to its intense floral odour and its skin sensitizing properties.

Phenyl acetaldehyde: RIFM Monograph (1976) FCT, **14**, 827, Special Issue III.

IFRA GUIDELINE

The Committee recommends the use of phenylacetaldehyde as a fragrance ingredient in conjunction with substances preventing sensitization, as for example equal weights of phenylethylalcohol or dipropyleneglycol.

This recommendation is based on results showing sensitizing potential for the individual ingredient, but absence of sensitizing reaction in a number of compounds (private communication to IFRA), as well as in mixtures with equal weights of phenylethylalcohol or dipropyleneglycol (D. L. J. Opdyke (1976), Food Cosmet. Toxicol., **14**, 197, D. L. J. Opdyke (1979), Food Cosmet. Toxicol., **17**, 377–80).

October 1975

PHENYLACETALDEHYDE DIMETHYLACETAL [101-48-4]

Chemical name	1,1-Dimethoxy-2-phenyl-ethane Viridine; Vetde lilas; PADMA $C_{10}H_{14}O_2$

Isolation	Produced from phenylacetaldehyde and methanol.
Appearance	Colourless liquid.
Odour	Intense, green-leafy odour.
Physical	b.p. 221 °C, s.g. 1.01.
Uses	Widely used in perfumery in floral compositions, being invaluable in lilac perfumes, and certain types of roses and lilies, exhibits excellent stability in hostile bases, strong alkalis, bleaches etc. RIFM Monograph (1975) *FCT*, **13**, 899, Special Issue II (Binder, p. 624).

PHENYLACETIC ACID [103-82-2]

Chemical name	Benzeneacetic acid alpha-Toluic acid $C_8H_8O_2$
Source	
NATURAL	Neroli oil.
Isolation	Obtained by treating benzyl cyanide with dilute sulphuric acid.
Appearance	White crystalline compound.
Odour	Sweet animalic-honey-like odour which becomes more honey-like on extreme dilution.
Physical	b.p. 266 °C, m.p. 77 °C. Slightly soluble in water and miscible with alcohol, chloroform or ether.
Uses	In small quantities in floral notes such as lilac, jasmine, acacia, trefle, neroli, and rose: forms the basis of many honey compounds. Also considered to be a very good fixative. RIFM Monograph (1975) *FCT*, **13**, 901, Special Issue II (Binder, p. 627).

PHENYLAMYL ALCOHOL [10521-91-2]

Chemical name	5-Phenylpentan-1-ol $C_6H_5(CH_2)_4CH_2OH$
Appearance	Colourless oil.
Odour	Light, citrus, sweet-herbaceous note.
Physical	b.p. 244 °C, s.g. 0.97.
Uses	In verbena and citrus cologne types.

PHENYL BENZOATE [93-99-2]

Chemical name	Benzoic acid phenyl ester $C_6H_5COOC_6H_5$
Appearance	Colourless crystalline substance.
Odour	Faint, balsamic, metallic green and geranium-like.
Physical	b.p. 314°C, m.p. 70 °C, s.g. 1.23. Insoluble in water, freely soluble in hot alcohol.
Uses	Rarely used due to unstable nature which produces hazardous phenols.

PHENYLBUTYL ALCOHOL [3360-41-6]

Chemical name	4-Phenylbutan-1-ol
	$C_6H_5(CH_2)_3CH_2OH$
Appearance	Colourless liquid.
Odour	Pleasant, rosaceous fragrance.
Physical	b.p. 238 °C.
Uses	In mimosa, cassia, rose and oriental types.

p-PHENYLENEDIAMINE [106-50-3]

Chemical name	1,4-Benzenediamine
CTFA name	p-Phenylenediamine
	Paraphenylenediamine; CI 76060; 1,4-phenylene-diamine
	$C_6H_4(NH_2)_2$
Isolation	Reduction of aminobenzene with tin and hydrochloric acid, forming aniline and p-phenylenediamine.
Appearance	The pure compound occurs as colourless crystals, which become slightly reddish on exposure.
Physical	m.p. 147 °C: slightly soluble in water, soluble in alcohol, chloroform or ether.
Chemical	Its use depends upon its oxidation to Bandrowski base, usually effected with hydrogen peroxide.
Uses	A well-known and successful hair-dye.
Comments	Should be stored in amber glass vessels and in coloured glass bottles when made into solution.

PHENYLETHYL ACETATE [103-45-7]

Chemical name	2-Phenylethyl acetate, beta-phenylethyl acetate
	$C_{10}H_{12}O_2$
Source	
NATURAL	Pandanus oil.
Isolation	Synthesized from phenylethyl alcohol and acetic acid.
Appearance	Colourless liquid.
Odour	Sweet fruity rose–honey note, suggestive of green rose leaves.
Physical	b.p. 232 °C, s.g. 1.05.
Uses	In perfumery in white rose and gardenia types.
	RIFM Monograph (1974) FCT, 12, 957, Special Issue I (Binder, p. 628).

PHENYLETHYL ALCOHOL [60-12-8]

Chemical name	2-Phenylethanol
CTFA name	Phenethyl alcohol
	beta-Phenylethyl alcohol; benzyl carbinol; beta-hydroxyethylbenzene
	$C_8H_{10}O$

Source	
NATURAL	The absolutes of rose and orange blossom, and in the oils of neroli, champaca, geranium, and rose.
Isolation	Synthesized from benzene and ethylene oxide by the Friedel–Crafts reaction.
Appearance	Colourless liquid.
Odour	Mild, 'honey–rose' odour.
Physical	b.p. 220 °C, s.g. 1.03: Partially soluble in water, miscible with alcohol or ether.
Uses	Indispensable in perfumery in rose compositions, and also employed in orange-blossom, neroli, jasmine and many other sweet floral perfumes.

RIFM Monograph (1975) *FCT*, **13**, 903, Special Issue II (Binder, p. 629).

PHENYLETHYL *iso*-AMYL ETHER [56011–02–0]

	iso-Amyl phenylethyl ether
Appearance	Colourless liquid.
Odour	Powerful rose–hyacinth-type.
Physical	s.g. 0.90: Soluble in 12 vols. 80% alcohol.
Uses	In perfume compounds for soaps and household products.

RIFM Monograph (1983) *FCT*, **21**, 873.

PHENYLETHYL ANTHRANILATE [133–18–6]

Chemical name	2-Phenylethyl *o*-aminobenzoate
	$NH_2C_6H_4COOCH_2CH_2C_6H_5$
Appearance	White crystalline solid.
Odour	Almost odourless when pure but imparts a sweet-floral character to fragrances.
Physical	b.p. 324 °C, m.p. 42 °C.
Uses	In perfumery in gardenia, honeysuckle and violet compounds, imparting a sweet persistence: odour blends well with methyl ionone and orris.

RIFM Monograph (1976) *FCT*, **14**, 831, Special Issue III (Binder, p. 631).

PHENYLETHYL BENZOATE [94–47–3]

Chemical name	2-Phenylethyl benzoate
	$C_6H_5COOCH_2CH_2C_6H_5$
Appearance	Colourless oily liquid.
Odour	Faint, floral balsamic
Physical	b.p. approx. 300 °C, s.g. 1.10.
Uses	In perfumery as a fixative in rose, carnation, neroli and oriental compositions.

RIFM Monograph (1975) *FCT*, **13**, 905, Special Issue II (Binder, p. 632).

PHENYLETHYL *iso*-BUTYRATE [103–48–0]

Chemical name	2-Phenylethyl 2-methyl propanoate
	$(CH_3)_2CHCOOCH_2CH_2C_6H_5$
Appearance	Colourless liquid.
Odour	Fresh, fruity, tea-rose-type.
Physical	b.p. 230°C, s.g. 1.00.
Chemical	More stable than the normal butyrate.
Uses	In perfumery in rose and floral compounds.
	RIFM Monograph (1978) *FCT*, **16**, 847, Special Issue IV.

PHENYLETHYL *n*-BUTYRATE [103–52–6]

Chemical name	2-Phenylethyl *n*-butanoate
	$CH_3CH_2CH_2COOCH_2CH_2C_6H_5$
Appearance	Colourless liquid.
Odour	Natural rosy floral-fruity note.
Physical	b.p. 238°C, s.g. 1.00.
Uses	In a wide range of floral compositions.
	RIFM Monograph (1979) *FCT*, **17**, 889, Special Issue V.

PHENYLETHYL CINNAMATE [103–53–7]

Chemical name	2-Phenylethyl 3-phenylacrylate
	$C_6H_5CH{=}CHCOOCH_2CH_2C_6H_5$
Appearance	Crystalline solid.
Odour	Faint, sweet balsamic odour, resembling purified Tolu balsam.
Physical	b.p. above 300°C, m.p. 58°C.
Uses	As a fixative in a wide range of floral compositions.
	RIFM Monograph (1978) *FCT*, **16**, 845, Special Issue IV.

PHENYLETHYL DIMETHYL CARBINOL [103–05–9]

Chemical name	2-Methyl-4-phenylbutan-2-ol
	Dimethyl phenylethyl carbinol
	$C_6H_5CH_2CH_2C(CH_3)_2OH$
Appearance	Colourless viscous liquid.
Odour	Soft, floral-green and lily-like.
Physical	b.p. 238°C.
Uses	As a blender in jasmine, rose, muguet and magnolia types.
	Dimethyl Phenylethyl Carbinol: RIFM Monograph (1974) *FCT*, **12**, 537 (Binder, p. 332).

PHENYLETHYL DIMETHYL CARBINYL ACETATE [103–07–1]

Chemical name	2-Methyl-4-phenylbutan-2-yl acetate
	Dimethyl phenylethyl carbinyl acetate
	$CH_3COOC(CH_3)_2CH_2CH_2C_6H_5$
Appearance	Colourless liquid.
Odour	Rose fragrance.
Physical	b.p. 244 °C.
Uses	Widely used to intensify delicate floral notes.
	Dimethyl Phenylethyl Carbinyl Acetate: RIFM Monograph (1978) *FCT*, **16**, 721, Special Issue IV.

PHENYLETHYL FORMATE [104–62–1]

Chemical name	Benzyl carbinyl formate
	$HCOOCH_2C_6H_5$
Appearance	Liquid.
Odour	Powerful, green herbaceous, rosy note.
Physical	b.p. 226°C, s.g. 1.03.
Uses	As a modifier in white rose muguet, lilac and hyacinth fragrances.
	RIFM Monograph (1974) *FCT*, **12**, 959, Special Issue I (Binder, p. 634)

PHENYLETHYL METHYL ETHYL CARBINOL [10415–87–9]

Chemical name	3-Methyl-5-phenylpentan-3-ol
	$C_6H_5CH_2CH_2C(C_2H_5)OH$
Appearance	Colourless liquid.
Odour	Mild, warm floral, rosy-muguet type.
Physical	b.p. 254 °C.
Uses	As a modifier for rose, muguet and lilac types.
	RIFM Monograph (1979) *FCT*, **17**, 891, Special Issue V.

PHENYLETHYL PHENYLACETATE [102–20–5]

Chemical name	2-Phenylethyl phenylacetate
	$C_6H_5CH_2COOCH_2CH_2C_6H_5$
Appearance	Crystalline solid.
Odour	Tenacious heavy, sweet balsamic odour.
Physical	b.p. 325 °C, m.p. 28 °C.
Uses	Extensively in perfumery in floral, balsamic, sweet-woody and oriental types.
	RIFM Monograph (1975) *FCT*, **13**, 907, Special Issue II (Binder, p. 637).

PHENYLETHYL PROPIONATE [122–70–3]

Chemical name	2-Phenylethyl propanoate
	$C_2H_5COOCH_2CH_2C_6H_5$

Appearance	Colourless liquid.
Odour	Herbaceous-rosy, deep-fruity note.
Physical	b.p. 244 °C, s.g. 1.02.
Uses	A modifier in floral and herbaceous types.
	RIFM Monograph (1974) *FCT*, **12**, 963, Special Issue I (Binder, p. 638).

PHENYLETHYL SALICYLATE [87–22–9]

Chemical name	2-Phenylethyl o-hydroxybenzoate
	$HOC_6H_4COOCH_2CH_2C_6H_5$
Appearance	Crystalline solid.
Odour	Balsamic rose–hyacinth-type.
Physical	b.p. above 300 °C, m.p. 44 °C.
Uses	As a fixative for bouquet perfumes.
	RIFM Monograph (1978) *FCT*, **16**, 849, Special Issue IV.

PHENYLETHYL iso-THIOCYANATE [2257–09–2]

Chemical name	2-Phenylethyl iso-thiocyanate
Source	
NATURAL	Oils of mustard, mignonette, and nasturtium.
Appearance	Colourless or pale yellow liquid.
Odour	Pungent, earthy–sweet, horseradish-like.
Physical	b.p. 256 °C.
Uses	Important note in reseda.

PHENYLETHYL n-VALERATE [7460–74–4]

Chemical name	2-Phenylethyl *n*-pentanoate
	Phenylethyl valerianate
	$CH_3(CH_2)_3COOCH_2CH_2C_6H_5$
Appearance	Colourless liquid.
Odour	Fruity rose, herbaceous, tobacco-like note.
Physical	b.p. 268 °C.
Uses	In jasmine, narcisse and tobacco types.

PHENYLETHYLIDENE ACETONE [10521–97–8]

Chemical name	5-Phenylpent-3-en-2-one
	$C_6H_5CH_2CH=CHCOCH_3$
Isolation	Condensation of phenylacetaldehyde and acetone.
Appearance	Oily liquid.
Odour	Powerful floral-green sweet-pea-type.
Physical	254 °C, s.g. 1.014.
Uses	In sweet-pea, honeysuckle and green floral types.

PHENYLGLYCOL BUTYRATE

Chemical name	2-Hydroxy-2-phenylethyl n-butyrate
	Phenylglycol monobutyrate
	$CH_3(CH_2)_2COOCH_2CH(OH)C_6H_5$
Appearance	Colourless liquid.
Odour	Fruity, lily–hyacinth-type.
Physical	b.p. 231 °C.

PHENYLGLYCOL DIACETATE [6270-03-7]

Chemical name	Phenylethylene glycol diacetate
	$C_6H_5CH(OCOCH_3)CH_2OCOCH_3$
Appearance	Colourless viscous oil.
Odour	Gardenia, muguet, and hyacinth.
Physical	b.p. 274 °C, s.g. 1.03.
Uses	In perfume compounds of the muguet, jasmine, hyacinth and gardenia types.

PHENYLGLYCOL PROPIONATE [25496-78-0]

Chemical name	2-Hydroxy-2-phenylethyl propanoate
	Phenylglycol monopropionate
	$C_2H_5COOCH_2CH(OH)C_6H_5$
Appearance	Colourless liquid.
Odour	Mild, ethereal-fruity, green-floral note.
Physical	b.p. 223 °C.
Uses	In perfumery, narcissus, gardenia and hyacinth type compounds.

PHENYLHEPTYL ALCOHOL [3208-25-1]

Chemical name	7-Phenylheptan-1-ol
	$C_6H_5(CH_2)_6CH_2OH$
Appearance	Colourless oil.
Odour	Pleasant; mild, oily-rose type.
Physical	b.p. 261 °C.
Uses	As a modifier for rose compositions.

PHENYLHEXYL ALCOHOL [2430-16-2]

Chemical name	6-Phenylhexan-1-ol
	$C_6H_5(CH_2)_5CH_2OH$
Appearance	Colourless liquid.
Odour	Rosy-lime, citrus-green note.
Physical	b.p. 258 °C.
Uses	Occasionally in rose and citrus bases.

PHENYL NAPHTHYL KETONE [644-13-3]

Chemical name	2-Naphthyl phenyl ketone
	2-Benzoyl-naphthalene
	$C_6H_5COC_{10}H_{17}$
Appearance	Crystalline solid.
Odour	Weak, tenacious, floral-amber type.
Physical	b.p. above 300 °C, m.p. 82 °C.
Uses	In perfumery in neroli, jasmine, and sweet-pea compounds.

PHENYLPROPYL ACETATE [122-72-5]

Chemical name	3-Phenylpropan-1-yl acetate
	Hydrocinnamyl acetate
	$CH_3COOCH_2CH_2CH_2C_6H_5$
Source	
NATURAL	Probably in cassia oil.
Appearance	Colourless liquid.
Odour	Sweet, floral-fruity and mildly balsamic odour.
Physical	b.p. 244 °C, s.g. 1.02.
Other	Resembles the alcohol.
Uses	In perfumery in reseda, lilac, rose, hyacinth and muguet types. RIFM Monograph (1974) *FCT*, **12**, 965, Special Issue I (Binder, p. 640).

3-PHENYLPROPYL ALCOHOL [122-97-4]

Chemical name	3-Phenylpropan-1-ol
	Hydrocinnamyl alcohol; hydrocinnamic alcohol; phenylethyl carbinol
	$C_6H_5CH_2CH_2CH_2OH$
Source	
NATURAL	Styrax and white Peru balsam.
Isolation	Prepared by the reduction of cinnamic alcohol.
Appearance	Colourless oily liquid.
Odour	Sweet, balsamic-floral.
Physical	b.p. 235 °C.
Uses	In perfumery in the preparation of fancy bouquets of the lilac–hyacinth type: a useful basis for mignonette and may be used in conjunction with phenylethyl phenylacetate for jonquille perfumes. RIFM Monograph (1979) *FCT*, **17**, 893, Special Issue V.

PHENYL PROPYL ALDEHYDE [104-53-0]

Chemical name	3-Phenyl propanal
	Benzylacetaldehyde; hydrocinnamic aldehyde
	$C_6H_5CH_2CH_2CHO$

Appearance	Colourless liquid.
Odour	Floral hyacinth–lilac, balsamic green type.
Physical	b.p. 222 °C, s.g. 1.02.
Uses	In the preparation of hyacinth, rose, lilac, heliotrope, sweet pea and in fougere and spice notes. RIFM Monograph (1974) *FCT*, **12**, 967, Special Issue I (Binder, p. 641).

PHENYLPROPYL *iso*-BUTYRATE [103–58–2]

Chemical name	3-Phenylprop-1-yl 2-methylpropanoate $(CH_3)_2CH_2COO(CH_2)_3C_6H_5$
Appearance	Colourless liquid.
Odour	Tenacious fruity-balsamic, sweet character.
Physical	b.p. 252 °C, s.g. 0.99.
Uses	Occasionally in jasmine compositions. RIFM Monograph (1978) *FCT*, **16**, 851, Special Issue IV.

PHENYLPROPYL *n*-BUTYRATE [7402–29–1]

Chemical name	3-Phenylprop-1-yl *n*-butyrate $COO(CH_2)_3C_6H_5$
Appearance	Colourless liquid.
Odour	Heavy, sweet-fruity, resembling plums.
Physical	b.p. 258 °C.
Uses	In jasmine, narcisse and oriental compositions.

PHENYLPROPYL CINNAMATE [122–68–9]

Chemical name	3-Phenylprop-1-yl 3-phenylacrylate $C_6H_5CH{=}CHCOO(CH_2)_3C_6H_5$
Source	
NATURAL	Storax.
Odour	Tenacious sweet balsamic odour with a slightly fruity-rose note.
Physical	b.p. over 300 °C, s.g. 1.08.
Uses	A fixative in a similar way to phenylethyl cinnamate. RIFM Monograph (1974) *FCT*, **12**, 969, Special Issue I (Binder, 2 Propenoic Acid, 3-Phenyl-, 3-Phenyl Propyl Ester: p. 642).

PHENYLPROPYL FORMATE [104–64–3]

Chemical name	3-Phenylprop-1-yl formate $HCOO(CH_2)_3C_6H_5$
Appearance	Colourless liquid.
Odour	Powerful, sweet, floral-herbaceous odour.
Physical	b.p. 238 °C.
Uses	In perfumery in narcissus–hyacinth compounds.

RIFM Monograph (1976) *FCT*, **14**, 835, Special Issue III (Binder, p. 643).

PHENYLPROPYL n-VALERATE

Chemical name	3-Phenylprop-1-yl *n*-pentanoate
	Phenylpropyl valerianate
	$CH_3(CH_2)_3COO(CH_2)_3C_6H_5$
Appearance	Colourless liquid.
Odour	Heavy, fruity, slight odour of strawberries.
Physical	b.p. 285 °C.
Uses	Occasionally in the reproduction of red rose notes.

PHENYL SALICYLATE [118-55-8]

Chemical name	2-Hydroxybenzoic acid phenyl ester
	Phenyl *o*-hydroxybenzoate; salol
	$HOC_6H_4COOC_6H_5$
Appearance	Crystalline solid.
Odour	Mild, sweet, fruity-balsamic notes.
Physical	b.p. 282 °C, m.p. 43 °C: almost insoluble in water, soluble in acetone, chloroform, ether or oils.
Uses	Rarely in perfumery, more commonly as a pharmaceutical, chemical and mild bactericide.

RIFM Monograph (1976) *FCT*, **14**, 837, Special Issue III.

PHENYL p-TOLYL ETHER [1706-12-3]

Chemical name	Phenyl cresyl oxide
	para-Cresol phenylether
	$C_6H_5OC_6H_4CH_3$
Appearance	Colourless liquid.
Odour	Powerful, heavy floral odour reminiscent of narcissus, rose and orange-blossom.
Physical	b.p. 278 °C, s.g. 1.06.
Uses	In perfumes for cosmetics and soaps of the above types.

PIMENTO OIL [8016-45-3]

Allspice Oil

Source	
GEOGRAPHIC	An evergreen tree native to the West Indies, but also found in Central America, India, and Réunion.
NATURAL	*Pimenta officinalis*, Lindley, N. O. Myrtaceae.

Isolation	The unripe berries are collected and dried in the sun. The oil is obtained by steam distillation of the unripe fruit, or from the leaves.
Yield	3.3–4.5% from the berries and 0.3–2.9% from the leaves.
Appearance	Yellowish-brown liquid.
Odour	Fragrant and balsamic resembling clove, cubeb, and nutmeg.
Chemical	Contains eugenol, methyl eugenol, caryophyllene, cineole, phellandrene, palmitic acid, and probably some terpene alcohols.
Uses	In perfumery to modify the odour of carnation oils, and may be employed in the soap industry.

Pimenta Berry Oil: RIFM Monograph (1979) *FCT*, **17**, 381
Pimento Leaf Oil: RIFM Monograph (1974) *FCT*, **12**, 971.

PINENE [80–56–8]

Chemical name	2-Pinene
	alpha-Pinene
	$C_{10}H_{16}$
Source	
NATURAL	Many essential oils.
Isolation	Obtained from turpentine by distillation.
Appearance	Colourless liquid.
Odour	Typical pine-type resinous, refreshing.
Physical	Practically insoluble in water, soluble in alcohol, chloroform or ether, b.p. 157 °C.
Chemical	Occurs as both alpha- and beta-pinene: both optically active and exist in both *l*-, *d*- and racemic forms.

Alpha-Pinene: RIFM Monograph (1978) *FCT*, **16**, 853, Special Issue IV. Beta-Pinene: RIFM Monograph (1978) *FCT*, **16**, 859, Special Issue IV.

PINE NEEDLE OIL [8002–09–3]

CTFA name	Pine needle extract
Source	
GEOGRAPHIC	A tree which grows principally in Russia, and all over Europe.
NATURAL	Several species of N. O. Coniferae.
Isolation	Steam distillation of the needles, twigs, and cones.
Yield	0.1–0.5%.
Appearance	Liquid.
Odour	Fresh, typical pine-type.
Chemical	Contains a high percentage of bornyl acetate.
Uses	In perfumes for bath preparations, and household products.

Pinus pumilio Oil: RIFM Monograph (1976) *FCT*, **14**, 843, Special Issue III (Binder, p 650).

PINE OILS [8002–09–3]

CTFA name	Pine oil
	Yarmor

Source	
GEOGRAPHIC	A tree cultivated in the USA, Finland, France, Portugal, and the USSR.
NATURAL	*Pinus palustris*, *P. ponderosa* and other species.
Isolation	Steam distillation of wood chips of the heartwood and stumpwood.
Yield	3–5%.
Appearance	There are two qualities; yellow pine oil and water white oil.
Odour	Fresh, harsh, pine-type.
Physical	Insoluble in water, soluble in the usual organic solvents.
Chemical	Mainly isomeric tertiary and secondary cyclic terpene alcohols.
Uses	In low cost fragrances for household and detergent products.
Comments	Irritating to the skin or mucous membranes.

RIFM Monograph (Yarmor Pine Oil) (1983) *FCT*, **21**, 875.

IFRA GUIDELINE

Oils derived from the Pinacea family

The Committee recommends that essential oils and isolates derived from the Pinacea family (e.g. Pinus and Abies genera) should only be used when the level of peroxides is kept to the lowest practicable level, for instance by adding antioxidants at the time of production. Such products should have a peroxide value of less than 10 millimoles peroxide per litre, determined according to the EOA method.

This recommendation is based on the published literature, mentioning sensitizing properties when containing peroxides (*Food Cosmet. Toxicol.*, **11**, 1053 (1973); **16**, 843 (1978); **16**, 853 (1978)).

May 1976, last amendment October 1979

POLYETHYLENE GLYCOL(S) [25322-68-3]

Macrogol
General formula $H(OCH_2CH_2)_nOH$

Source	
CHEMICAL	Liquid or solid polymers.
Isolation	Polymerization of ethylene oxide in the presence of traces of water.
Appearance	PEG 300 and PEG 400 are clear, colourless liquids: viscosity increases with increasing molecular weight; thus, PEG 1000 is a white, waxy solid and PEG 1540 is a polymer wax with a similar consistency to beeswax.
Odour	None.
Physical	PEG 1000, m.p. 35–40 °C, PEG 1540, m.p. 42–46 °C: soluble in water (the solid polyethylene glycols form clear solutions): not sticky or greasy.
Chemical	The number after each abbreviation indicates the molecular weight.
Other	Properties vary with their molecular weight.
Uses	In cosmetic creams as humectants or lubricants and in hair lotions and creams, shaving creams, skin creams and lotions.

POLYETHYLENE GLYCOL ESTERS

Isolation	Esterification of polyethylene glycols, of selected molecular weight, with fatty acids.
Appearance	Change from liquids to solids as the chain length of the fatty acid increases.
Chemical	The solid esters are non-reactive to electrolytes.
Other	Properties vary with the molecular weight of the polyglycol and chain length.
Uses	The monooleates, monolaurates, and monostearates of polyethylene glycols with molecular weight 400 are the most useful in cosmetic and toilet preparations. The liquid esters are used in perfumery as solubilizers and may be employed as foam stabilizers and viscosity builders in liquid preparations. The solid esters are used in hair dressings and conditioners, low viscosity creams and lotions.

POLYETHYLENE GLYCOL ETHERS

Isolation	Condensation products of ethylene oxide with a high molecular weight fatty alcohol.
Physical	Those marketed under the name *Volpo* are soluble in alcohols, glycols, and most chlorinated and aromatic solvents: lower members of the series are soluble in mineral oil and other non-polar oils, insoluble in water: higher members with 10, 15 and 20 mol. ethylene oxides are generally insoluble in mineral oil, soluble in water.
Chemical	Cetomacrogol 1000 is prepared from cetyl alcohol and contains 20–24 ethylene oxide groups. Condensates of lauryl, cetyl, stearyl, and oleyl alcohols each have a specific number of oxyethylene groups and are marketed under the name *Brij*.
Other	Properties vary depending upon the fatty alcohol used.
Uses	Variable, but generally used as emulsifying agents for cosmetic creams and lotions, as solubilizing agents, and in the formulation of clear gels with mineral oils, fatty acid esters, and other oily minerals.

POLYOXYETHYLENE SORBITAN FATTY ACID ESTERS

	Trade name *Tweens*: when mixed with non-ethoxylated fatty acid esters of sorbitol they are known as *Spans*.
Isolation	Esterification of condensates of ethylene oxide and sorbitol, generally with one or three molecules of a fatty acid.
Appearance	Varies (according to the grade) from a thin liquid to a viscous oil.
Physical	Generally soluble in water depending upon amount of ethylene oxide present.
Uses	Solubilizers and general purpose emulsifiers.

POLYVINYL PYRROLIDONE (PVP) [9003-39-8]

Chemical name	1-Ethenyl-2-pyrrolidone polymers
CTFA name	PVP
	Povidone
	$(C_6H_9NO)_n$
Source	
CHEMICAL	A film-forming linear polymer of 1-vinyl-2-pyrrolidone monomers.
Isolation	Available as a range of products.
Appearance	White or faintly yellow free-flowing powder resembling albumen.
Physical	Soluble in water, alcohol, glycols, and a wide range of organic solvents.
Chemical	Average molecular weight of the polymer used in the cosmetic industry is 40 000.
Uses	In the formulation of hair sprays, hair setting lotions and gels: also as a thickening and dispersing agent in eye liners, and as a protective colloid in make-up preparations and skin creams.

POTASSIUM HYDROXIDE [1310-58-3]

Chemical name	Potassium hydroxide
CTFA name	Potassium hydroxide
	Caustic potash; potassa
	KOH
Isolation	Electrolytic decomposition of potassium chloride.
Appearance	Sticks or lumps which contain about 80–85% of KOH.
Physical	Dissolves in water with an exothermic reaction.
Chemical	Rapidly absorbs moisture and CO_2 from the air.
Uses	A saponifying agent: in cosmetics in the preparation of vanishing creams.
Comments	Extremely corrosive.

PROFARNESOL

	$C_{14}H_{26}O$
Isolation	Synthesized from methyl heptenone.
Appearance	Colourless to pale yellow liquid.
Odour	Delicate floral note.

iso-PROPANOLAMINE [78-96-6]

Chemical name	Monoisopropanolamine
	$CH_3CHOHCH_2NH_2$
Appearance	Viscous liquid.
Odour	Slightly ammoniacal.
Physical	s.g. 0.96: water soluble.
Uses	In cosmetic preparations as an emulsifying agent and generally to prepare liquid emulsions.

PROPENYL HEPTANOATE [142-19-8]

Allyl heptanoate, allyl heptylate
$C_{10}H_{18}O_2$

Isolation	By direct esterification of allyl alcohol with heptanoic acid.
Appearance	Colourless liquid.
Odour	Sweet, fruity, slightly pungent banana-like.
Physical	b.p. approx. 210 °C.
Uses	As a topnote ingredient in fruit blossom types such as apple blossom and peach blossom. Also blends well in lavender and herbaceous fragrance types.

RIFM Monograph (Allyl heptylate) (1977) *FCT*, **15**, 619.

p-iso-PROPYLACETOPHENONE

$CH_3COC_6H_4CH(CH_3)$
Acetocumene; cumyl methyl ketone

Appearance	Colourless oil.
Odour	Powerful, dry-woody. Reminiscent of orris.
Physical	b.p. 252 °C, s.g. 0.97.
Uses	Rarely in cassie and mimosa bases.

iso-PROPYL ALCOHOL [67-63-0]

Chemical name	Propan-2-ol
CTFA name	Isopropyl alcohol
	2-Propanol; isopropanol; dimethyl carbinol; petrohol.
	$CH_3CHOHCH_3$
Source	
CHEMICAL	A by-product of the petroleum and natural gas industry in America. Obtained from acetone in Britain.
Isolation	In America, made by absorbing olefin gases, containing propylene, in sulphuric acid and hydrolysing the resulting alkyl sulphuric acids: the crude alcohol is distilled and purified by chemical treatment and rectification.
Appearance	Colourless liquid.
Odour	Sour, ethereal, acetone-like.
Physical	b.p. 82 °C: miscible with water, alcohol, ether or chloroform: may be salted out on saturating the solution with sodium chloride.
Other	More toxic than ethyl alcohol.
Uses	In cosmetic and toilet preparations such as perfumes, hair washes, dentifrices, and toilet waters: as a solvent for essential oils, and as a gum-resin solvent in preparing resin-odours from olibanum, galbanum, and ginger.

iso-PROPYL BENZYL CARBINOL [705-58-8]

Chemical name	3-Methyl-1-phenylbutan-2-ol
	$C_6H_5CH_2CH(OH)CH(CH_3)_2$

Appearance	Colourless liquid.
Odour	Fresh earthy-green, foliage-like.
Physical	b.p. 238 °C.
Uses	Commonly used in sweet pea, gardenia-like reseda and lilac types.

n-PROPYL FUROATE [615–10–1]

Chemical name	n-Propyl furan-2-carboxylate
	$C_4H_3OCOOC_3H_7$
Appearance	Colourless oily liquid.
Odour	Oily, herbaceous-earthy note.
Physical	b.p. 211 °C.
Uses	In perfumery as a fixative but mainly as a flavour ingredient.

PROPYL GALLATE [121–79–9]

Chemical name	Propyl 3:4:5-trihydroxy benzoate
CTFA name	Propyl gallate
	Benzoic acid, 3,4,5-trihydroxy, propyl ester; 3,4,5-trihydroxy-benzoic acid, propyl ester; progallin
	$C_{10}H_{12}O_5$
Appearance	White crystalline solid.
Odour	None: slightly bitter taste.
Physical	Soluble 1 part by weight in 1000 of water, and 1 part in 3 parts of propylene glycol: also soluble in fats and oils, more readily so using heat and agitation.
Chemical	A phenolic compound.
Uses	As an antioxidant in cosmetic formulations to inhibit the auto-oxidation of fats and fixed oils, preventing rancidity and instability.
Comments	Should be protected from exposure to air, heat, and light: discolours in the presence of iron.

iso-PROPYL LINOLEATE [22882–95–7]

CTFA name	Isopropyl linoleate
	$C_{21}H_{38}O_2$
Isolation	Obtained from an essential fatty acid.
Appearance	Low viscosity liquid.
Physical	s.g. 0.867.
Uses	In the formulation of cosmetic preparations for dry and rough skin conditions: also in lipsticks.
Comments	Should be used with an antioxidant to prevent rancidity.

iso-PROPYL MYRISTATE [110–27–0]

CTFA name	Isopropyl myristate
	1-Methylethyl tetradecanoate
	$C_{17}H_{34}O_2$

Appearance	Colourless, mobile liquid.
Odour	Virtually odourless.
Physical	Insoluble in water and glycerine: soluble 1 to 3 vols. 90% alcohol.
Other	iso-Propyl palmitate and iso-propyl stearate have similar properties.
Uses	In cosmetic creams and lotions, especially as an emollient in face creams and hand lotions and in perfumery as a diluent for some low-cost fragrances.

RIFM Monograph (1976) FCT, 14, 323.

p-iso-PROPYLPHENYLACETALDEHYDE [4395–92–0]

Cortexal
$(CH_3)_2CHC_6H_4CH_2CHO$

Appearance	Colourless liquid.
Odour	Powerful, green, woody, sap-like character. Reminiscent of the peeled cortex of fresh branches.
Physical	b.p. 243 °C.
Uses	In modern fragrance types in combination with citrus oils, aldehydes, oakmoss.

p-isopropyl Phenylacetaldehyde: RIFM Monograph (1979) FCT, 17, 533.

iso-PROPYLQUINOLINE [135–79–5]

Chemical name	6-iso-Propylquinoline
	$CH(CH_3)_2C_9H_6N$
Appearance	Pale amber-coloured liquid.
Odour	Intense, diffusive, earthy root-like character.
Physical	b.p. 260 °C.
Uses	Mainly in woody, fougere, chypre and oriental types.

RIFM Monograph (1975) FCT, 13, 821.

PROPYLENE GLYCOL [57–55–6]

Chemical name	1,2-Propanediol
	1,2-Dihydroxypropane; methylethylene glycol; methyl glycol
	$CH_3CH(OH)CH_2OH$
Appearance	Colourless, viscous, hygroscopic liquid.
Physical	b.p. 189 °C, s.g. 1.040: miscible with water, alcohols, acetone, and chloroform in all proportions: will dissolve in many essential oils, but not in fixed oils.
Uses	In cosmetics as a humectant in foundation creams, moisture creams, and in skin tonics: also as a solvent for the esters of p-hydroxybenzoic acid.
Comments	*l*-, *d*- and *dl*- forms exist.

PROPYLENE GLYCOL MONOLAURATE [142–55–2]

CTFA name	Propylene glycol laurate
	2-Hydroxypropyl dodecanoate
	$C_{15}H_{30}O_3$
Appearance	Clear, oily liquid.
Physical	s.g. 0.915: miscible with mineral and vegetable oils: soluble in alcohol, and dispersible in water.
Uses	In cosmetics in lipsticks and other stick make-up preparations: also in hairsprays as a plasticizer for resins and film-forming polymers.

PROPYLENE GLYCOL MONOMYRISTATE [29059–24–3]

Chemical name	2-Hydroxypropyl myristate
CTFA name	Propylene glycol myristate
	Tetradecanoic acid, monoester with 1,2-propanediol
	$C_{17}H_{34}O_3$
Appearance	Soft creamy-coloured wax.
Physical	s.g. 0.906.
Chemical	Similar to the laurate.
Uses	In cosmetics, particularly lipsticks, as a co-solvent for the eosins: also in rouges and coloured make-up preparations.
Comments	Used in preference to the laurate when a firmer consistency and superior odour are required.

PROPYLENE GLYCOL MONOSTEARATE [1323–39–3]

CTFA name	Propylene glycol stearate
	Octadecanoic acid, monoester with 1,2-propanediol
	$C_{21}H_{42}O_3$
Appearance	White to creamy wax.
Physical	m.p. 50 °C.
Chemical	Has a monoester content of 50–70%: the rest consists of distearate, small quantities of free fatty acid, and free propylene glycol.
Uses	In cosmetics to control the viscosity in cleansing lotions and liquid foundation creams: the self-emulsifying form, which is water dispersible, may be used in the preparation of oil-in-water creams and lotions of a softer consistency.

iso-PULEGOL [89–79–2]

Chemical name	2-iso-Propenyl-5-methylcyclohexanol
	$C_{10}H_{17}OH$
Isolation	Cyclization of citronellal.
Appearance	Colourless liquid.
Odour	Minty-herbaceous, reminiscent of the first smell of tuberose.

Physical	b.p. 201 °C.
Chemical	Mixture of four isomers, the predominant corresponding to natural menthol.
Uses	Occasionally in rose, geranium and oriental notes.

PULEGONE [89-82-7]

Chemical name	5-Methyl-2-(1-methylethylidene)cyclohexanone 4(8)-*p*-Menthen-3-one; 1-methyl-4-isopropylidene-3-cyclo-hexanone $C_{10}H_{16}O$
Source	
NATURAL	Oil of pennyroyal.
Odour	Herbaceous, minty, resinous character.
Physical	b.p. 224 °C, s.g. 0.94.
Uses	In dental preparations. *d*-Pulegone: RIFM Monograph (1978) *FCT*, **16**, 867, Special Issue IV.

iso-PULEGYL ACETATE [89-49-6]

Chemical name	2-iso-Propenyl-5-methylcyclohexyl acetate $CH_3COOC_{10}H_{17}$
Odour	Fresh, green, minty, leafy note.
Physical	b.p. 232 °C, s.g. 0.93.
Uses	Adds fresh character to rose, geranium, oriental, lavender and fougere types. iso Pulegol Acetate: RIFM Monograph (1976) *FCT*, **14**, 327 (Binder, p. 474).

iso-PULEGYL FORMATE

Chemical name	2-iso-Propenyl-5-methylcyclohexyl formate $HCOOC_{10}H_{17}$
Physical	b.p. 226 °C, s.g. 0.96.
Chemical	Similar to the alcohol.
Odour	Fresh, green, minty and earthy character.

QUASSIA [76-78-8]

CTFA name	Quassin Bitter wood; bitter ash
Source	
GEOGRAPHIC	Trees which grow in Jamaica, Colombia, Panama, and the West Indies.
NATURAL	Obtained from the wood of the trunk and the branches of *Picroena excelsa*, *Quassia amara*, or Surinam quassia.
Appearance	White crystals.
Odour	None.

Physical	Soluble in alcohol, methanol and chloroform; slightly soluble in water or ether.
Uses	As a denaturant for alcohol.
Comments	The bitter principle in the wood: the extract is used for fly poison on flypaper and is an anthelmintic.

QUATERNARY AMMONIUM COMPOUNDS

	Includes the range of alkyl trimethylammonium bromides, and the benzylalkylammonium chlorides.
Physical	Freely soluble in water forming stable solutions.
Chemical	Have bactericidal and germicidal properties.
Uses	Cationic surface-active agents which are used in cosmetics and some medications. The stearyl compound is valuable in the formulation of hair conditioning preparations. The bactericidal and germicidal properties control dandruff. Fungal and bacterial skin infections are treated with the mixed benzylalkylammonium chloride: in a low concentration (0.01–0.02%) may be used for the treatment of nappy rash.

QUERCITRON [522-12-3]

	Quercitrin; quercitrocide; quercimelin; quercitin-3-L-rhamnoside; thujin
	$C_{21}H_{20}O_{11}$
Source	
GEOGRAPHIC	A tree which grows in South America.
NATURAL	The bark from *Quercus tinctoria*, Olivier, or *Q. discolor*, N. O. Cupuliferae.
Chemical	Flavine shade yellow is prepared by extracting quercitron bark.
Uses	Dyeing.

QUILLAJA [68990-67-0]

CTFA name	Quillaja
	Quillaja bark; soap bark; quillay bark; Panama bark; China bark; Murillo bark
Source	
GEOGRAPHIC	A tree grown in South America.
NATURAL	The inner bark of *Quillaja saponaria*, Molina, N. O. Rosaceae.
Physical	A tincture may be produced of 1 in 20 of 60% alcohol.
Chemical	Contains saponin.
Uses	In place of soap to produce a froth or foam.

QUINCE [90106-03-9]

CTFA name	Quince extract
COLIPA name	Quince seed gum; semen cydonia; golden apple seed; cydonia seed

Quinine

Source
GEOGRAPHIC   A tree cultivated in Europe.
NATURAL   *Pyrus cydonia*, L., N. O. Rosaceae.
Isolation   The mucilaginous principle is extracted from the seeds.
Chemical   The needs contain 20% of a mucilaginous principle called cydonin.
Uses   In cosmetics as a basis to hair and setting lotions.

QUININE [130-95-0]

Chemical name   6'-Methoxycinchonan-9-ol
CTFA name   Quinine
                   Cinchonan-9-ol, 6'-methoxy
Source
NATURAL   Cinchona bark.
Physical   m.p. 177 °C (decomposes).
Appearance   Colourless or white crystals turning reddish-brown on exposure to air.
Uses   In carbonated beverages at low level.

QUINO-QUINO

Source
GEOGRAPHIC   A tree which grows in Florida, Bolivia, and Peru.
NATURAL   *Myroxylon balsamum*, var. *punctatum*, Klotzsch, N. O. Leguminosae.
Isolation   The balsam is obtained from the bark.
Appearance   Resembles Peru and Tolu balsam.
Chemical   Contains vanillin, benzyl benzoate, and benzyl cinnamate.
Uses   A varnish and also in the preparation of incense.

RESORCINOL [108-46-3]

Chemical name   1,3-Benzenediol
                   Resorcin; *m*-dihydroxybenzene
                   $C_6H_4(OH)_2$
Appearance   Colourless crystalline solid.
Odour   Practically odourless.
Physical   m.p. 111 °C: soluble in water, alcohol, and glycols: changes to a pink colour on exposure to air, light, or on contact with iron.
Other   Has antiseptic properties.
Uses   In cosmetics, especially in hair lotions and shampoos for the treatment of seborrhoea: with salicylic acid in the preparation of exfoliating creams.
Comments   May discolour fair or grey hair, and resorcinol monoacetate is preferred in hair lotions: resorcinol is irritant to the skin and mucous membranes.

RESORCINOL DIMETHYLETHER [151–10–0]

	$C_8H_{10}O_2$
Isolation	Synthesized from resorcinol and dimethyl sulphate.
Appearance	Colourless liquid.
Odour	Slightly medicinal, powerful, hazelnut-like.
Physical	b.p. 215 °C.
Uses	Mainly as a flavouring ingredient.

RHODINOL [6812–78–8]

	Laevo-citronellol (principal ingredient)
	$C_{10}H_{20}O$
Isolation	Originally an impure compound isolated from rose otto: may be prepared by the fractionation of Bourbon geranium oil, producing a mixture of geraniol and *l*-citronellol: claimed to be a pure alcohol isomeric with alpha-citronellol.
Appearance	Colourless oily liquid.
Odour	Natural, sweet, rosy-floral, the best approximates to a somewhat sweet odour of roses.
Physical	b.p. 115 °C: slightly soluble in water, miscible with alcohol or ether. The commercially available esters vary according to source and compositon of their basic rhodinol.
Uses	In perfumery in conjunction with phenylethyl alcohol and the esters of geraniol and citronellol for high quality fragrances.

RHODINYL ACETATE [141–11–7]

Isolation	Produced by acetylating rhodinol.
Appearance	Colourless liquid.
Odour	Fresh, sweet, rose, muguet type.
Chemical	A mixture of esters, consisting mainly of *l*-citronellol and geraniol.
Uses	In good quality rose, muguet, carnation, freesia etc.
	RIFM Monograph (1974) *FCT*, **12**, 975, Special Issue I (Binder, p. 657).

RICE STARCH [9005–25–8]

Chemical name	Amylum
CTFA name	Rice starch
Source	
NATURAL	*Oryza sativa*, L., N.O. Graminaceae.
Isolation	Polysaccharide granules from rice.
Appearance	Polygonal, nearly uniform granules, with no evidence of hilum or striae.
Uses	In cosmetic powders.

ROBINIA OIL

Source
NATURAL — Flowers of *Robinia pseudo-acacia*, L., N.O. Leguminosae.
Isolation — The absolute is obtained by extraction with volatile solvents.
Odour — Possesses an intense odour of the blossoms.
Chemical — Contains indole, methyl anthranilate, linalol, benzyl alcohol, heliotropin, and terpineol, with traces of aldehydes and ketones of a peach odour.

RONDELETIA

Cinchonaceae
History — The name given to an extensive West Indian and tropical American genus of the many-seeded division of Cinchonaceae, by Plumier, in memory of William Rondelet, a famous physician and natural historian.
Source
NATURAL — Mainly shrubs, but some grow to the size of trees.
Odour — Possess fragrant flowers.
Comments — The perfumes prepared and sold under this name are not usually from the natural flowers.

ROSE CONCRETE/ROSE ABSOLUTE [8007–01–0]

Rose de Grasse: the name used to describe the oil collected from rose water which is prepared on a large scale. It is used in exclusive perfumes. Rose de Mai: obtained in southern France.
Source
GEOGRAPHIC — A plant which grows in southern France, Algeria, Morocco, Egypt, and Bulgaria.
NATURAL — *Rosa centifolia*, and other species, and from *Rosa damascena* in Bulgaria.
Isolation — Rose de Mai concrete is obtained from the petals by enfleurage; further extraction with volatile solvents yields the absolute. Extraction may also be from the pomade made by maceration.
Yield — 0.25% as rose de Mai concrete of which 67% is obtained as absolute.
Appearance — Oil.
Odour — Resembles the fresh flower.
Chemical — Rose absolute contains phenylethyl alcohol, together with geraniol, citronellol, nerol, and farnesol.
Uses — In perfumery in fine fragrances.
RIFM Monograph (1975) *FCT*, **13**, 911, Special Issue II (Binder, p. 661).

ROSE OIL [8007–01–0]

CTFA name — Rose oil
Rose otto, attar of roses

Source	
GEOGRAPHIC	Roses cultivated mainly in Bulgaria, but also in France, Italy, Morocco, and Turkey.
NATURAL	A red rose, *Rosa damascena*, varieties of *Rosa centifolia*, and *Rosa alba*, a white rose.
Isolation	The hand-picked petals are steam distilled and the otto floats to the top of the distillate. The rose water is then redistilled (cohobated) and the oil which separates is mixed with that previously collected. The absolute may be obtained by the volatile solvent process.
Yield	0.02–0.05%.
Appearance	Oil.
Odour	Intensely rose-like.
Chemical	Principal constituent is *l*-citronellol (40–65%): also contains geraniol, nerol, *l*-linalol, and their esters, eugenol, farnesol, citral, nonyl aldehyde, and stearoptenes: phenylethyl alcohol may occur in traces, but is usually lost in the distillation waters.
Uses	In perfumery in fine fragrances.
Comments	May be subject to adulteration with dimethyl octanol, rhodinol, geraniol, citronellol, etc. It is not possible to reproduce this odour from synthetics and natural isolates alone. RIFM Monograph (1974) *FCT*, **12**, 979 (Rose Oil, Bulgarian), (1974) *FCT*, **12**, 981 (Rose Oil, Moroccan), (1975) *FCT*, **13**, 913 (Rose Oil, Turkish).

ROSE–GERANIUM OIL

Source	
GEOGRAPHIC	Mainly prepared at Grasse in the south of France.
Isolation	Prepared from the distillation of French geranium oil with approximately 50 times its weight of fresh rose petals: may also be blended from African and Bourbon oils.
Yield	80% of the original oil.

ROSEMARY OIL [8000-25-7]

CTFA name	Rosemary oil *Rosmarinus officinalis* Oil
Source	
GEOGRAPHIC	A herb which grows around the Mediterranean, mainly in the south of France, Spain, Tunisia, and Yugoslavia.
NATURAL	*Rosmarinus officinalis*, L., and possibly other species of N.O. Labiatae.
Isolation	Steam distilled from the flowers and leaves, principally in the south of France and Spain, throughout the year. The stills are set into the ground and heated by direct fire. The process takes 3–4 h. In England, distillation of this oil takes place on a comparatively small scale, and the product is more expensive.

Rosewood oil

Yield	1–2% depending upon source of the plant material.
Appearance	Oil.
Odour	Varies depending upon source and the part of the plant used: woody, herbaceous, lavender-like odour: generally, the French oil is very fine and superior to the Spanish: the best product is obtained from dried leaves.
Chemical	Principal constituents are borneol, camphor, cineole, pinene, and camphene.
Uses	In cheap perfumery, especially hair washes and soaps, blending well with lavender and spike. The French and English oils may be employed for blending in eau-de-colognes.
Comments	Liable to adulteration with turpentine, sage, and spike oils. RIFM Monograph (1974) *FCT*, **12**, 977, Special Issue I (Binder, p. 665).

ROSEWOOD OIL *See* **BOIS DE ROSE FEMELLE**

ROSE OXIDE [16409–43–1]

$C_{10}H_{18}O$

Source	
NATURAL	Rose and geranium oils.
Isolation	Synthesized from citronellol by photo-oxidation with subsequent reduction and cyclization.
Appearance	Colourless liquid.
Odour	Reminiscent of rose and geranium.
Chemical	An oxide mixture containing many isomers. Rose oxide laevo: RIFM Monograph (1976) *FCT*, **14**, 855, Special Issue III (Binder, p. 666).

RUE OIL [8014–29–7]

CTFA name	Rue oil *Ruta graveolens* oil
Source	
GEOGRAPHIC	A plant which grows in France and Algeria.
NATURAL	Different species of Ruta, N.O. Rutaceae: in France, comes from *R. graveolens*, and in Algeria from *R. montana*, L., which yields the summer oil, and *R. bracteosa*, L., which yields the winter oil.
Isolation	Distillation.
Appearance	Oil.
Chemical	The principal constituents (about 90%) are methyl nonyl ketone and methyl heptyl ketone.
Uses	The oil from *R. montana* is highly appreciated as the raw material in the preparation of methyl-*n*-nonylacetic aldehyde by Darzen's glycidic reaction: may also be employed at low levels in perfumery, in sweet-pea compounds.

RIFM Monograph (1975) *FCT*, **13**, 455 (Binder, p. 667).
IFRA GUIDELINE
The Committee recommends for applications on areas of skin exposed to sunshine, excluding bath preparations, soaps and other products which are washed off the skin, to limit rue oil to 3.9% in the compound (see remark on phototoxic ingredients in the introduction).

This recommendation is based on the results of RIFM on the phototoxicity of rue oil (*Food Cosmet. Toxicol.*, **13**, 455 (1975)) and on the observed no-effect level of a sample in a test using the hairless mouse (private communication to IFRA).

November 1974, amended October 1978

SACCHARIN [81-07-2]

Chemical name	1,2-Dihydro-2-ketobenzisosulfonazole
CTFA name	Saccharin
	o-Benzoyl sulphimide; glucide; sacharin insoluble; benzosulfimide; o-sulfobenzimide; benzoic sulfimide; o-sulfobenzoic acid imide $C_7H_5NO_3S$
Appearance	White, crystalline, or powdery solid.
Odour	None, an intensely sweet taste.
Physical	m.p. 226–230 °C: soluble 1 in 290 parts of water, 1 in 30 parts of alcohol, 1 in 12 parts of acetone, 1 in 50 parts of glycerine.
Other	Considered to have 550 times the sweetening power of sucrose.
Uses	In toothpaste as a sweetening agent. The sodium salt, saccharin sodium or soluble saccharin may be used in the same way as saccharin: may also be used in the preparation of tablets containing 15, 30, or 60 mg saccharin sodium.
Comments	Listed as a carcinogen by the EPA.

SAFFRON

Source	Crocus or *Crocus sativus*, L., N.O. Iridaceae
GEOGRAPHIC	A plant cultivated principally in Spain, but also in South-West Asia, and India.
NATURAL	*Crocus sativus*, L., N.O. Iridaceae.
Isolation	Dried stigmas and tops of styles are used to obtain the chief constituent (crocin): saffron oil may also be produced.
Appearance	Saffron oil is a limpid slightly yellow-coloured liquid.
Odour	Intense odour of saffron.
Chemical	Principal constituent is crocin, a glucoside easily soluble in water.
Uses	Saffron oil was employed as a colouring agent, but has now been replaced by tartrazine yellow: may be employed in oriental-type perfumes or to produce original bouquets.

SAFROLE [94-59-7]

Chemical name	5-(2-Propenyl)-1,3-benzodioxole
	3,4-Methylenedioxyallyl benzene; 4-allyl-1,2-methylenedioxybenzene; allylcatechol methylene ether
	$CH_2O_2C_6H_3CH_2CH=CH_2$
Source	
NATURAL	Oils of sassafras, star anise, nutmeg, ylang-ylang, cinnamon leaf, and camphor.
Isolation	Obtained commercially by separation from camphor oil.
Appearance	Colourless or slightly yellow liquid; becomes a white crystalline solid at lower temperatures.
Odour	Warm, spicy, woody-floral note.
Physical	m.p. 11 °C, b.p. 235 °C: insoluble in water, very soluble in alcohol.
Uses	Principally in the synthesis of heliotropin; not used as such as a fragrance ingredient.
Comments	Listed as a carcinogen by the EPA.

RIFM Monograph (1974) FCT, **12**, 983, Special Issue I (Binder, p. 668).

IFRA GUIDELINE

The Committee recommends that safrole as such should not be used as a fragrance ingredient; essential oils containing safrole should not be used at a level such that the total concentration of safrole exceeds 0.05% in fragrance compounds. Examples of essential oils with a high safrole content are Sassafras oil (*Sassafras officinale*, Nees & Eberm.), Ocotea Cymbarum oil (*Ocotea pretiosa*, Metz) and certain qualities of Camphor oils.

The Committee considers it prudent to recommend that the total concentration of safrole, isosafrole and dihydrosafrole should not exceed 0.05% in fragrance compounds.

These recommendations are based on the conclusions of the Scientific Committee on Cosmetology of the EEC on safrole and on the similarity of the biological activity of these substances (Scientific Committee on Cosmetology of the EEC, opinion reached on September 2, 1980; Communication to the EEC Commission ENV/521/79 and IARC Monograph (1976), **10** 231–44).

October 1976, amended July 1987

iso-SAFROLE [120-58-1]

Chemical name	5-(1-Propenyl)-1,3-benzodioxole
	Isosafrole; 1,2-(methylenedioxy)-4-propenylbenzene
	$C_{10}H_{10}O_2$
Source	
NATURAL	Ylang-ylang oil.
Isolation	Obtained from safrole by treatment with alcoholic potash.
Appearance	Colourless liquid.

Odour	Warm, soft floral, anisic and slightly spicy.
Physical	b.p. 248 °C; m.p. 7 °C.
Chemical	Isomeric with safrole; *cis-* and *trans-* forms.
Uses	Mainly in the synthesis of heliotropin (via safrole): not used as a fragrance ingredient.

Isosafrole: RIFM Monograph (1976) *FCT*, **14**, 329 (Binder, p. 475).

IFRA GUIDELINE

The Committee recommends that safrole as such should not be used as a fragrance ingredient; essential oils containing safrole should not be used at a level such that the total concentration of safrole exceeds 0.05% in fragrance compounds. Examples of essential oils with a high safrole content are Sassafras oil (*Sassafras officinale*, Nees & Eberm.), Ocotea Cymbarum oil (*Ocotea pretiosa*, Metz) and certain qualities of Camphor oils.

The Committee considers it prudent to recommend that the total concentration of safrole, isosafrole and dihydrosafrole should not exceed 0.05% in fragrance compounds.

These recommendations are based on the conclusions of the Scientific Committee on Cosmetology of the EEC on safrole and on the similarity of the biological activity of these substances (Scientific Committee on Cosmetology of the EEC, opinion reached on September 2, 1980; Communication to the EEC Commission ENV/521/79 and IARC Monograph (1976), **10** 231–44).

October 1976, amended July 1987

SAGE OIL DALMATIAN [84082-79-1]

CTFA name	Sage oil
Source	
GEOGRAPHIC	A shrub-like herb found throughout south-eastern Europe, especially in Yugoslavia.
NATURAL	*Salvia officinalis*, N.O. Labiatae.
Isolation	Steam distillation of the dried leaves.
Yield	1–2%.
Appearance	Yellow liquid.
Odour	Powerful and camphoraceous.
Uses	In soap perfumery.

RIFM Monograph (1974) *FCT*, **12**, 987, Special Issue I (Binder, p. 672).

SAGE OIL SPANISH [90106-49-3]

Source	
GEOGRAPHIC	Wild herb which grows throughout the Mediterranean, especially in Spain.
NATURAL	*Salvia lavandulaefolia*.
Isolation	Steam distillation of the dried herb.

Yield	0.8–1.0% depending upon the raw material.
Odour	Fresh and camphoraceous.
Uses	In herb-type perfumes.
Comments	May be used as an adulterant for rosemary and spike lavender oils. RIFM Monograph (1976) *FCT*, **14**, 857, Special Issue III (Binder, p. 673).

SALICYLALDEHYDE [90-02-8]

Chemical name	2-Hydroxybenzaldehyde o-Hydroxybenzaldehyde; salicylic aldehyde $C_6H_4(OH)CHO$
Source	
NATURAL	Some essential oils, e.g. those from *Spiraea ulmaria*, L., and other species of N.O. Rosaceae.
Isolation	Distilled from some essential oils or prepared synthetically.
Appearance	Clear colourless oily liquid.
Odour	Pungent, irritating, bitter-almond-like; burning taste.
Physical	b.p. 196 °C, m.p. 2 °C: slightly soluble in water, soluble in alcohol or ether.
Uses	No longer used as a perfumery ingredient due to its skin irritant properties. RIFM Monograph (1979) *FCT*, **17**, 903, Special Issue V.

SALICYLIC ACID [69-72-7]

Chemical name	2-Hydroxybenzoic acid o-Hydroxybenzoic acid $C_6H_4OHCOOH$
Appearance	White crystalline solid.
Physical	b.p. 211 °C, m.p. 156 °C: soluble in water and alcohol: slightly soluble in fixed oils, melted fats and waxes.
Other	Has antiseptic properties.
Uses	In toilet preparations as an antiseptic and fungicide and in hair preparations as a treatment for dandruff: also in concentrations up to 25% for the removal of corns and warts.

SANDALWOOD OIL EAST INDIAN [8006-87-9]

CTFA name	Sandalwood oil Oil of santal; arheol
Source	
GEOGRAPHIC	An evergreen tree indigenous to the mountainous regions of India, and Indonesia: produced in Macassar in Indonesia and the wood may also be known as Macassar wood: very small quantities of fine oil are distilled in England.

NATURAL	*Santalum album*, L., N.O. Santalaceae.
Isolation	Trees between 18 and 25 years old, fully mature and showing signs of natural death, are used. Most of the oil is produced in Mysore: is obtained by lengthy steam distillation of the powdered heartwood and roots.
Yield	4–6%.
Appearance	Volatile, viscid, pale yellow oil.
Odour	Faint, but persistent balsamic odour; slightly urinous.
Chemical	Has a very high boiling point: contains at least 90% santalol, which is a mixture of two alcohols.
Other	Good fixative properties.
Uses	Considerably in perfumery, being one of the oldest and most expensive raw materials available: in violet, cassie, rose and reseda compositions: also blends well with heliotropin, coumarin, vetivert, and musk ambrette: widely used in soap compounds. RIFM Monograph (1974) *FCT*, **12**, 989, Special Issue I (Binder, p. 675).

SANDALWOOD OIL WEST INDIAN [8015-65-4]

Amyris oil

Source	
GEOGRAPHIC	A tree found in Indonesia, and the West Indies.
NATURAL	*Amyris balsamifera*.
Isolation	Steam distillation of the wood chips.
Yield	1.5–3.5%.
Odour	Mild and cedarwood-like.
Uses	In low-cost perfumes.

SANDALWOOD OIL WESTERN AUSTRALIAN [92875-02-0]

Source	
GEOGRAPHIC	Trees widely distributed in Western Australia.
NATURAL	*Fusanus spicatus*, R.Br., and other species of N.O. Santalaceae.
Isolation	After felling, the logs are transported to Perth for distillation: the nearer the habitat of the tree to the equator, the finer the quality of oil.
Appearance	Volatile, viscid, pale yellow liquid.
Odour	Similar to that of sandalwood oil East Indian, but lacks the balsamic note.
Chemical	Contains upwards of 90% sesquiterpene alcohol.
Uses	In sandalwood soap perfumery.
Comments	About two-thirds the price of sandalwood oil East Indian: the wood may also be exported to China, where it is made into joss-sticks.

SANDARAC [9000-57-1]

CTFA name	Sandarac gum
Source	
GEOGRAPHIC	A tree indigenous to the mountainous regions of north-west Africa.
NATURAL	*Callitris quadrivalvis*, Ventenal, N.O. Coniferae.
Isolation	Obtained by incisions into the stems of the trees.
Appearance	Small, pale yellow resinous tears.
Odour	Slight terebinthinaceous.
Physical	Insoluble in water, soluble in alcohol, ether and acetone.
Uses	In tooth cements, lacquers and varnishes, and as an incense.

SANTALOL [alpha-Santalol, 115-71-9; beta-santalol, 77-42-9]

Source	
NATURAL	Sandalwood oil.
Isolation	Obtained from sandalwood oil East Indian.
Appearance	Colourless liquid.
Odour	Soft, sandalwood odour.
Physical	Alpha form, b.p. 302 °C, beta form, b.p. 309 °C.
Chemical	The commercial product is a mixture of alpha and beta isomers.
Uses	In perfumery in oriental, modern woody, chypre and floral fragrance types.
	Alpha-santalol: RIFM Monograph (1974) *FCT*, **12**, 991, Special Issue I (Binder, p. 676).

SAPONINS [11006-75-0]

CTFA name	Saponins
	Sapogenin glycosides
Source	
NATURAL	A type of glycoside widely distributed in plants, e.g. in quillaia bark.
Odour	Bitter taste.
Physical	Immediately produce a foam when shaken with water.
Chemical	Mixtures of two glucosides, e.g. quillaic acid, and quillaia-sapotoxin: these are similar to the polygalic acid and senegin of senega root.

SASSAFRAS OIL [8006-80-2]

CTFA name	Sassafras oil
Source	
GEOGRAPHIC	A tree indigenous to North America.
NATURAL	*Sasafras officinale*, Nees., N.O. Lauraceae. Australian oil comes from *Atherosperma moschatum*, Lab., N.O. Monimiaceae.
Isolation	Steam distillation of the shavings of the inner root bark and wood.

Yield	About 2% from the wood and 8% from the inner root bark.
Chemical	Contains 80% safrole, together with eugenol, camphor, safrene, phellandrene, and pinene.

RIFM Monograph (1982) *FCT*, **20**, 825, Special Issue VI.

IFRA GUIDELINE

The Committee recommends that safrole as such should not be used as a fragrance ingredient; essential oils containing safrole should not be used at a level such that the total concentration of safrole exceeds 0.05% in fragrance compounds. Examples of essential oils with a high safrole content are Sassafras oil (*Sassafras officinale*, Nees & Eberm.), Ocotea Cymbarum oil (*Ocotea pretiosa*, Metz) and certain qualities of Camphor oils.

The Committee considers it prudent to recommend that the total concentration of safrole, isosafrole and dihydrosafrole should not exceed 0.05% in fragrance compounds.

These recommendations are based on the conclusions of the Scientific Committee on Cosmetology of the EEC on safrole and on the similarity of the biological activity of these substances (Scientific Committee on Cosmetology of the EEC, opinion reached on September 2, 1980; Communication to the EEC Commission ENV/521/79 and IARC Monograph (1976), **10** 231–44).

October 1976, amended July 1987

SAVORY OIL [84775-98-4]

CTFA name	Savory extract
Source	
GEOGRAPHIC	A herb cultivated in Yugoslavia, France, and the USA.
NATURAL	*Satureja hortensis*.
Isolation	Steam distillation of the twigs and leaves.
Yield	0.2–0.3% depending upon source of the plant material.
Odour	Fresh and medicinal, slightly reminiscent of thyme and origanum.
Uses	In perfumery in low concentrations in herbal notes.

Savory Oil (Summer Variety): RIFM Monograph (1976) *FCT*, **14**, 859, Special Issue III (Binder, p. 678).

SEA HOLLY OIL [85116-77-4]

Source	
NATURAL	A herb, *Eryngium campestre*, L., N. O. Umbelliferae.
Isolation	Distillation.
Appearance	Yellow oil.
Odour	Musky fragrance.

SESAME OIL [8008-74-0]

CTFA name	Sesame oil
	Gingilli oil; Benne oil; gingelly oil; teel oil

Shellac

Source	
GEOGRAPHIC	A plant grown in India, and other tropical countries.
NATURAL	*Sesamum indicum*, L.
Isolation	Expression from the seeds.
Yield	50–57%.
Appearance	Commercial samples are limpid, pale yellow liquids.
Odour	Pleasant and grain-like, with a bland taste.
Physical	Solidifies at about $-5\,°C$: soluble in chloroform or ether, insoluble in water.

SHELLAC [9000–59–3]

CTFA name	Shellac
	Lacca; lac
Source	
NATURAL	Obtained from the stems of various plants as deposits by a scale insect, *Laccifer lacca*.
Isolation	The insect feeds on the sap of various plants which become encrusted with its resinous secretion: this is washed and dried to prepare 'seed-lac': shellac is obtained by melting and straining, or by solvent extraction.
Appearance	Commercial samples appear as pale yellow to orange–brown hard transparent scales.
Odour	None.
Physical	m.p. 115–120 °C: insoluble in water, dissolves readily in warm alcohol: is almost completely soluble in weak alkali solutions such as sodium carbonate, borax, or ammonia.
Chemical	Contains about 6% wax and 70% resin.
Uses	In the preparation of hair sprays, after bleaching and dewaxing.

SHIU OIL [8022–91–1]

Ho oil

Source	
GEOGRAPHIC	A plant which grows in Japan belonging to N.O. Lauraceae.
Isolation	Distillation: comes in two forms; the crude state and the state from which the camphor content has been separated.
Chemical	Contains 50–65% linalol, linalyl acetate, eugenol, safrole, cineole, pinene, camphene, dipentene, and formaldehyde.
Uses	Commercially as a source for linalol and its esters: in perfumery in the production of artificial bergamot oil.

RIFM Monograph (1974) *FCT*, **12**, 917 (Ho leaf oil).

SILICONES

The simplest composition is $[-(CH_3)_2SiO-]_n$

History	The name was originated by Kipping; now used to designate any

	organosilicon oxide polymer where there is a common structural unit.
Appearance	Low viscosity fluids, viscous liquids, or waxy solids, which are colourless and tasteless.
Odour	None.
Physical	Stable to heat: solubility depends upon composition.
Chemical	Properties depend upon composition and the organic groupings present: generally they consist of alternate silicon and oxygen atoms with various organic groups attached to silicon.
Other	Water-repellent properties.
Uses	In cosmetics, particularly in hand creams, shampoos, hairsprays and lipsticks: also in barrier creams.

SILVER FIR NEEDLE OIL [8021-27-0]

Source	
GEOGRAPHIC	Central Europe.
NATURAL	Silver spruce, or white spruce, *Abies alba*, Mill.
Isolation	Steam distillation of the needles and twigs.
Yield	0.2–0.6%.
ODOUR	Balsamic and herbaceous reminiscent of fir needle oil.
Uses	Similar to that of fir needle oil.
	RIFM Monograph (1974) *FCT*, **12**, 809 (hom cones)
	FCT, **12**, 811 (hom needles).

SKATOLE [83-34-1]

Chemical name	3-Methyl-1*H*-indole
	C_9H_9N
Source	
NATURAL	Civet absolute, and in the wood of *Celtis reticulosa* and *C. Durandii*, Engl., N. O. Urticaceae.
Isolation	Obtained from high-temperature coal tar: may also be produced when sugar is extracted from molasses by the strontia method: may be synthesized from propylidene phenylhydrazide.
Appearance	White crystals which turn brown on exposure to air and light.
Odour	Powerful diffusive, disagreeable, animalic, slightly medicinal: becomes warm and reminiscent of rotting leaves on dilution.
Physical	b.p. 266 °C, m.p. 96 °C: soluble in hot water, alcohol, benzene, chloroform or ether.
Chemical	Related to indole (the beta-methyl derivative).
	RIFM Monograph (1976) *FCT*, **14**, 863, Special Issue III.

SODIUM ALGINATE [9005-38-3]

Chemical name	Alginic acid sodium salt
CTFA name	Algin
	Sodium polymannuronate

Source	
NATURAL	Native algae.
Isolation	A powder manufactured from the algae to produce a standard mucilage: the viscosity may be controlled and standardized by the addition of calcium salts.
Appearance	Greyish-white fibrous powder which produces a transparent, water-white mucilage.
Odour	None.
Physical	Soluble in water, insoluble in chloroform or ether: viscosity and solubility of sodium alginate in water may be affected by other chemicals such as alcohols.
Uses	In cosmetics in hand jellies, greaseless hair creams and hair-setting lotions: also in toothpastes.

SODIUM ALUMINIUM CHLORHYDROXY LACTATE [8038-93-5]

CTFA name	Sodium aluminium chlorhydroxy lactate
Isolation	Prepared by complexing aluminium chlorhydrate with sodium lactate.
Appearance	Commercial samples appear as a colourless, syrupy solution in water.
Physical	pH approx. 8.5: the solution contains about 40% solids.
Uses	In cosmetics and toiletries based on sodium stearate, such as antiperspirant sticks.

SODIUM CARBONATE MONOHYDRATE [497-19-8; 5968-11-6]

Thermonatrite
$Na_2CO_3 \cdot H_2O$

Source	
NATURAL	Can be produced from lake brines or seawater by electrolytic processes.
CHEMICAL	Produced by the ammonia–soda or Solvay process.
Appearance	White crystalline solid.
Odour	Alkaline taste.
Physical	Occurs in crystals of various sizes, the smaller ones having comparatively slow solubility: insoluble in alcohol.
Uses	For bath crystals.

SODIUM CARBOXYMETHYLCELLULOSE [9004-32-4]

Chemical name	Carboxymethylcellulose sodium
CTFA name	Cellulose gum Sodium cellulose glycolate; CMC
Appearance	White fibrous solid or granules: marketed under various grades of viscosity.
Physical	Dissolves in hot and cold water forming a mucilage: the viscosity is not affected by the addition of metal salts.

Chemical	The mucilage is stable between pH 2 and 10.
Uses	As a thickening agent in cosmetics and as an auxiliary emulsifier in toothpaste.

SODIUM HEXAMETAPHOSPHATE [10124-56-8]

Chemical name	Sodium polymetaphosphate
CTFA name	Sodium hexametaphosphate
	Metaphosphoric acid, hexasodium salt; Graham's salt; glassy sodium metaphosphate
	$H_6O_{18}P_6Na_6$
Appearance	White hygroscopic powder.
Chemical	Composed of sodium metaphosphate, and a buffered mixture of anhydrous phosphates.
Uses	As a water softener; increases the lathering properties in toilet soaps.

SODIUM HYDROXIDE [1310-73-2]

CTFA name	Sodium hydroxide
	Caustic soda; soda lye; sodium hydrate
	NaOH
Isolation	Hydrolysis of sodium chloride.
Appearance	Lumps or powder.
Physical	Absorbs carbon dioxide and water from the air rapidly on exposure.
Uses	In cosmetics in the preparation of vanishing creams.

SODIUM PERBORATE [7632-04-4]

CTFA name	Sodium perborate
	Perboric acid, sodium salt
	$NaBO_3 \cdot 4H_2O$
Isolation	Obtained by treating a saturated solution of borax with an equal quantity of sodium hydroxide and about twice the amount of solution of hydrogen peroxide: on cooling the sodium perborate crystallizes out.
Appearance	White crystalline powder.
Odour	None: has a saline taste.
Chemical	Should contain about 85% of available oxygen.
Uses	In dentifrices and in bath salts.
Comments	Should be kept well closed and in a cool place.

SODIUM SESQUICARBONATE [533-96-0]

CTFA name	Sodium sesquicarbonate
	Urao; trona; carbonic acid, sodium salt
	$Na_2CO_3 \cdot NaHCO_3 \cdot 2H_2O$

Appearance	Crystalline solid or white powder.
Chemical	Stable in air.
Uses	May be coloured and perfumed for bath salts.

SODIUM SULPHOCARBOLATE [1300-51-2]

Chemical name	Sodium phenolsulfonate
	Hydroxybenzenesulfonic acid sodium salt
	$C_6H_4(OH)SO_2ONa \cdot 2H_2O$
Isolation	p-Phenol sulphonic acid, produced from phenol and an excess of sulphuric acid, is converted into the sodium salt.
Appearance	Colourless rhombic prisms.
Odour	None, but has slightly bitter taste.
Physical	The aqueous solution is neutral.
Uses	In cosmetics in skin and sunburn lotions.

SORBIC ACID [110-44-1]

Chemical name	2,4-Hexadienoic acid
CTFA name	Sorbic acid
	Hexa-2:4-dienoic acid; 2-propenyl-acrylic acid
	$CH_3(CH=CH)_2COOH$
Appearance	White crystalline solid.
Physical	Slightly soluble in water, alcohol, ether, and propylene glycol.
Other	Inhibits the growth of moulds and yeasts.
Uses	In cosmetics as a preservative: may be used with non-ionic emulsifiers.
Comments	Should be stored in a cool dark place.

SORBITOL [50-70-4]

Chemical name	d-Glucitol
CTFA name	Sorbitol
	Hexan-1,2,3,4,5,6-hexol; d-sorbite; sorbol; hexahydric alcohol; sorbit; l-gulitol; d-sorbitol
	$CH_2OH(CHOH)_4CH_2OH$
Appearance	White crystalline solid.
Odour	None.
Physical	m.p. 110–111 °C (anhydrous), 75 °C (hydrate), s.g. 1.45: soluble in water and hot alcohol.
Chemical	Available as sorbitol syrup (a 70% solution).
Other	Sweet taste.
Uses	As a humectant, or with glycerine in moisturizing creams and lotions.

SOUTHERNWOOD [89957-58-4]

'Old man'

Source	
NATURAL	A small shrub, *Artemisia abrotanum*, N.O. Compositae.
Appearance	Has feathery, fragrant foliage and bears small yellow flowers.

SPEARMINT OIL [8008-79-5]

CTFA name	Spearmint oil Mentha spicata oil; oil of spearmint
Source	
GEOGRAPHIC	England, Russia, Germany, Brazil, Japan and the USA.
NATURAL	The flowering spearmint plant, *Mentha viridis*, L., and other species: known as green mint in the USA.
Isolation	Obtained from the dried, flowering tops of the plant (the most fragrant oil being obtained from the fruiting plants) by steam distillation: the largest quantities are distilled in the USA.
Yield	0.5–2% of the fresh plant, or about 20 lb (9 kg) oil per acre.
Odour	Typical warm and herbaceous.
Chemical	Contains about 60% carvone, phellandrene, limonene, and either dihydrocarveol acetate or dihydrocuminic acetate.
Uses	Has a limited use in perfumery: employed mainly as flavouring agent in dental hygiene products and chewing gum. RIFM Monograph (1978) *FCT*, **16**, 871, Special Issue IV.

SPICEWOOD OIL

	Wild allspice
Source	
GEOGRAPHIC	A bush which grows in North America, where it is known as 'spice bush'.
NATURAL	*Laurus benzoin*, L., N.O. Lauraceae.
Odour	Fragrant.
Chemical	Contains methyl salicylate.
Uses	In perfumery in lavender bouquets to impart a spicy note.

SPIKE-LAVENDER OIL

Source	
GEOGRAPHIC	A plant which grows wild and is cultivated in the lower Alps of France, on the mountains of Spain and also in Yugoslavia.
NATURAL	*Lavendula spica*, D.C., N.O. Labiatae.
Isolation	The plant is taller than true lavender and the blossoms appear about 3 weeks later. The weight may be increased by mixing the cut plants with another plant such as *Calamintha officinalis*, Manch. The oil is obtained by steam distillation of the flowering tops.
Yield	0.5–1.0% depending upon source of plant material.
Appearance	Pale yellow oil.

Spikenard oil

Odour	Camphoraceous, reminiscent of rosemary and lavender. The French oil has a finer odour than the Spanish.
Chemical	Contains camphor, borneol, linalol, camphene, cineol, and probably terpineol.
Uses	Mainly in the soap industry in masculine or lavender-type perfumes.
Comments	May be adulterated with rosemary, sage, and turpentine oils. RIFM Monograph (Spike Lavender Oil) (1976) FCT, 14, 453.

SPIKENARD OIL [90064-28-1]

Source	
GEOGRAPHIC	A perennial herb native to Nepal, Bhutan, and other parts of the Himalayan mountains: also found in parts of Japan.
NATURAL	*Nardostachys Jatamansi*, D.C., N.O. Valerianaceae.
Isolation	Distillation of the root.
Appearance	Pale yellow liquid.
Odour	Faintly musky, reminiscent of patchouli, and valerian.
Uses	The roots are used as a perfume for the hair in India.
Comments	Practically unobtainable and is sometimes replaced by the valeric acid esters of citronellol and rhodinol.

SQUALANE [111-01-3]

Chemical name	2,6,10,15,19,23-Hexamethyltetracosane
CTFA name	Squalane
	Perhydrosqualene; dodecahydrosqualene; spinacane
	$C_{30}H_{62}$
Source	
NATURAL	Obtained from squalene, which is prepared from fish liver oils: is also found in human sebum.
Isolation	Catalytic hydrogenation of squalene.
Appearance	Colourless liquid oil.
Odour	Odourless and tasteless.
Physical	Solidifies at about $-38\,°C$: miscible with vegetable and mineral oils and other lipophilic materials.
Chemical	Stable in air/oxygen.
Uses	As a lubricant and emollient in cosmetics: also as an ingredient in the oil phase in other creams and lotions.
Comments	Squalene itself is unsuitable for use in cosmetic preparations as it readily oxidizes in air forming a viscous mass.

STANNIC OXIDE [18282-10-5]

CTFA name	Tin oxide
	White tin oxide; tin dioxide; stannic anhydride; CI 77861
	SnO_2

Appearance	Heavy white to grey or slightly yellowish powder.
Physical	Insoluble in water or alcohol.
Uses	Was used for frictional nail polishes.

STEARIC ACID [57-11-4]

Chemical name	Octadecanoic acid
CTFA name	Stearic acid
	$CH_3(CH_2)_{16}COOH$
Source	
NATURAL	Tallow and other fats.
Isolation	Saponification of the fats with lime or magnesia, in the presence of high pressure steam. The resulting stearate is then treated with acid and purified.
Appearance	White crystalline solid.
Odour	None.
Physical	m.p. 69–70 °C: very slightly soluble in water and partially soluble in alcohol, benzene, chloroform or acetone.
Chemical	There are several varieties according to purity.
Uses	Cosmetics in the preparation of vanishing creams.
	RIFM Monograph (1979) FCT, **17**, 383.

STEARYL ALCOHOL [112-92-5]

Chemical name	1-Octadecanol
CTFA name	Stearyl alcohol
	Stenol
	$CH_3(CH_2)_{16}CH_2OH$
Source	
CHEMICAL	A mixture of solid alcohols, but mainly stearyl alcohol.
Appearance	White flakes or granules.
Odour	Faint but characteristic.
Physical	m.p. 56–60 °C: soluble in alcohol, ether, benzene or acetone.
Uses	In the stabilization, and to control the viscosity, of oil-in-water and water-in-oil emulsions: as an alternative to cetyl alcohol.

STILBENE [588-59-0]

Chemical name	*trans*-1,2-Diphenylethylene, *trans-sym*-diphenylethylene (stilbene)
	Toluylene
	$C_6H_5CH=CHC_6H_5$
Appearance	Colourless to yellow crystalline solid.
Physical	b.p. 307 °C, m.p. 124 °C: practically insoluble in water, soluble in hot alcohol.
Uses	In sunscreen preparations.

STOCK

Source	*Matthiola tristis*, N.O. Cruciferae
NATURAL	The natural oil is unknown.
Isolation	Artificial compounds may be prepared from rose alcohols, heliotropin, orange flower synthetic, jasmine, anisic aldehyde, iso-eugenol, terpineol, and hydroxy citronellal.
Odour	The natural odour is reminiscent of wallflower and mimosa.
Uses	In perfumery.

STORAX [8024-01-9]

Styrax resinoid/oil; sweet oriental gum

Source	
GEOGRAPHIC	A tree which grows in Asia Minor, Persia, Honduras, and Guatemala.
NATURAL	*Liquidamber orientalis*, and *Liquidamber styraciflua*.
Isolation	A natural balsam formed when the tree is bruised or beaten: it exudes into the inner bark, which is stripped and boiled with water causing the balsam to separate and float to the top: the resinoid is obtained by extraction of the balsam followed by neutralization: the oil is produced by steam distillation.
Yield	Approx. 50% as resinoid and 4–5% as oil.
Appearance	Greyish-brown opaque balsamic resin: after purification in alcohol, followed by filtration and evaporation of the solvent, it becomes transparent: the essential oil is a yellowish-brown liquid.
Odour	Balsamic and recalling naphthalene: on dilution it has a bouquet reminiscent of hyacinth, jonquille, and tuberose. The essential oil has a pleasant odour.
Physical	Insoluble in water, almost completely soluble in alcohol, ether or acetone.
Chemical	Contains about 1% volatile oil together with resin: the essential oil contains styrene, the ethyl, benzyl, cinnamyl, and phenyl propyl esters of cinnamic acid, phenyl propyl alcohol, cinnamyl alcohol, vanillin, and probably naphthalene.
Uses	The resinoid and oil are used in perfumery as fixatives and widely employed in soap perfumery: may be used in hawthorn, hyacinth, jonquille, lavender, magnolia, night-scented stock, tuberose, and verbena perfumes: the use of styrax gum is now obsolete due to their skin sensitizing nature.
Comments	The outer bark of the tree is sold for fumigation: the exhausted inner bark, after drying, may be sold as *Cortex thymiamatis*.

Styrax IFRA GUIDELINES

Styrax American

The Committee recommends that crude gums of American styrax should not be used as fragrance ingredients. Only preparations of

American styrax (the exudation from *Liquidamber styraciflua*, L., var. *macrophylla*) produced by methods which give products not showing a potential for sensitization should be used (e.g. steam distillation, vacuum distillation, extraction with ethanol or hexane, neutralization with aqueous alkali followed by solvent extraction).

These recommendations are based on test results of RIFM on the sensitizing potential of gums and resinoids of American styrax, and on the absence of sensitizing reactions in tests with samples obtained by steam distillation, vacuum distillation, extraction with ethanol, hexane or benzene or by treatment with aqueous alkali and solvent extraction (private communication to IFRA).

November 1977, last amendment May 1980

Styrax Asian

The Committee recommends that crude gums of Asian styrax should not be used as fragrance ingredients. Only preparations of Asian styrax (the exudation from *Liquidamber orientalis*, Mill.) produced by methods which give products not showing a potential for sensitization should be used (e.g. vacuum distillation, extraction with ethanol).

These recommendations are based on test results of RIFM on the sensitizing potential of gums and resinoids of Asian styrax, and on the absence of sensitizing reactions in tests with samples obtained by vacuum distillation, extraction with ethanol or benzene (private communication to IFRA).

November 1977, last amendment May 1980

STORAX (LIQUID) [8024-01-9]

Styrax gum and rectified oil

Source	
GEOGRAPHIC	A tree which grows in the mountains of Honduras and Thailand.
NATURAL	*Liquidamber styraciflua*, L., *Orientalis*, L.
Isolation	Found in slight excrescences on the bark.
Appearance	Viscous, transparent liquid.
Odour	Aromatic and balsamic.
Uses	In perfumery.

STYRENE [100-42-5]

Chemical name	Ethenylbenzene
	Phenylethylene; styrol; styrolene; cinnamene; cinnamol; vinylbenzene
	$C_6H_5CH=CH_2$
Source	
NATURAL	Storax and several other balsamic exudations.
Isolation	May be synthesized.

Appearance	Colourless mobile liquid.
Odour	Penetrating and fragrant, reminiscent of naphthalene.
Physical	b.p. 146 °C: sparingly soluble in water, soluble in most organic solvents.
Chemical	On exposure to air it slowly polymerizes and oxidizes with the formation of peroxides.
Uses	In traces in perfumery to enhance topnote effects, especially in floral compounds.

SUCROSE OCTAACETATE [126-14-7]

Appearance	Fine, white, hygroscopic, crystalline solid.
Physical	Soluble in alcohol, acetone, and benzene, only slightly soluble in water.
Other	Very bitter taste.
Uses	In certain countries is permitted as a denaturant for alcohol. RIFM Monograph (1982) *FCT*, **20**, 827, Special Issue VI.

SULPHONATED FATTY ALCOHOLS

	Alkali metal and amine salts of the sulphuric acid esters of higher aliphatic monohydric alcohols.
Isolation	May be prepared from a fatty alcohol which contains lauryl alcohol with smaller proportions of myristyl, cetyl, and stearyl alcohols.
Appearance	Sodium lauryl sulphate is a white or yellow powder. Sodium lauryl ether sulphates are viscous liquids. The ammonium, triethanolamine, and monoethanolamine alkyl sulphates are pale amber-coloured liquids.
Odour	Sodium lauryl sulphate has a faint odour.
Chemical	Ionic surface-active agents.
Other	Have wetting and cleansing properties.
Uses	In cosmetics and toiletries. Sodium lauryl sulphate may be used as a powder in the preparation of shampoos and for toothpaste. The ammonium, triethanolamine, and monoethanolamine alkyl sulphates are milder detergents and degreasing agents than sodium alkyl sulphates and may be used in the preparation of hair shampoos and bubble baths. Sodium lauryl ether sulphates have high foaming properties and may be used in the preparation of shampoos and bubble baths.

SUMBUL OIL [90028-71-0]

History	The botanical source was discovered in 1869 by Fedschenko in the mountains of Maghian.
Source	
GEOGRAPHIC	A plant grown in India, Russia, and Turkestan.

NATURAL Isolation	*Ferula sumbul*, Hooker filius, N.O. Umbelliferae. Obtained with difficulty by distillation of the roots: however, a strong tincture of the root, 1 in 5, in 90% alcohol, will replace the oil, and a resin may be extracted.
Yield	Approx. 0.5%: it is possible to extract about 6% resin from the tincture.
Appearance	Dark viscous liquid.
Odour	Musky, characteristic of the root, also known as musk root.
Chemical	Contains a mixture of cyclic and aliphatic hydrocarbons with nine carbon atoms, a sesquiterpene, sumbulene, a dextrorotatory sesquiterpene like that found in W.I. santal, a mixture of tertiary alcohols and an ester.
Uses	In perfumery as a fixative.
Comments	The root imported today has an odour of angelica and would appear to be a product from another species, possibly *Ferula suaveolens*.

SWEET PEA [90604–48–1]

Lathyrus odoratus, L., N.O. Leguminosae

Source GEOGRAPHIC	Native to Sicily.
Isolation	
Odour	Resembles orange blossom and hyacinth with a suggestion of rose.
Uses	Sweet-pea compounds are used in perfumery.

SYRINGA [94279–93–3]

History	This plant is known as 'syringa', although this is the botanical name for the common lilac, *Syringa vulgare*, L., N.O. Oleaceae.
Source GEOGRAPHIC	There are two varieties, the inodorous, noted in Carolina in 1738, and the large-flowered 'Gordon' syringa which was introduced from California in 1838, and is strongly perfumed.
NATURAL Isolation	The mock orange, *Philadelphus coronarius*, L., N.O. Saxifragaceae. The concrete and absolute may be prepared by extraction of the flowers with petroleum ether: an essential oil may be produced by distillation and cohobation.
Appearance	The concrete essence is red–brown and yields a thick red–brown liquid absolute, giving a white deposit: it produces a yellowish essential oil.
Odour	The concrete has an odour of jonquille and narcissus, and the absolute on dilution produces a finer odour of the flower.
Chemical	Mock orange perfumes are generally mixtures of synthetic muguet and neroli.
Uses	The perfume is seldom obtained from the flower, but the synthetic mixtures are used in perfumery.

TAGET(ES) OIL [8016-84-0]

Source
GEOGRAPHIC A plant cultivated in Italy, Spain and South Africa.
NATURAL *Tagetes glandulifera* and *T. patula*, N.O. Compositae.
Isolation Obtained from the flowers by steam distillation; performed by manufacturers in the south of France.
Yield 0.1–0.5%.
Appearance Reddish liquid.
Odour Intense, somewhat fruity and cloying; on extreme dilution resembles apples.
Uses In traces in fine perfumery, especially in floral fragrances and cosmetics.
RIFM Monograph (1982) *FCT*, 20, 829, Special issue VI.
IFRA GUIDELINE
The Committee recommends for applications on areas of skin exposed to sunshine, excluding bath preparations, soaps and other products which are washed off the skin, to limit Marigold oil and absolute (oils and absolutes obtained from *Tagetes minuta*, L., syn., *Tagetes glandulifera*, Schrank and *Tagetes patula*, L.) to 0.25% in the fragrance compound (see remark on phototoxic ingredients in the introduction).

This recommendation is based on test results of RIFM on the phototoxicity of Marigold oil and absolute obtained from *Tagetes minuta* and *Tagetes patula* from Egypt and South Africa indicating similar phototoxic potential. A no-effect level of 0.05% for phototoxicity was determined on humans using Egyptian *Tagetes minuta* absolute (private communication to IFRA).

October 1986

TALC, PURIFIED [14807-96-6]

CTFA name Talc
Purified french chalk; talcum
$Mg_6(Si_2O_5)_4(OH)_4$
Source
GEOGRAPHIC The finest quality comes from Italy.
CHEMICAL A purified native hydrated magnesium silicate, which should have a good 'slip'.
Appearance White, slightly lustrous powder: inferior qualities may be greyish.
Physical Insoluble in water, cold acids, or alkalis.
Uses In cosmetics as a base in talcum powders, dusting powders and face powders: prevents chaffing and irritation: micronized talc, giving better covering power, is used in aerosol preparations.
Comments May be sterilized by heating to 160 °C and maintaining the temperature for 1 h.

TANGERINE OIL [8008-31-9]

History	In use since the end of the Nineteenth Century.
Source	
NATURAL	It is found in the peels of *Citrus reticulata*.
Isolation	Expression of the peel of ripe fruit.
Other	Tangerine oil components include d-limonene, n-octaldehyde, n-decylaldehyde, citral, linalool, citronella and cadinene. Also present are terpenes, aldehydes, alcohols and esters.

TANSY OIL [84961-64-8]

Source	
GEOGRAPHIC	A herb which grows in Europe and the USA.
NATURAL	*Tanacetum vulgare*, L., N.O. Compositae.
Isolation	Obtained by steam distillation of the leaves and tops.
Yield	0.2%.
Appearance	Yellow oil.
Physical	Turns brown with age and on exposure to air.
Chemical	Contains thujone with some camphor.
	RIFM Monograph (1976) *FCT*, 14, 869, Special Issue III (Binder, p. 688).

TARRAGON OIL [8016-88-4]

Estragon

Source	
NATURAL	The plant *Artemisia dracunculus*, L., N.O. Compositae, generally used for culinary purposes.
Isolation	Obtained by distillation.
Yield	0.5%.
Appearance	Yellowish-green oil.
Odour	Powerful and persistent, reminiscent of anise.
Chemical	Contains methyl chavicol, phellandrene, an aliphatic hydrocarbon resembling ocimene, and probably also *p*-methoxy-cinnamic aldehyde.
Uses	In perfumery, up to 1.5% in fancy bouquets, and colognes.
	Estragon oil: RIFM Monograph (1974) *FCT*, 12, 709 (Binder, p. 340).

TEA ABSOLUTE [68647-73-4]
[84650-60-2]

Source	
GEOGRAPHIC	A shrub native to China, but cultivated in India.
NATURAL	*Thea chinensis*, Sims, N.O. Termstremiaceae.
Isolation	The absolute may be obtained by extraction with volatile solvents

	from the tea leaves: on distillation an essential oil may be obtained after fermentation.
Yield	0.006% essential oil.
Appearance	Dark semi-fluid substance: may also be obtained in a colourless form.
Odour	Warm and reminiscent of tobacco.
Chemical	Principal constituents are methyl salicylate and phenylethyl alcohol, together with iso-valeric aldehyde, citronellol, hexanal and iso-butyric aldehyde.
Uses	In perfumery in leather, chypre, fougere and hay types.

IFRA GUIDELINES

The Committee recommends that the following materials should not be used as fragrance ingredients:

– Allylisothiocyanate
– Chenopodium oil
– 3,7-Dimethyl-2-octen-1-ol (6,7-Dihydrogeraniol)
– Furfurylideneacetone
– Methyl methacrylate
– Phenylacetone (Methyl benzyl ketone)
– Esters of 2-octynoic acid, except those covered elsewhere in these Guidelines i.e. methyl and allyl heptine carbonate
– Esters of 2-nonynoic acid, except methyl octine carbonate
– Thea sinensis absolute

These recommendations are based on the absence of reports on the use of these materials as fragrance ingredients and inadequate evaluation of possible physiological effects resulting from their use in fragrances.

March 1988, last amendment July 1990

TEA TREE OIL [68647-73-4]

CTFA name	Tea tree oil
	Melaleuca oil
Source	
GEOGRAPHIC	A tree found in Australia.
NATURAL	*Melaleuca alternifolia*, N.O. Myrtaceae.
Isolation	Obtained by distillation of the leaves.
Yield	Almost 2%.
Chemical	Principal constituents are pinene, terpinene, cymene, terpineol and cineole (8%).
Other	Has germicidal properties.
Uses	In perfumery in deodorants and disinfectants: also in medicated soaps and dentifrices.
Comments	No toxicity or irritation associated with this oil.
	RIFM Monograph (1988) *FCT*, **26**, 407, Special Issue VII.

alpha-Terpineol   321

TEA TREE OIL – LEMON SCENTED

Source
GEOGRAPHIC   New South Wales and Queensland.
NATURAL   Obtained from leaves and twigs of the tree *Laptospermum citratum*, N.O. Myrtaceae.
Isolation
Yield   1.5–2%.
Appearance   Bright yellow liquid.
Odour   Strong lemon-type.

TERPENELESS OILS

                  Concentrated oils; sesqui-terpeneless
Source
NATURAL   Manufactured from volatile oils.
Isolation   The concentrated oils are manufactured from the natural oils by removing the non-oxygenated constituents (hydrocarbons, terpenes and resins) by fractional distillation *in vacuo*.
Appearance   Oils containing more 'body' than the synthetic.
Odour   Characteristic odour of the natural oil.
Physical   More soluble in dilute alcohol and do not decompose so readily with age.
Chemical   Consist only of the odiferous ingredients of the natural oil.
Uses   One of the most useful raw materials in perfumery and are used in preference to the commercial essential oils.
Comments   Should not be confused with synthetics or natural isolates.

alpha-TERPINENE   [99–86–5]

Chemical name   1-methyl-4-isopropylcyclohexadiene-1,3 p-menthadiene-1,3; terpilene; 1-3-cyclohexadiene, 1-methyl-4-(l-methylethyl)-; p-mentha-1,3-diene
Source
NATURAL   Found in cardamom and a number of other essential oils: also in orange, coriander and eucalyptus.
Isolation   Acid isomerization of terpinolene.
Appearance   An oil with a lemon odour.
Physical   Practically insoluble in water but miscible with alcohol or ether. GC RIFM no. 72–227; IRC RIFM no. 72–227.

alpha-TERPINEOL   [98–55–5]

Chemical name   *p*-Menth-1-en-8-ol
CTFA name   Terpineol
                  $C_{10}H_{18}O$
Source
NATURAL   Several essential oils including bergamot, neroli, magnolia and petitgrain.

Isolation	Synthesized on a large scale from alpha-pinene by terpin hydrate: may also be obtained from pine oil by distillation.
Appearance	Colourless liquid.
Odour	Delicately floral, reminiscent of lilac.
Physical	b.p. 219 °C, m.p. 35 °C (*d* or *l*), 40–41 °C (*dl*).
Chemical	Occurs commercially as alpha-terpineol with a mixture of other isomers.
Uses	Extensively in perfumery as a basis for lilac perfumes and as a principal ingredient in many varieties of pine fragrances. RIFM Monograph (1974) *FCT*, **12**, 997, Special Issue I (Binder, p. 695).

TERPINOLENE [586-62-9]

Chemical name	*para*-Mentha-1,4(8)-diene Terpadiene, Terpinene $C_{10}H_{16}$
Isolation	As a by-product in the manufacture of terpineol from pinene.
Appearance	Colourless oily liquid.
Odour	Sweet-piny with a pleasant citrus characteristic when pure.
Physical	b.p. 184 °C.
Uses	Widely used in pine and citrus compositions for household products, airfresheners and industrial purposes. RIFM Monograph (1976) *FCT*, **14**, 877, Special Issue III (Binder, p. 697).

TERPINYL ACETATE [80-26-2]

Chemical name	*p*-Menth-1-en-8-yl acetate $C_{12}H_{20}O_2$
Source	
NATURAL	Many essential oils including cardamom, cypress and melaleuca.
Isolation	May be synthesized from terpineol and acetic anhydride.
Appearance	Colourless liquid.
Odour	Fresh, sweet, slightly herbaceous, bergamot, lavender character.
Physical	b.p. 220 °C.
Uses	Extensively in lavender, fougere, citrus and pine fragrances for soaps, detergents and household products. RIFM Monograph (1974) *FCT*, **12**, 999.

alpha-TERPINYL ANTHRANILATE

Chemical name	*p*-Menth-1-en-8-yl *o*-aminobenzoate $NH_2C_6H_4COOC_{10}H_{17}$
Isolation	Obtained from anthranilic acid and terpineol.
Appearance	Colourless or straw-coloured oily liquid.
Odour	Very tenacious, heavy, sweet, floral, neroli type.

RIFM Monograph (1974) *FCT*, **12**, 999, Special Issue I (Binder, p. 699).

TERPINYL iso-BUTYRATE

Chemical name p-Menth-1-en-8-yl 2-methylpropanoate
 $(CH_3)_2CH_2COOC_{10}H_{17}$
Appearance Colourless oily liquid.
Odour Mild, sweet, fruity floral herbaceous note.
Physical b.p. 242 °C.
Uses Occasionally in perfumery in rose, lavender and geranium fragrances.
 RIFM Monograph (1988) *FCT*, **26**, 409, Special Issue VII.

TERPINYL n-BUTYRATE [2153-28-8]

Chemical name p-Menth-1-en-8-yl n-butanoate
 $CH_3CH_2CH_2COOC_{10}H_{17}$
Source
GEOGRAPHIC Oil of the mountain pine, which grows in New South Wales.
Isolation May be synthesized.
Appearance Colourless oily liquid.
Odour Mild, sweet, balsamic, floral note.
Physical b.p. 246 °C.
Uses In perfumes as a modifier in oriental, fougere and rose compositions.

alpha-TERPINYL CINNAMATE

Chemical name p-Menth-1-en-8-yl 3-phenylacrylate
 $C_6H_5CH=CHCOOC_{10}H_{17}$
Appearance Colourless or pale straw-coloured liquid.
Odour Persistent, mild, sweet, floral, balsamic resembling that of Peru balsam.
Uses Useful in perfumery in the preparation of new-mown hay, fougere, oriental and certain muguet fragrances.

TERPINYL FORMATE [2153-26-6]

Chemical name p-Menth-1-en-8-yl formate
 $HCOOC_{10}H_{17}$
Source
NATURAL Ceylon cardamon oil.
Isolation May be synthesized.
Appearance Colourless liquid.
Odour Fresh floral, citrus-like, and slightly herbaceous in character.
Physical b.p. 213 °C.

Uses	Useful in perfumery in the compounding of citrus colognes, neroli notes and lavender compositions.

RIFM Monograph (1976) *FCT*, **14**, 879, Special Issue III.

TERPINYL PROPIONATE [80-27-3]

Chemical name	*p*-Menth-1-en-8-yl propanoate
	$C_2H_5COOC_{10}H_{17}$
Appearance	Colourless oily liquid.
Odour	Mild, fruity, sweet-herbaceous character.
Physical	b.p. 240 °C.

RIFM Monograph (1974) *FCT*, **12**, 1001.

TERPINYL *n*-VALERATE [14481-55-1]

Chemical name	*p*-Menth-1-en-8-yl *n*-pentanoate
	Terpinyl valerianate
	$CH_3(CH_2)_3COOC_{10}H_{17}$
Source	
NATURAL	Oil of cypress.
Isolation	May be synthesized.
Appearance	Colourless liquid.
Odour	Sweet-herbaceous, piny and amber-like.
Physical	b.p. 256 °C.
Uses	In tobacco flavours.

TESU

Source	
GEOGRAPHIC	A plant which grows in the East.
NATURAL	*Butea frondosa*, Roxb., N.O. Leguminosae.
Isolation	Obtained from the dried flowers.
Appearance	An acid decoction of this substance will dye materials a brilliant yellow.
Chemical	Excess alkali turns the colour a reddish-orange.
Uses	As an Indian dye.

TETRAHYDROFURFURYL ACETATE [637-64-9]

CTFA name	Tetrahydrofurfuryl acetate
	Tetrahydro-2-furanmethanol acetate
	$C_4H_7OCH_2OOCCH_3$
Appearance	Colourless liquid.
Physical	Soluble in water, alcohol, ether and chloroform; b.p. 196 °C.
Other	An excellent solvent.
Uses	Generally used with other esters of tetrahydrofurfuryl alcohol to dissolve eosins and make them miscible with fats and waxes used in make-up preparations such as lipsticks, and cream rouges.

4,4a,5,9b-TETRAHYDROINDENO[1,2:d]-m-DIOXIN [18096-62-3]

$C_{11}H_{12}O_2$

Isolation — Synthesized from indene and formaldehyde.
Appearance — White crystalline substance.
Odour — Intense, faecal odour, which becomes floral and jasmine-like on dilution.

TETRAHYDROLINALOL [57706-88-4]

Chemical name — 3,7-Dimethyloctan-3-ol
$(CH_3)_2CH(CH_2)_3C(CH_3)(OH)CH_2CH_3$
Appearance — Colourless liquid.
Odour — Fine delicate floral character superior to linalol.
Physical — b.p. 197 °C.
Uses — As a major component of many delicate floral types such as muguet, lilac, honeysuckle etc.
Tetrahydrolinalool: RIFM Monograph (1979) *FCT*, **17**, 909, Special Issue V.

TETRAHYDRO-2-METHYLQUINOLINE [1780-19-4]

Chemical name — Tetrahydroquinaldine, hydroquinaldine
$CH_3C_9H_9NH$
Odour — Powerful, sweet-floral odour resembling lilac.

TETRAHYDRO-4-METHYLQUINOLINE [19343-78-3]

Civettal crystals
$CH_3C_9H_9NH$
Appearance — Colourless or pale amber-coloured crystalline solid.
Odour — Powerful, tenacious sweet animalic note with honey-like undertones.
Physical — m.p. 38 °C, b.p. 256 °C.
Uses — Most frequently in artificial civet bases and in oriental, amber, tabac and honey fragrances.

TETRAHYDROMUGUOL [57706-88-4]

Chemical name — Commercial mixture of 2,6-dimethyl-2-octanol and 3,7-dimethyl-3-octanol
Isolation — By hydrogenation of allo-ocimenol via allo-ocimene.
Appearance — Colourless liquid.
Odour — Fresh sweet floral with citrusy-lime overtones.
Uses — Finds extensive use in perfumes for soap and detergents forming the basis of many floral types such as muguet, lilac, rose and honeysuckle notes. Blends well with almost all floral ingredients. RIFM Monograph (1976) *FCT*, **14**, 881, Special Issue III (Binder, p. 702).

TETRAHYDROQUINOLINE [635-46-1]

Chemical name	1,2,3,4-Tetrahydroquinoline $C_9H_{10}NH$
Appearance	Colourless or pale yellow liquid.
Odour	Harsh honey-like with an animalic note.
Physical	b.p. 248 °C, m.p. 22 °C.

THANAKA [91722-91-7]

Source	
GEOGRAPHIC	Burma.
NATURAL	The chief constituent is obtained from the bark and root of a flowering shrub, *Murraya paniculata*, N.O. Aurantiaceae.
Appearance	The chief constituent is a yellow powder.
Odour	The blossoms from the tree smell of jasmine and the red fruit possesses a strong odour of gooseberry.
Uses	An accepted toilet cream in Burma.

THIOGLYCOLLIC ACID [68-11-1]

Chemical name	Mercaptoacetic acid
CTFA name	Thioglycolic acid $HSCH_2COOH$
Appearance	Liquid.
Odour	Unpleasant.
Physical	b.p. 104–106 °C at 11 mm Hg, m.p. −16.5 °C: miscible with water and most organic solvents.
Uses	The salts of thioglycollic acid are used in aqueous solution in cold waving preparations. It breaks down the disulphide linkage of cystine while the hair is on curlers, to cysteine, and is then oxidized by substances such as potassium bromate or hydrogen peroxide to reconstitute the cystine molecule. The hair is then left in its altered form.
Comments	Can cause severe burns and blistering of the skin.

THIOTROPE BASE

Chemical name	4-Aldehydophenyl thiocarbimide
Isolation	Prepared by the interaction of thiocarbonyl chloride and 4-aminobenzaldehyde under suitable conditions.
Appearance	White crystalline solid.
Odour	Strong and resembling heliotropin.
Physical	m.p. 32 °C, to clear liquid.

THUJA-LEAF OIL [8007-20-3]

Thuja oil, Cederleaf oil

Source	
GEOGRAPHIC	A tree which grows in the Pennsylvania district of the USA, and Canada.
NATURAL	*Thuja occidentalis*, L., N.O. Coniferae: also known as arbor vitae, and in North America as swamp cedar and white cedar, but is not a true cedar.
Isolation	Steam distillation from the fresh leaves and twigs: the best oil comes from those trees which have been well exposed to sunlight.
Yield	The highest yield is obtained when the steam pressure in the still is greatest (0.4–0.6%).
Appearance	Colourless to yellowish oil.
Odour	Peculiar and reminiscent of bornyl acetate and male fern.
Chemical	Contains thujone, fenchone, pinene and borneol.
Uses	In perfumery as a fixative. Combines well with lavender, pine and citrus notes, especially in masculine fragrances.

Cedar Leaf Oil (*Thuja occidentalis*, L.): RIFM Monograph (1974) FCT, **12**, 843, Special Issue I (Binder, p. 197).

THYMENE [91745-13-0]

Source	
NATURAL	Ajowan oil.
Isolation	A by-product in the isolation of thymol from ajowan oil.
Chemical	Consists principally of cymene.
Uses	In soap perfumery.

THYME OIL [8007-46-3]

CTFA name	Thyme oil
Source	
GEOGRAPHIC	About 40 species of the plant, from Spain and the mountainous districts of southern France.
NATURAL	The French oil is obtained from the herb *Thymus vulgaris*, L., and the Spanish from *T. zygis*, var. *gracilis*, Boiss, *T. vulgaris*, L., and *T. capitatus*, Hoff et Link.
Isolation	Obtained from the flowering herb by distillation.
Appearance	Two forms; 'red' and 'white': the former owes its colour to imperfect distilling apparatus. The French oil is considered the finest. Spanish oils may be greenish in colour due to the source of the raw material.
Odour	Spicy.
Chemical	Those obtained from the Sevilla district contain carvacrol, while those from the Murcia province are characterized by their thymol content. French oils may contain either thymol or its isomer, while Spanish oils have a higher phenol content.
Uses	Widely employed in soap perfumery to impart a fresh smell: the French red thyme oil is favoured: may be used occasionally in eau-de-colognes and in flavouring liquid dentifrices.

Thyme Oil, Red: RIFM Monograph (1974) *FCT*, **12**, 1003, Special Issue I (Binder, p. 704).

THYMOL [89-83-8]

Chemical name	5-Methyl-2-isopropyl-1-phenol
	3-*p*-Cymenol; 3-hydroxy-*p*-cymene; thyme camphor
	$C_{10}H_{14}O$
Source	
NATURAL	Thyme and ajowan oil.
Isolation	Synthesized from *p*-cymene, spruce, turpentine, and *m*-cresol.
Appearance	Colourless to whitish crystalline solid.
Odour	Intense sweet, medicinal, warm, herbaceous reminiscent of thyme oil.
Physical	b.p. 233 °C, m.p. 51 °C: slightly soluble in water, soluble in alcohol, chloroform or ether.
Chemical	Liquefies when triturated with menthol and/or camphor.
Other	Has fungicidal and antiseptic properties.
Uses	In shampoos and dental preparations as a fungicide and in discrete amounts in herbaceous and piny fragrance types for soaps and household products.

TILLEUL [90063-55-1]

Source	
GEOGRAPHIC	A tree which grows in southern Europe and northern Asia and may be cultivated for ornamental purposes.
NATURAL	The linden tree, *Tilia europeatree*, has particularly fragrant flowers and belongs to the lime tree family, Tiliaceae.
Chemical	The perfume may be imitated by mixtures of hydroxycitronellal, geraniol, linalol, methyl anthranilate, anisic aldehyde, ionone and benzyl acetate, fixed with musk ketone, benzoin or Peru balsam.
Uses	The synthetic is used in perfumery.

TITANIUM DIOXIDE [13463-67-7]

	Unitane; CI 77891
	TiO_2
Appearance	White pigment.
Odour	Odourless and tasteless.
Uses	In cosmetics to give covering power and opacity to talcum powders, foundation creams, eye make-up and stick make-up preparations.

TOBACCO FLOWERS

	Nicotiana affinis, N.O. Solanaceae
Source	
GEOGRAPHIC	Frequently found in English gardens.

NATURAL	A perennial with a sweet-scented flower: another variety, *N. persica* constitutes the raw material for tobacco of Shiraz.
Odour	Reminiscent of clary sage, tonka, rose, carnation and honey; more noticeable towards the evening.
Chemical	This perfume may be compounded and varied with labdanum, olibanum, methyl ionone, and the iso-butyrates of phenylethyl and phenylpropyl alcohols.

TOBACCO LEAF ABSOLUTE/RESINOID [8037–19–2]

Source	
GEOGRAPHIC	Plants native to America, but cultivated throughout the world.
NATURAL	*Nicotiana tabacum*.
Isolation	The concrete is obtained by extraction of the leaves and the absolute is made from the concrete: a resinoid is prepared by the leading Grasse manufacturers by extraction of the leaves with volatile solvents.
Yield	Varies according to source of the plant material.
Appearance	The resinoid is a colourless liquid.
Odour	Warm, aromatic, intense odour of tobacco.
Uses	In perfumery in certain masculine perfumes: the absolute is essential to obtain the genuine tobacco notes required. RIFM Monograph (1978) *FCT*, **16**, 875, Special Issue IV.

TOLU BALSAM RESINOID/TOLU BALSAM OIL [9000–64–0]

Source	
GEOGRAPHIC	A tree which grows in Central and South America.
NATURAL	*Myroxylon toluiferum*, H.B. and K., N.O. Leguminosae.
Isolation	The balsam is obtained by slashing the trunks of the trees and collecting the exudation from the schizogenous ducts. Several cuts are made and the trees are bled for about 8 months of the year. The resinoid is obtained by extraction and the oil by steam distillation.
Yield	Approx. 90% resinoid and 7% oil.
Appearance	A natural oleo-resinous secretion which becomes hard and brittle on exposure to air: a colourless tolu balsam is also available.
Odour	Warm, balsamic and hyacinth-like.
Chemical	Contains about 80% resin, together with benzoic and cinnamic acids, vanillin, benzyl benzoate, benzyl cinnamate, and a small quantity of volatile oil.
Uses	A useful raw material in perfumery, being employed as a fixative and blending well with cinnamyl alcohol, coumarin, musk etc.: also used in soap perfumery. Balsamtolu (Gum): RIFM Monograph (1976) *FCT*, **14**, 689, Special Issue III.

TONKA BEAN ABSOLUTE [8046-22-8]

Source	
GEOGRAPHIC	A tree cultivated in Africa and Eastern Venezuela.
NATURAL	Two species of Dipteryx; *D. odorata*, Willd., and *D. oppositifolia*, Willd.: the beans from the former are known as 'angostura beans' while those from the latter, which are smaller and less valuable, are known as 'para' beans.
Isolation	The beans are collected, and the seeds removed and dried in the sun: they may be crystallized on the exterior by immersion in rum, placed in casks and covered in strong alcoholic liquor for 24 h: the beans are dried in the air, which causes them to shrink and become covered in small crystals of coumarin: the dried comminuted beans may be extracted to yield the concrete from which the absolute is obtained.
Yield	Varies according to source of the plant material and production procedure.
Odour	Pleasant and herbaceous with slight caramel and coumarin undertones.
Chemical	The beans contain about 3% coumarin.
Uses	In perfumery as a fixative: the beans may also be used as a tobacco flavour.
Comments	Has largely been replaced by synthetic coumarin. Tonka Absolute: RIFM Monograph (1974) *FCT*, **12**, 1005, Special Issue I (Binder, p. 710).

TORMENTILLA [85085-66-1]

CTFA name	Tormentil extract
Source	
GEOGRAPHIC	Plants which grow in Europe and northern Asia.
NATURAL	Rhizomes of *Potentilla sylvestris*, *P. repens*, *P. tormentilla* and other species of N.O. Rosaceae.
Isolation	An extract may be obtained by infusion or decoction of the rhizomes.
Appearance	Deep reddish colour.
Chemical	The colour is due to tormentilla red, a decomposition product of tannin.
Other	The presence of tannic acid produces remarkable astringent properties.

TRAGACANTH [9000-65-1]

CTFA name	Tragacanth gum Gum tragacanth
Source	
GEOGRAPHIC	A shrub grown in Asia Minor.

NATURAL	*Astragalus gummifer*, Lab., and other species of N.O. Leguminosae.
Isolation	Obtained by an incision into the stems of the plants, causing a gummy exudation.
Appearance	Commercial samples are thin, ribbon-like flakes, or tears.
Odour	Odourless and tasteless.
Physical	Its molecular weight is about 840 000: miscible with alcohol, partly soluble in water, swelling to a homogeneous gelatinous mass: should be stored in a dry place.
Uses	An emulsion stabilizer.

TREEMOSS, CONCRETE [68648–41–9]
Source
NATURAL	Lichens (Evernia furfuracae and Usnea barbata) found on the bark of fir and pine trees.
Isolation	Extraction of the moss, twigs and needles with volatile solvents and evaporation under vacuum.
Appearance	A grey-black to brownish semi-solid or viscous liquid.
CHEMICAL	The main constituents are lichen acids (atranorin, furfuracinic acid and chloroatranorin).

TREFLE [89997–78–4]
Source
GEOGRAPHIC	A plant common in England; there is also a variety found in Italy.
NATURAL	There are many species of clover, the most fragrant being *Trifolium incarnatum* and *T. odoratum*.
Odour	Sweet.
Chemical	Amyl salicylate, iso-butyl salicylate, or a mixture of methyl salicylate and vanillin recall the odour of trefle.
Uses	In perfumery: blends with ylang-ylang, clary sage and oakmoss resin: it is fixed with musk ambrette or benzyl iso-eugenol.

2,4,4'-TRICHLORO-2'-HYDROXYDIPHENYL ETHER [3380–34–5]
	Irgasan (trade name)
Appearance	White powder.
Odour	Faintly aromatic.
Physical	Soluble in alcohol and most organic solvents, insoluble in water.
Other	An effective bactericide.
Uses	In cosmetics in soaps, deodorants, creams and lotions, and in detergents.

TRICHLOROMETHYL PHENYL CARBINYL ACETATE [90–17–5]
Chemical name	1-Phenyl-2,2,2-trichloroethyl acetate Rosacetol, rosone, roseacetone and many other names $C_{10}H_9Cl_3O_2$

Triethanolamine

Isolation	Synthesized from benzene and chloral by the Friedel–Crafts reaction, with subsequent acetylation.
Appearance	White or colourless crystalline solid.
Odour	Balsamic, tenacious green-rosy type.
Physical	b.p. 282 °C, m.p. 88 °C.
Uses	Widely in perfumery as a fixative in rose and geranium types for soaps, detergents and other powder perfumes.

RIFM Monograph (1975) *FCT*, **13**, 919, Special issue II (Binder, p. 713).

TRIETHANOLAMINE [102-71-6]

CTFA name	Triethanolamine TEA; trihydroxytriethylamine $N(CH_2CH_2OH)_3$
Appearance	Pale yellow, viscous and very hygroscopic liquid.
Odour	Slightly ammoniacal.
Physical	b.p. 335.4 °C: completely soluble in water, acetone, and alcohols: slightly soluble in hydrocarbons.
Chemical	Combines with free fatty acids in molecular proportions to form almost neutral soaps: contains approximately 5% mono- and 18% di-ethanolamine.
Comments	Turns brown on exposure to air.

TRIETHANOLAMINE STEARATE [4568-28-9]

CTFA name	TEA-stearate
Isolation	A fatty acid is dissolved with the oil to produce the 'oil phase' and the amine is dissolved with an appropriate amount of water to give the 'aqueous phase'. Emulsification occurs when the two solutions are heated to 70–75 °C and mixed.
Appearance	A hard wax-like soap: the oleate is a semi-liquid resembling petroleum jelly.
Physical	b.p. 277–279 °C at 150 mm Hg, m.p. 21.2 °C.
Chemical	The chemical requirements for emulsification are: 2–4% triethanolamine and 5–15% oleic or stearic acids, based upon the weight of oils to be emulsified.
Uses	Extensively in cosmetics.

TRIETHYL CITRATE [77-93-0]

Chemical name	Triethyl 2-hydroxypropan-1,2,3-carboxylate
CTFA name	Triethyl citrate Ethyl citrate; 2-hydroxy-1,2,3-propanetricarboxylic acid, triethyl ester $C_{12}H_{20}O_7$

Appearance	Colourless slightly viscous liquid.
Odour	Very mild sweet note, virtually odourless.
Physical	b.p. approx. 294 °C.
Uses	A blender, diluent and solvent for perfumes and flavours.

RIFM Monograph (1979) FCT, **17**, 389.

TRIMETHYL CYCLOHEXANOL [116–02–9]

Chemical name	3,5,5-Trimethyl cyclohexanol
	Cyclonol
	C_9H_8O
Isolation	By hydrogenation of phorone.
Appearance	Colourless crystalline mass.
Odour	Powerful camphoraceous menthol-like note.
Physical	m.p. 59 °C, b.p. approx. 228 °C.
Uses	Widely used to provide radiance to low cost fragrances for household products, detergents, airfresheners etc.

RIFM Monograph (1974) FCT, **12**, 1007, Special Issue I (Binder, p. 715).

TRIMETHYL HEXANAL [5435–64–3]

Chemical name	3,5,5-Trimethyl hexanal
	iso-Nonylaldehyde, Inonanal
	C_9H_{18}
Isolation	By oxidation of di-iso-butylene.
Appearance	Colourless oily liquid.
Odour	Powerful grassy green with a somewhat gassy characteristic.
Uses	Most commonly used in low cost masking fragrances or in airfreshener compositions where it blends well with citrus and other terpenes.

RIFM Monograph (1982) FCT, **20**, 841, Special issue VI.

TRIMETHYL UNDECYLENIC ALDEHYDE [141–13–9]

	Adoxal, Farenal
	$C_{14}H_{26}O$
Isolation	Synthesized from partially reduced pseudo-ionone by the glycidic ester.
Appearance	Colourless or pale yellow liquid.
Odour	Tenacious sweet, waxy note reminiscent of rose and lily of the valley.
Uses	In perfumery for its power and tenacity in many floral, balsamic, woody and oriental compositions.

TRIPHENYL PHOSPHATE [115-86-6]

	$(C_6H_5O)_3PO$
Appearance	Crystalline solid.
Odour	None.
Physical	b.p. 245 °C at 11 mm Hg, m.p. 49–50 °C.
Uses	A valuable plasticizer.

TRUFFLES [85085-76-3]

Tuber cibarium, T. griseum, T. melanosporum

History	These were mentioned by Theophrastus and Pliny.
Source	
GEOGRAPHIC	Cultivated in the districts of Riez and Digne in France, where they are found growing under nut trees.
NATURAL	A fungus of the subgroup Oscomycetes.
Isolation	Unearthed by dogs or pigs and packed in alcohol: may be extracted with alcohol and petroleum ether.
Appearance	Fleshy fungi.
Odour	Mild and curious.
Chemical	Lauryl propionate has an odour reminiscent of truffles.
Uses	Consumed in France as a delicacy, but may prove successful in chypre perfumes.

TUBEROSE CONCRETE/ABSOLUTE [8024-05-3]

Source	
GEOGRAPHIC	A plant native to Central America, and cultivated in France, India, Morocco and Egypt.
NATURAL	*Polianthes tuberosa*, N.O. Amaryllidaceae.
Isolation	The concrete is obtained from the flowers by cold enfleurage and the absolute by extraction with volatile solvents.
Yield	0.1% concrete from which 30% absolute may be obtained.
Odour	Heavy and honey-like.
Chemical	The enfleurage product is very rich in methyl anthranilate: the essential oil contains methyl benzoate, methyl anthranilate, benzyl benzoate, methyl salicylate and benzoic acid, and also a ketone (tuberone): may also contain benzyl alcohol, geraniol, nerol, eugenol and farnesol.
Uses	Widely used in modern perfumery.

TULIP

Tulipa suaveolens

Source	
GEOGRAPHIC	Cultivated in both England and Holland.
NATURAL	The bulbous plant known as 'sweet tulip'.

Odour	The only variety which possesses a marked odour, recalling Marechal Niel roses.
Uses	Tulip perfumes are seldom compounded.

TURMERIC OIL *See* CURCOMA OIL

TURPENTINE [8052-14-0]

CTFA name	Turpentine Gum thus
Source	
NATURAL	A mixture of terpene hydrocarbons (mainly alpha-pinene) obtained from various species of Pinus.
Appearance	Yellowish opaque sticky mass.
Physical	An oleo-resin: insoluble in water, soluble in alcohol, chloroform, ether or glacial acetic acid.
Uses	The source of turpentine oil: as a constituent of stimulating ointments.
Comments	Is absorbed through the skin, lungs and intestines and can cause skin and mucous membrane irritation.

TURTLE OIL

History	Has been used for centuries by the Mayas and Mexican Indians as a body skin beautifier: is also reputed to have been used by Cleopatra.
Source	
GEOGRAPHIC	The Seychelles, southern California, Mexico, Brazil and Panama.
NATURAL	Turtle livers and particularly the green Atlantic and Pacific turtles, known as the loggerhead and leatherback varieties: may also come from the giant sea turtle, *Chelonian athecae*, Sp., N. O. Sphargidae.
Isolation	There are three methods of production: (1) by steam distillation of the turtle fat and liver: (2) as a by-product in the production of turtle soup: (3) by boiling the fat in kettles and separating the oil when it floats to the surface. The crude oil is refined by washing, bleaching and filtration.
Appearance	Varies according to source and the refining method employed. The best quality is greyish-white, but others may vary from yellow to golden-brown: may be solid or liquid at room temperature.
Odour	The better grades have an odour reminiscent of cod-liver oil: others are nauseating and may develop rancidity.
Physical	m.p. 20 °C: those oils liquid at room temperature deposit high melting stearins on standing. The oil extracted from turtle eggs is thought to have the same nutritive value as cod-liver oil.
Other	The oil extracted from the giant sea turtle is thought to have astringent properties.
Uses	In the cosmetic industry as an emollient and for its vitamin content.

UMBURANA SEEDS

Source
GEOGRAPHIC   Large trees which grow in Brazil.
NATURAL   *Amburana claudii*, Schwacke et Taube, N. O. Leguminosae.
Isolation   Consignments of the seeds sometimes appear on the English market.
Odour   Resemble coumarin.
Uses   In Brazil as a tobacco perfume.

gamma-UNDECALACTONE [104-67-6]

Chemical name   1,4-Hendecanolide
                Peach aldehyde, aldehyde C_{14} (so called) peach lactone, tetradecyl aldehyde
                $C_{11}H_{20}O_2$
Isolation   Synthesized by the cyclization of undecanoic acid through the reaction of *n*-octanol with acrylic acid.
Appearance   Colourless liquid.
Odour   Sweet, oily, peach-like odour.
Uses   Widely in perfumery in compounds such as jasmine, rose, tuberose and lilac.
                RIFM Monograph (1975) *FCT*, **13**, 921 (Binder, p. 718).

UNDECENOIC ACID [112-38-9]

                Undecylenic acid
                $CH_2{=}CH(CH_2)_8CO_2H$
Appearance   Colourless leafy crystals.
Odour   Acrid, sour fatty note.
Physical   b.p. 275 °C, m.p. 24 °C: miscible with fixed and volatile oils and with alcohol: almost insoluble in water.
Chemical   Consists mainly of undec-10-enoic acid.
Other   Effective fungicide and bactericide.
Uses   As a fungicide in concentrations of 2–5% in dusting powders, creams and lotions.
                Undecylenic Acid: RIFM Monograph (1978) *FCT*, **16**, 883, Special issue IV.

UNDECYL ALCOHOL [112-42-5]

                Undecylic alcohol, Undecanol-1, Alcohol C_{11} (saturated)
                $C_{11}H_{24}O$
Isolation   Synthesized by the catalytic hydrogenation of methyl undecylerate.
Physical   m.p. 16 °C, b.p. 241 °C.
Appearance   Colourless liquid.
Odour   Floral–fruity, slightly fatty character.

Alcohol C-11, Undecylic: RIFM Monograph (1978) *FCT*, **16**, 641, Special Issue IV.

UNDECYLENIC ALCOHOL [112-43-6]

Chemical name	n-Hendec-10-en-1-ol
	$CH_2\!=\!CH(CH_2)_8CH_2OH$
Source	
NATURAL	Oil from the leaves of *Litsea odorifera* (Oil of Trewas).
Odour	Fresh but rather fatty odour.
Physical	b.p. 245 °C, m.p. -2 °C, s.g. 0.85.
Uses	In small quantities in perfumery as a modifier and to give lift to rose, jasmine, and other floral fragrances and in citrus and fruity topnotes.

Alcohol C-11: RIFM Monograph (1973) *FCT*, **11**, 107 and (1973) *FCT*, **11**, 1079 (Binder, p. 43).

UNDECYLENIC ALDEHYDE [112-45-8]

Chemical name	n-Hendec-10-en-1-al
	Aldehyde C_{11} (unsaturated)
	$CH_2\!=\!CH(CH_2)_8CHO$
Appearance	Colourless liquid.
Odour	Powerful, mild, waxy, rosy-citrus note.
Physical	b.p. 235 °C, m.p. 7 °C.
Uses	Widely in perfumery with other aldehydes in rose, amber, moss and many other fragrance types.

Aldehyde C-11, Undecylenic: RIFM Monograph (1973) *FCT*, **11**, 479 and (1973) *FCT*, **11**, 1080 (Binder, p. 54).

UNDECYLIC ALDEHYDE [112-44-7]

Chemical name	n-Hendecanal
	Aldehyde C_{11} (saturated)
	$CH_3(CH_2)_9CHO$
Isolation	Synthesized by the oxidation of alpha-hydroxylauric acid.
Appearance	Colourless liquid.
Odour	Powerful, waxy, floral odour, with fruity overtones reminiscent of rose and incense.
Physical	b.p. 223 °C, m.p. -4 °C.
Chemical	Belongs to the saturated series.
Uses	Widely in perfumery in traces as a modifier.

RIFM Monograph (1973) *FCT*, **11**, 479, (1973) *FCT*, **11**, 481 and (1973) *FCT*, **11**, 1080.
Aldehyde C-11, Undecylic: RIFM Monograph (1973) *FCT*, **11**, 479, (1973) *FCT*, **11**, 481 and (1973) *FCT*, **11**, 1080 (Binder, p. 55), (1976) *FCT*, **14**, 609.

VALERIAN OIL [8008-88-6]

CTFA name	Valerian
Source	
GEOGRAPHIC	A plant cultivated in Europe, Japan, China and the USSR.
NATURAL	*Valeriana officinalis*, L., N. O. Valerianaceae.
Isolation	Steam distillation of the dried and comminuted rhizomes.
Yield	0.2–2% depending upon quality of the plant material.
Appearance	Brown liquid.
Odour	Characteristic and spicy–balsamic.
Chemical	Constituents: valeric acid, bornyl acetate, formate, butyrate and valerate, pinene and camphene.
Uses	In low concentrations in medicine and as a tobacco flavour.

VANILLA RESINOID/ABSOLUTE [8024-06-4]

CTFA name	Vanilla
Source	
GEOGRAPHIC	A plant native to tropical America, but also cultivated in Indonesia, Madagascar and in the West Indies.
NATURAL	*Vanilla planifolia*, Andr., and other species of N. O. Orchidaceae.
Isolation	The plants are grown as climbers on young trees: artificial fertilization is performed as this rarely takes place naturally: a plantation may last 10 years, and will yield 4 or 5 crops: the beans are harvested when still unripe, sorted and cured. They are first immersed in water at about 63 °C for 2–3 min, packed into barrels lined with wool, and allowed to sweat for 24 h. During the next 7 days they are exposed to sunlight for about 6 h a day. The last stage of the curing process is slow drying in the shade, for 4–6 weeks. When the pods are sufficiently shrunken and hard they are packed into parchment-lined tins. The finest variety is grown in Mexico, but these are replaced in Europe by pods imported from Bourbon and Madagascar. The resinoid is produced by extraction of the pods. A fine absolute may be prepared in Grasse. A very valuable natural product.
Yield	Approx. 1 cwt (50 kg) per acre: may vary according to the plant material.
Appearance	The cured pods are a dark brown colour, about 20 cm in length and filled with small, black, shiny seeds. A whitish bloom, which is vanillin, crystallizes on the surface of each bean.
Odour	None when picked: after fermentation an intensely sweet, warm, balsamic odour is displayed.
Chemical	Contains about 2% vanillin, together with anisic aldehyde, anisyl alcohol, and free anisic acid.
Uses	A very valuable product in perfumery and used extensively in all types of fine fragrances: largely replaced by vanillin in many flower compounds: used in the preparation of face powders: may

be used in the form of an alcoholic tincture as a flavouring: is also used as a tobacco flavour.
RIFM Monograph (1982) *FCT*, **20**, 849 (Vanilla Tincture).

VANILLIN [121-33-5]

Chemical name	4-Hydroxy-3-methoxybenzaldehyde
COLIPA name	Vanillin
	Methylprotocatechuic aldehyde; vanillic aldehyde; 3-methoxy-4-hydroxybenzaldehyde
	$C_8H_8O_3$

Source
GEOGRAPHIC   Produced on a large scale in Europe and America.
NATURAL   Vanilla pod, Peru and Tolu balsams, and gum benzoin.
Isolation   Manufactured from the lignin present in sulphite liquors: may also be synthesized from guaiacol by the introduction of an aldehyde group and from protocatechuic aldehyde by methylation: can be made from clove oil: is prepared from iso-eugenol by oxidation on a large scale.
Appearance   White to cream crystalline solid.
Odour   Intense, sweet, typical of vanilla.
Physical   m.p. 83 °C, s.g. 1.056 (liq.): partially soluble in water, freely soluble in alcohol, ether, chloroform and other organic solvents.
Uses   Extensively in perfumery as a fixer, modifier and blender: combines well with heliotropin, coumarin, and benzyl iso-eugenol: of limited use in soap perfumery as it tends to turn soap brown after some time.
Comments   Slowly oxidizes in moist air.
RIFM Monograph (1977) *FCT*, **15**, 633.

VASSOURA OIL [94406-01-6]

Source
GEOGRAPHIC   A bush which grows in Brazil.
NATURAL   *Baccharis dracunculifolia.*
Isolation   Steam distillation of the fresh green leaves.
Yield   Approx. 0.3–0.5%.
Odour   Spicy and grassy.
Uses   In perfumery to impart a harsh spiciness.

VERBENA OIL [8022-79-5]

Source
GEOGRAPHIC   A small bush native to South America, but now cultivated in Algeria, Tunisia, and the south of France.
NATURAL   *Lippia citriodora*, Kunth, known as verveine citronelle in the south of France.

Isolation	Steam distillation with cohobation of the leaves: a synthetic may be prepared.
Yield	0.1–0.7%.
Appearance	The volatile oil is a yellow liquid.
Odour	Fresh, fragrant citrus odour.
Chemical	Genuine verbena oil contains traces of acetic acid, 33% aldehydes and ketones as citral, methyl heptenone, carvone, and furfural, 4% oxide as cineole, 22% terpene as l-limonene, 15% sesquiterpenes, dipentene and d-beta-caryophylene, 20% alcohols as linalol, borneol, nerol, geraniol, nerolidol, and cedrol, together with traces of nitrogenous substances such as pyrrol: a synthetic substance which closely resembles verbena oil is cymylacetic aldehyde.
Uses	No longer used in perfumery due to its skin sensitizing and phototoxic properties.
Comments	Should not be confused with Spanish Verbena oil.

IFRA GUIDELINE

The Committee recommends that Verbena oil obtained from *Lippia citriodora*, Kunth., should not be used as a fragrance ingredient.

This recommendation is based on test results of RIFM showing sensitizing and phototoxic potential for this material (private communication to IFRA).

Commercial compositions of the Verbena type should meet the requirements of the IFRA Guidelines.

December 1981

VETIVER(T) OIL [8016-96-4]

Oil of vetiver; oleum andropogonis muricati; vetyver oil; khas khas oil; khus oil; cus cus oil; vetiver oil Java; vetiver oil Haiti; vetiver oil Réunion

History	In India the roots from vetiver are known as khus-khus and are woven into mats.
Source	
GEOGRAPHIC	A wild, perennial, tufted grass which grows in India and is cultivated in Indonesia, Jamaica, Java, Brazil, and China.
NATURAL	*Vetiveria zizamoides*, Stapf., N. O. Graminaceae.
Isolation	Grown in volcanic sand or ash and propagated by root division. The roots are cleaned, chopped and digested with water: steam distillation yields the oil in two fractions. A proportion of the raw material may be sent to Europe for distillation, but good oils may be distilled in the country of origin. A resinoid, which is a good substitute for the oil, may be obtained by extraction of the roots with volatile solvents.
Yield	2–3% depending upon quality of the plant material and method of distillation.

Appearance	Viscous dark brown liquid.
Odour	Persistent, heavy, and earthy: English oils have a finer odour and this may be due to their being single distillates. The Javanese and Réunion oils differ in odour and are cheaper.
Physical	Soluble in all proportions in most fixed oils, and in 1–3 vols. 80% alcohol.
Chemical	Contains vetiverol, vetivene, an alcohol and hydrocarbon.
Other	An excellent fixative.
Uses	Widely used in perfumery as a fixative, especially for rose and opoponax perfumes, and in larger quantities with patchouli as a basis for oriental-type perfumes: used in soap perfumery as a fixer for violet odours.
	RIFM Monograph (1974) *FCT*, **12**, 1013, Special Issue I (Binder, p. 728).

VETIVEROL [68129-81-7]

	Vetivenol
	$C_{15}H_{26}O$
Isolation	A mixture of alcohols obtained from various types of vetiver oil.
Appearance	Pale yellow to pale olive-green viscous liquid.
Odour	Sweet balsamic, similar to vetiver.
Physical	b.p. 264 °C.
Chemical	Principal components are bicyclo-vetiverol and tricyclo-vetiverol.
Uses	In perfumery; blends well with all santal products.
	RIFM Monograph (1979) *FCT*, **17**, 923, Special Issue V.

VETIVERYL ACETATE [68917-34-0]

Chemical name	Vetivenyl acetate
	Vetacety, Acetivenol
Isolation	Synthesized by the acetylation of vetiver oil with acetic anhydride.
Appearance	Pale yellow viscous liquid.
Odour	Sweet, dry woody of the santal type.
Physical	b.p. 285 °C.
Uses	In perfumery as a base for chypre and modern aldehyde perfumes.
	Vetiver Acetate: RIFM Monograph (1974) *FCT*, **12**, 1011, Special Issue I (Binder, p. 726).

VETIVERYL n-BUTYRATE

Chemical name	Vetivenyl butyrate
Appearance	Pale straw-coloured or pale olive-green viscous liquid.
Odour	Similar but more fruity than vetiveryl acetate.
Physical	b.p. 300 °C.

VETIVERYL n-VALERATE

Chemical name	Vetivenyl valerianate
Appearance	Pale straw-coloured or amber-coloured viscous liquid.
Odour	Tenacious, dry and sweet but mildly herbaceous character.
Physical	b.p. 310 °C.

VIOLET LEAF CONCRETE/ABSOLUTE [8024-08-6]

Source
GEOGRAPHIC   A perennial herb native to Great Britain, but largely cultivated in Italy and the south of France.
NATURAL   *Viola odorata*, N. O. Violaceae.
Isolation   The concrete is obtained from the leaves by extraction with volatile solvents: further extraction yields the absolute: is imitated artificially by methyl octine carbonate.
Yield   Approx. 0.1% concrete from which about 30% absolute may be obtained.
Appearance   A concrete or a colourless liquid.
Odour   Intense, rather unattractive, green, and peppery.
Chemical   The main constituent of an ethereal oil from green violet leaves is an aldehyde $C_9H_{14}O$ containing an open unbranched chain of nine carbon atoms.
Uses   In perfumery as a toner in concentrations from 0.5–1%: may be employed in the finest violet perfumes which also contain ionone or methyl ionone as a base, together with modifiers and blenders.
RIFM Monograph (1976) *FCT*, **14**, 893, Special Issue III (Binder, p. 729).

WALLFLOWER OIL [89997-45-5]

Source   The plant from which this is obtained is also known as giroflee.
NATURAL   The fragrant perennial plant, *Cheiranthus cheiri*, N. O. Cruciferae, of which there are many varieties.
Isolation   An essential oil may be obtained by extraction of the flowers with volatile solvents, followed by steam distillation: may also be prepared synthetically from anisic aldehyde and the rose alcohols with traces of *p*-cresyl methyl ether.
Yield   Approx. 0.06% of essential oil.
Uses   In perfumery.

WAX – CARNAUBA WAX [8015-86-9]

CTFA name   Carnauba
Brazil wax

Source	
NATURAL	*Copernica cerifera*, N. O. Palmae.
Isolation	Extracted from the leaves.
Appearance	Hard solid material.
Odour	When solid, is odourless and tasteless; has a sharp, characteristic and not unpleasant odour on melting.
Physical	m.p. 82–85.5 °C: sparingly soluble in fat solvents.
Chemical	Contains cerotic acid and myricyl alcohol.
Uses	In the cosmetic industry.

WAX – CHINESE WAX

Insect wax

Source	
NATURAL	The insect, *Coccus ceriferus*.
Isolation	Secreted by the insect.
Appearance	Similar in appearance to cetaceum (white to yellowish white).
Odour	Almost none.
Physical	m.p. approx. 92 °C: insoluble in water, freely soluble in benzene, hexane and chloroform.
Chemical	The chief constituent is ceryl cerotate.
Uses	In the manufacture of cosmetics and toiletries.

WAX – JAPAN WAX

Vegetable wax; surmach wax; Japan tallow

Source	
NATURAL	*Rhus succedanea*, N. O. Anacardeaceae.
Isolation	A by-product of the lacquer industry, being obtained from the fruits by extraction with volatile solvents or with heat and pressure.
Appearance	Hard, brittle substance.
Physical	m.p. 50–54 °C: insoluble but readily emulsifies in water: insoluble in cold alcohol.
Chemical	Contains tripalmitin and its acid.
Uses	In the preparation of pomades.

WAX – MONTAN/MONTANA WAX [8002-53-7]

CTFA name	Montan wax
	Lignite wax
Source	
NATURAL	Obtained by the extraction of lignite.
Appearance	Dark brown in its crude form.
Physical	m.p. 100 °C: when super-heated it becomes white: insoluble in water, soluble in organic solvents.
Uses	In cosmetics and toiletries.

WAX – MYRTLE [8038-77-5]

CTFA name	Bayberry wax
Source	
GEOGRAPHIC	Imported from America.
NATURAL	*Myrica cerifera*, N. O. Myrtaceae.
Isolation	Obtained when the berries are boiled in water.
Appearance	Hard green solid.
Chemical	Its chief constituent is tripalmitin.
Uses	In the manufacture of toiletries.

WILD PIMENTO OIL

Source	
GEOGRAPHIC	A plant grown in America.
NATURAL	*Amonis jamarcensis*, N. O. Scitaminae.
Isolation	Obtained from the leaves by distillation.
Odour	Resembles spike lavender.
Chemical	Contains 15% cineole, 38% alcohols, chiefly *l*-linalol and some geraniol, 1.5% linalyl acetate, 17% terpenes, and about 0.1% phenols and aldehydes: there are also sesquiterpenes with traces of acetic and caproic acids.

WOOL ALCOHOLS [8027-33-6]

Chemical name	Alcoholia lanae
Source	
NATURAL	The wool-fat of sheep.
Isolation	Obtained by saponification of the wool-fat followed by removal of the unwanted fatty acid, and extraction and purification of the remainder.
Appearance	Brittle golden-brown solid.
Odour	Faint and characteristic.
Physical	Insoluble in water, soluble in most organic solvents, and partially soluble in alcohol.
Chemical	Contains at least 20% cholesterol, together with other sterols and triterpene alcohols.
Uses	As a water-in-oil emulsifier and as a stabilizer for oil-in-water emulsions.

WORMWOOD OIL [8008-93-3]

	Absinthe oil
Source	
GEOGRAPHIC	A plant which grows wild in northern Africa and southern Europe: widely cultivated in the USA.
NATURAL	*Artemisia absinthium*, L., N. O. Compositae.
Isolation	Steam distillation of the dried leaves and flowering tops.

Yield	0.5–1%.
Appearance	Greenish-blue liquid.
Odour	Unpleasant and intense characteristic odour of the herb: reminiscent of cedar leaf oil.
Uses	In perfumery as a toner in masculine notes: also in the preparation of the liqueur absinthe.

Artemisia Oil (Wormwood). RIFM Monograph (1975) FCT, **13**, 721, Special Issue II (Binder, p. 109).

XOLISMA ABSOLUTE

Xolisma is also known as Sour Wood and Sorrel Tree

Source	
GEOGRAPHIC	A plant which grows in the Florida jungle, extending north as far as South Carolina.
NATURAL	*Xolisma ferruginea*, N. O. Ericaceae.
Isolation	Obtained by extraction with volatile solvents.
Appearance	The plant resembles sweet myrtle and forms flowers in clusters.
Odour	The flowers emit a characteristic fragrance reminiscent of hydroxy-citronellal.
Uses	In perfumery.

YLANG-YLANG OIL [8006-81-3]

Source	
GEOGRAPHIC	A tree native to Indonesia and the Philippines, and cultivated on the Comoro Islands.
NATURAL	*Cananga odorata*, Hook, and other species of *Cananga latifolia*.
Isolation	Obtained by slow steam distillation with fresh water of the freshly picked flowers. There are four different qualities of oil according to the geographical distribution of the trees. In Manila the best quality oil comes from the fully developed yellow blossoms. In Réunion the oil is fractionated and the second runnings are rectified to produce a good second quality oil. An artificial oil may be produced which is much stronger than the natural oil.
Yield	Approx. 1.5–2.5%: 350–400 kg of flowers are required to produce 1 kg of the finest oil. A fully developed tree will produce 20–30 kg blossoms per season.
Appearance	Oil.
Odour	Narcotic, floral, and jasmine-like.
Chemical	The first runnings contain oxygenated bodies and esters, whereas a large proportion of sesquiterpenes come over with the second fraction. The following constituents have been identified: p-cresyl methyl ether, geraniol, linalol (and probably their acetic and benzoic acid esters), cadinene, pinene, benzyl alcohol, iso-eugenol, eugenol, eugenyl methyl ether, methyl anthranilate, methyl

Uses	benzoate, methyl salicylate, benzyl benzoate, benzyl acetate, benzaldehyde and iso-safrole: there may be nerol, farnesol, and valeric acid present in some oils. Widely used in high-class perfumery, blending well with bois de rose oil, amyl salicylate, phenylethyl cinnamate and vetivert oil, especially in face powders: used as a modifier in artificial violets and lilacs.
Comments	May be replaced in cheaper perfumes and soaps by cananga oil. RIFM Monograph (1974) *FCT*, **12**, 1015, Special Issue I (Binder, p. 731)

ZDRAVETS OIL [92347-05-2]

Source	
GEOGRAPHIC	A plant indigenous to the northern shores of the Mediterranean, and is cultivated in central and southern Europe.
NATURAL	*Geranium macrohizum*, L.
Isolation	Distillation of the leaves: an absolute may be obtained by extraction with petroleum ether.
Appearance	Semi-crystalline mass.
Odour	Persistent and reminiscent of clary sage: the absolute has a fuller and more persistent odour which may be retained on absorbent paper for at least 10 days.
Chemical	The principal constituent is germacrol ($C_{15}H_{22}O$).
Uses	In perfumery; has been taken for medicinal purposes and as an aphrodisiac for many years: also grown for ornamental purposes.

ZEDOARY OIL [84961-49-9]

Source	
GEOGRAPHIC	A plant cultivated in Ceylon.
NATURAL	*Curcuma Zedoaria*, Roscoe, N. O. Zingiberaceae.
Isolation	Distillation from the rhizomes of the plants.
Appearance	Greenish-black or reddish viscid liquid.
Odour	Recalls ginger and camphor.
Chemical	Contains cineole.

ZINC OXIDE [1314-13-2]

Chemical name	Zinc white Flowers of zinc; philosopher's wool; zinc white; CI Pigment White 4; CI 77947 ZnO
Appearance	White or yellowish-white powder.
Odour	None.
Physical	Insoluble in water and alcohol, soluble in dilute acids.
Other	Has mild astringent properties.

Uses	In cosmetics in talcum and face powders: has good covering power and adhesiveness.

ZINC PHENOLSULPHONATE [127–82–2]

Chemical name	p-Hydroxybenzenesulfonic acid zinc salt
CTFA name	Zinc phenolsulfonate
	Zinc sulphocarbolate; 1-phenol-4-sulfonic acid zinc salt; zinc p-hydroxybenzenesulfonate; zinc sulfophenate
	$Zn(C_6H_5SO_4)_2 8H_2O$
Appearance	Colourless efflorescent crystalline solid or a white powder.
Odour	None.
Physical	Soluble in alcohol and water: may become slightly pink when exposed to air or light.
Uses	In toiletries, in antiperspirant and deodorant powders and lotions.

ZINC PYRIDINETHIONE [13463–41–7]

CTFA name	Zinc pyrithione
	Marketed under the name *Zinc Omadine*
Appearance	White to yellow crystalline solid.
Physical	Very low solubility in water (6 ppm).
Other	Has anti-bacterial and fungicidal properties.
Uses	In toiletries as a dandruff-controlling agent in cream or lotion-type shampoos: also in soap or other hair dressings, rinses or lotions.

ZINC STEARATE [557–05–1]

Chemical name	Octadecanoic acid zinc salt
CTFA name	Zinc stearate
	$Zn(C_{18}H_{35}O_2)_2$
Appearance	Light, smooth, unctuous white powder.
Physical	m.p. approx. 120 °C: has good adherent properties; repels water (in which it is insoluble): insoluble in alcohol or ether, soluble in benzene.
Other	Has mildly antiseptic properties.
Uses	In coloured cosmetics, in creams, lotions, face and talcum powders.

Appendix

MATERIALS WITH IFRA GUIDELINES, BUT NOT LISTED IN POUCHER'S

* = materials which IFRA recommends should not be used as fragrance ingredients

    Acetylated vetiveroil
* Acetyl ethyl tetramethyl tetralin (AETT)
    5-Acetyl-1,1,2,3,3,6-hexamethyl indan
* Acetyl isovaleryl
    Allylesters – (present in manuscript under individual names)
    Allyl heptine carbonate
    Amylcyclopentenone (2-pentyl-2-cyclopenten-1-one)
* Anisylidene acetone 4-(4-methoxyphenyl-3-buten-2-one)
* p-tert-Butylphenol
    Carvone oxide
    Cinnamic aldehyde – methyl anthranilate, Schiff's base
    Cyclamen alcohol 3-(4-iso-propylphenyl)-2-methylpropanol
* Diethyl maleate
* 2,4-Dihydroxy-3-methyl-benzaldehyde
* 4,6-Dimethyl-8-t-butyl coumarin
* Dimethylcitraconate
* Ethyl acrylate
* Fig leaf absolute
* trans-2-Heptenal
* Hexahydrocoumarin
* trans-2-Hexenal diethyl acetal
* trans-2-Hexenal dimethyl acetal
    alpha-Hexylidene cyclopentanone
* Hydroabietyl alcohol
* Hydroquinone monoethylether
* Hydroquinone monomethylether
* 6-Isopropyl-2-decalol
    Menthadienyl formate
    7-Methoxycoumarin
* alpha-Methyl anisylidene acetone 1-(4-methoxyphenyl)-1-penten-3-one
* 7-Methyl coumarin
* Methylcrotonate
* 4-Methyl-7-ethoxycoumarin
    Methyl heptadienone (6-methyl-3,5-heptadienone)
    Methyl N-methyl anthranilate (Dimethyl anthranilate)
    3-Methyl-2(3)-nonenenitrile
    1-Octen-3-yl acetate (amyl vinyl carbinyl acetate)
* Pentylidene cyclohexanone
    Perilla aldehyde

Propylidene phthalide
Pseudoionone   (2,6-dimethylundeca-2,6,8-trien-10-one)
Pseudomethylionones
Savin oil
Sclareol
Verbena Absolute
Versalide